Food Authentication

Food Authentication

Edited by

P.R. ASHURST
Dr P.R. Ashurst and Associates
Kingstone
Hereford
UK

and

M.J. DENNIS
CSL Food Science Laboratory
Norwich
UK

BLACKIE ACADEMIC & PROFESSIONAL
An Imprint of Chapman & Hall
London · Weinheim · New York · Tokyo · Melbourne · Madras

Published by
Blackie Academic & Professional, an imprint of Chapman & Hall,
2–6 Boundary Row, London SE1 8HN

Chapman & Hall, 2–6 Boundary Row, London SE1 8HN, UK

Chapman & Hall GmbH, Pappelallee 3, 69469 Weinheim, Germany

Chapman & Hall USA, Fourth Floor, 115 Fifth Avenue, New York
NY 10003, USA

Chapman & Hall Japan, ITP-Japan, Kyowa Building, 3F, 2-2-1
Hirakawacho, Chiyoda-ku, Tokyo 102, Japan

DA Book (Aust.) Pty Ltd, 648 Whitehorse Road, Mitcham 3132, Victoria, Australia

Chapman & Hall India, R. Seshadri, 32 Second Main Road, CIT East, Madras 600 035, India

First edition 1996

Softcover reprint of the hardcover 1st edition 1996
Typeset in 10/12pt Times by Cambrian Typesetters, Frimley, Surrey

ISBN-13: 978-1-4612-8426-0 e-ISBN-13: 978-1-4613-1119-5
DOI: 10.1007/978-1-4613-1119-5

A catalogue record for this book is available from the British Library
Library of Congress Catalog Card Number: 95–83679

∞ Printed on permanent acid-free text paper, manufactured in accordance with ANSI/NISO Z39.48-1992 (Permanence of Paper).

Preface

The issue of food authenticity is not new. For centuries unscrupulous farmers and traders have attempted to 'extend', or othewise alter, their products to maximise revenues. In recent years the subject has reached new prominence and there even have been situations where food authenticity has featured as a newspaper headline in various countries.

Food legislation covering the definition, and in some cases composition, of various commodities has been in place in developed countries for many years and paradoxically it is the legislative trend away from emphasis on composition and more on accurate and truthful labelling that has been one driving force for the authenticity issue.

Another, and many would speculate as the more potent, driving force is the move towards fewer and larger supermarket chains in many countries. Such trading companies with their images of quality products, buying power and commercial standing, exercise considerable commercial power which has been claimed as a significant source of financial pressure on food prices and food commodity product quality.

For whatever reason, recent food authenticity issues have become news and consumers, the media and enforcement authorities are showing more interest than ever before in the subject.

This book, which it is hoped will be the first in a series on the subject of food authenticity, deals with the major commodity items that feature in the food industry. After an opening review written by the editors the next two chapters deal with fruit juice and wines – both areas where there have been frauds of substantial proportions in recent years. The issues of authenticity in these subject areas is particularly complex and the relevant chapters provide a useful starting point for those wishing to learn more.

Chapters on meat, fish and cereals then follow. Many of the methods used are common to more than one commodity and here, whilst there have been no known commercial scandals of the same proportions as those in wine and juice markets, the potential for fraud remains.

Vegetable oils are covered in a comprehensive review as is honey and, particularly topical at the time of writing, is a chapter on coffee. Eggs and egg products, milk and a number of commercially less important items are dealt with in chapters at the end of the book.

The authors are well respected experts in their fields and the editors are grateful to them for the contributions to this book. It is not easy for busy authors to place a high priority on the preparation of a book chapter and satisfy the publisher's need for a deadline. Nevertheless, we believe that they have helped to produce a valuable work with wide interest. We trust that it will appeal to commodity processors, food manufacturers as well as to those in academia and the enforcement bodies responsible for ensuring compliance with food laws.

The book, as always, remains the work of its authors and the editors take full responsibility for its shortcomings and weaknesses. It is to be hoped that the reader will find it a useful companion in seeking to learn more of the subject of food authenticity.

Finally, the editors wish to thank Mrs Sue Bate for her invaluable assistance in preparing some of the manuscripts.

P.R.A.
M.J.D.

Contributors

Dr P.R. Ashurst	Dr P.R. Ashurst & Associates, Kingstone, Hereford, UK
Mr P. Brereton	CSL Food Science Laboratory, Norwich Research Park, Colney, Norwich NR4 7UQ, UK
Dr M.J. Dennis	CSL Food Science Laboratory, Norwich Research Park, Colney, Norwich NR4 7UQ, UK
Dr D. Firestone	Center for Food Safety and Applied Nutrition, Department of Health and Human Sciences, Food and Drug Administration, Washington, DC 20204, USA
Mr M. Hall	Campden & Chorleywood Food Research Association, Chipping Campden, Gloucestershire GL5 6LD, UK
Dr D.A. Hammond	Reading Scienctific Services Ltd, The Lord Zuckerman Research Centre, The University, Whiteknights, PO Box 234, Reading RG6 2LA, UK
Dr I. Lumley	Laboratory of the Government Chemist, Teddington, Middlesex TW11 0LY, UK
Dr I.M. Mackie	Food Science Laboratory Torry, PO Box 31, 135 Abbey Road, Aberdeen AB1 8DG, UK
Dr B. Médina	Laboratoire de la DGCCRF, Ministère de L'Economie et des Finances, 351 Cours de la Liberation, 33450 Talence, France
Dr P.C. Molan	University of Waikato, Te Whare Wananga o Waikato, Private Bag 3105, Hamilton 2020, New Zealand
Dr B.G. Osborne	Bread Research Institute of Australia, PO Box 7, Delhi Road, North Ryde, NSW 2113, Australia

Dr J. Prodolliet Nestlé Research Center Lausanne, Vers-Chez-les-Blancs, 1000 Lausanne 26, Switzerland

Mr R.J. Reina Winchester Engineering and Analytical Center, Department of Health and Human Services, Food and Drug Administration, Winchester, MA 01890, USA

Contents

1 An introduction to food authentication

M.J. DENNIS and P.R. ASHURST

1.1 The issue of authenticity

It is perhaps a reflection of the human condition that fraud in one form or another has accompanied commercial transactions over many centuries. The principle of '*caveat emptor*' (let the buyer beware) has become established but this maxim has now become clouded with the advent of written contracts tied to detailed product specifications and methods of analysis. Despite such a trend, the past few decades have seen large-scale fraud in commercial transactions involving commodities.

The final loser in most such situations has been the consumer who has to rely on product labelling and advertising to provide details of the nature of a product. It has become the responsibility of public enforcement authorities to police the claims made and to bring to justice fraudulent operators.

This book is concerned with the issue of authenticity of single food commodities, which still form an important sector of the food industry. It will be apparent immediately that, unless authenticity can be addressed in such commodities, there is little chance, if any, of ever tackling the problem of formulated foodstuffs. In authenticity issues it is not only the final consumer who suffers, food processors are increasingly part of large conglomerate business operations and a successful prosecution can effectively spell ruin for a food manufacturer.

The policing of food authenticity usually becomes the responsibility of the appropriate public enforcement body. For example, in the UK most food authenticity issues are likely to relate to prosecutions under food labelling regulations where a product does not conform to its description. This would fall within the scope of Trading Standards authorities. Issues of public health are, at present, within the area of operations of Environmental Health departments.

It is usually the case that food authenticity issues do not normally pose a threat to public health. However, this assertion relies heavily on the morality of those involved in illegal activities and, to an extent, their understanding of food chemistry. If they have little regard for human health in their desire to gain money serious consequences can result. The example provided by contaminated Spanish cooking oil provides an unfortunate demonstration of the need for vigilance over food authenticity.

This issue of authenticity is thus of great significance to enforcement bodies who need to have knowledge of not only how issues arise but how they can be monitored and evaluated.

1.2 The significance of authenticity to food processors

In many countries, legislation has become progressively more onerous, leaving little scope for a fraudulent food-processing operation. The enforcement authorities invariably have such wide areas of responsibility that they cannot usually carry out more than routine sampling of the sale of food commodities. It is, for example, typical in the UK for a Trading Standards department to cover such diverse areas of trade as Weights and Measures, petroleum products, toys and used cars, as well as all types of foodstuffs.

Different parts of food industries frequently have their own trade associations, which deal with problems pertinent to that sector. These could include oils and fats, fruit juices, meat, cereals, dairy products, and so on. Historically, these associations have dealt principally with the interests and commercial well-being of their members; however, increasingly they have become the focal point for technical problems of the industry concerned. Within the European Union, such trade bodies have played a vital part in advising individual member governments of the key issues that affect their industry, its suppliers and end users.

Trade bodies typically comprise both large and small manufacturers and a powerful group lobby can be exercised over any member who knowingly indulges in producing non-authentic or adulterated product. Probably the greatest risk within a trade body is that its databases, which are used for testing, may become corrupted. Thus product that is believed to be authentic – perhaps originating from a particular source – can have been subject to wholesale faud, which is then undetectable against a corrupted database. The issue of databases is thus vital to the control of authenticity. Each commodity area has its own special problems but the database must be sufficiently flexible to take account of varietal and seasonal variations, as well as different sources of origin. It will normally be the responsibility of the industry concerned to ensure it has access to adequate testing facilities, which, in turn, access independent, updated databases.

It is important that all parties, whether governments, enforcement agencies or the industry itself, have access to the same methodology and use the same authentic databases, so that all parties operate on a level playing field. In the UK, a number of government laboratories have been launched as executive agencies. These laboratories are now able to undertake work for industrial customers and this has played an important role in ensuring industry has access to the same data and advice that is used by government and enforcement authorities.

1.3 Authenticity testing

A processor producing or packaging a food commodity is often faced with the issue of deciding the relative importance of authenticity testing compared with other types of analysis. A central issue is the overall quality of the commodity; there is no point in producing an authentic product that is of such poor quality as to be rejected by or likely to cause illness to the consumer.

Typically, raw-materials (commodity) testing before the product is packaged will address four main issues:

1. 'Value for money' physico-chemical testing: typically this area ensures that a buyer has obtained what he ordered (e.g. the solids level in fruit juice, freedom from contaminants, conformity to specification etc.).
2. Organoleptic testing: does the product look, smell and taste as the consumer would expect?
3. Biological/microbiological testing: is the product free from enzyme activity or the presence of micro-organisms that would otherwise cause product spoilage or, at worst, harm the consumer?
4. Any special testing that may be required by the nature of the product (e.g. food safety criteria such as lead and cadmium content, nutritional labelling etc.).

From the above, it may be seen that authenticity can be reasonably described as part of the value-for-money testing. A processor then has to decide whether to test every batch of incoming raw material not only against the routine quality parameters but also by the best current practice for authenticity. The decision to carry out the latter may be influenced by retailers. For example, major supermarket chains in the UK have, since the major fruit juice scandal of 1991, insisted on positive release testing of every batch of juice packed for them, irrespective of the cost involved.

Decisions on frequency of authenticity testing will more typically be decided by the processor and will be influenced by factors such as cost, complexity and time involved.

Authenticity is however, more than just testing, and will usually involve a processor in a thorough understanding of the supplying sources, likely adulterants and where in the supply chain they could be added. It is said that the best defence against purchasing non-authentic product is to establish a long-term business relationship with a reputable supplier who can demonstrate an adequate audit trail of good-quality raw material, its processing and delivery.

However, it is important not to rely on trust alone, and continued vigilance by random but statistically significant authenticity testing to minimise risk of other forms and sources of adulteration should be performed. It is suggested that sampling to achieve a confidence level of at least 95% and preferably 99% should be established.

1.4 Compositional standards

Authenticity issues also relate to the composition of the commodity concerned. In most cases, a thorough understanding of the commodity is required by the food processor: those intent on adulterating a product usually have a better knowledge than most processors.

Although various types of authenticity fraud can be perpetrated, the objective in all types of fraud is usually to sell a cheap product as a more expensive type.

1.4.1 Species/variety

This area is perhaps the most obvious source of non-authentic product. This may take the form of complete substitution of one species for another (e.g. Fish – see chapter 5). More often it involves mixing authentic product with a similar but less expensive alternative; for example, orange juice adulterated with apple juice, olive oil blended with another vegetable oil. Even relatively low levels of adulteration (e.g. 5–10%) can, given the large volumes involved in most commodities, produce very large cash rewards for those involved.

This type of adulteration is usually relatively easy to detect as there are often natural markers present in the adulterating material that may be monitored. The most frequent problem is that relatively few components occur uniquely in one variety and not another and it is the level in the final mixture that is often the key to whether adulteration has occurred.

1.4.2 Region of origin

Much food labelling, particularly in the EU, places great emphasis on the regional name of a product (wines in particular are associated with a specific region). Products from one region may attract a considerably greater value than similar products from another area and such labelling can only be supported by analysis that confirms the region of origin.

1.4.3 Process

It is often the case that a commodity commands a premium price because it has been prepared by a specified process. Thus 'olive oil' may contain refined olive oil but virgin olive oil must be obtained by mechanical or physical means (see section 7.7). Similarly, orange juice should be obtained by squeezing the fruit and the product produced by the process called 'pulpwash' is not permitted in pure orange juice in the UK although 'in-line pulpwash' is permitted in the USA (see section 2.3.9).

In many countries, there is a legal restriction on the use of a name unless a particular process has been followed. Since certain processes are more efficient or cheaper to operate, there can often be a significant advantage in producing by one method but labelling the commodity as being produced by another. It is essential to be able to detect such types of fraud.

1.4.4 Industry frauds

Certain commodities possess properties that make them vulnerable to types of faud that are special to the particular industry concerned. For example, orange or other fruit juices can be diluted with sugar solutions. Apple juice has a greater value if its acidity is higher rather than lower. Such types of adulteration can yield especially high rewards to those operating them and very elegant techniques have been developed for their detection.

1.4.5 Brands

The 'passing off' of a commodity as a more expensive brand is a well-known type of fraud but generally relates more to manufactured products (e.g. Whisky) than to commodities. Nevertheless, individual commercial operations may wish to protect their products by developing specific tests.

1.5 Labelling and legislation

Authenticity problems are directly associated with the labelling of a product. For example, product described as 'vegetable oil' could properly be any edible oil of vegetable origin, whereas olive oil must be edible oil originating from olives. There are incidentally further reserved descriptions (e.g. 'virgin') that can apply to olive oil in countries of the European Union.

Certain countries have strict compositional standards that must apply before a product may be appropriately labelled. In Germany, for example, orange juice must meet the Richtwerte und Schwankungsbreiten bestimmter Kennzahlen – guide values and ranges of specific reference numbers – (RSK) compositional standard before it can be described as a fruit juice. By contrast, in the UK, there is no compositional standard for orange juice but it must be obtained from the fruit (oranges) by expression. Reserved descriptions apply in many countries to specific commodities and there is frequently copious legislation that describes the product in one way or another.

The issue of place or region of origin of certain commodities has already been referred to and most countries would seek to protect their own products. There are countless examples all over the world of products that bear a regional description and the legal status of such descriptions in each country will determine whether there is a potential authenticity issue (e.g. quality French wines). Determination of geographic origin is a particularly challenging task. However, it is an area of current research interest and some progress has been made in a few areas. Methods for the identification of Basmati rice represent one example (see chapter 6).

Similarly, the ability of a country to take legal action to protect these specific descriptive names or products will be determined by the availability of appropriately briefed enforcement authorities armed with valid methods and databases.

In the European Union, USA, Canada and most other first world countries commodities are, for the most part, legally and scientifically defined such that use on a food label of a particular name obliges the food processor to meet all implied descriptions.

Infringement of such labelling regulations can in many cases leave a food manufacturer and/or retailer open to criminal charges in the courts and penalties can be severe in many countries. Successful prosecution in the criminal courts can also leave the offender open to civil action for damages by the aggrieved party.

In summary, many countries have the legal framework and have shown determination to deal firmly with issues of authenticity.

1.6 Databases

It is virtually impossible to define food commodities in terms of a single all-encompassing scientific description. There are few commodities that are pure chemical substances (e.g. sucrose) and, for the most part, they are mixtures containing a majority of clearly defined principal components together with an admixture of many minor ingredients. The same product will show minor variations depending on its botanical variety, where and when it was grown and, to some extent at least, how it is processed.

To be able to confirm a commodity as authentic (or not as the case may be) there needs to be available a database of all possible variants of the minor and major ingredients of the commodity. It is readily apparent that such data requires continual updating to ensure that it remains contemporary and relevant to current agricultural and processing practice. New genetically engineered crops or annual climatic changes may have a considerable influence. It is also necessary that the data is available in a form that is readily comparable with test samples. The other essential requirement for such a database is that it must be collected by impartial and disinterested individuals who can verify the origin and method of processing where applicable.

There have been examples of authenticity testing schemes for commodities that have failed because the database became corrupted despite being supplied by well-meaning participants. The databases of testing laboratories are frequently regarded as highly confidential and this is usually much disliked by those seeking information.

There are still those intent on achieving adulteration or falsification of commodities because of the high commercial rewards and it is clear that access by such individuals or companies to testing databases could give major advantages.

There is, therefore, something of a paradox to be resolved in releasing database information. It is clearly necessary that industry and enforcement authorities work to the same standards. It is, therefore, essential that authenticity methods and the databases on which they rely are freely available to both parties. In some cases, compositional data on which food standards are based is published, for example fatty acid composition of vegetable oils is available through *Codex Alimentarius*. However, it is often seen as desirable that this basic information does not fall into the hands of those who would seek to adulterate food. Hence, many commercial laboratories guard jealously their database information. This may have the added advantage of preventing other commercial laboratories offering a competing service.

Testing data is available in two forms. There is the raw physico-chemical data such as the chemical composition or the content (per millilitre) of ^{13}C or ^{2}H in isotopic work. Unless the recipient has the base data available, such information means little and individuals must make their own judgement on authenticity. Most commercial laboratories offer the comparative service whereby the raw data is used to evaluate the product's authenticity.

This latter form of presentation suits commercial customers but can leave the testing laboratory open to criticism and potential legal action.

It can be very difficult for commercial customers of authenticity tests to evaluate how good a service they are getting from the laboratory they use. Timely results or a low cost analysis might hide fundamental weaknesses in the methods being used. The secrecy attached to authentic databases might well hide inadequate information on the range of products available. Later in this chapter we discuss the need for companies to exercise 'due diligence' in establishing that commodities they purchase are as described by the seller. At present we are aware of little guidance that is available to assist companies in meeting due diligence criteria when selecting a laboratory to perform authenticity tests. The consequences of selecting an inferior laboratory on the basis for instance of price are therefore difficult to judge. One possibility is to use laboratories that are accredited under a national accreditation scheme; in the UK, a scheme termed NAMAS is available. In general, this type of scheme details how tests should be carried out and ensures that these meet internationally recognised standards. Proficiency schemes such as FAPAS (Food Analysis Performance Assessment Service) are also available; these enable laboratories to check the accuracy of their measurements. Both these schemes are aimed at ensuring that sample measurements are carried out to the highest standard. As yet there is no internationally recognised scheme to deal with how the data so collected should be evaluated in terms of the authenticity of the sample.

The establishment and maintenance of a commodity database in parallel with running a commercial authenticity testing scheme is demanding and expensive. It places the testing laboratory in a powerful position when industry credibility is frequently at stake. Such operations must be managed with considerable skill and tact but are vital in ensuring authenticity of commodities.

1.7 Factors affecting choice of method for authenticity determination

The process of altering or substituting a food material in such a way as to alter its composition while permitting it to be described as ‘pure’ is sometimes referred to as an ‘economic adulteration’. To make money out of such a procedure, it is necessary to add one of the major components of foods. Water is such a major component and its addition to food products, like milk, make it the subject of one of the oldest authenticity issues. Any drink is therefore a target for such an adulteration. However, the addition of water is not confined to drinks and is also an issue in other products; for example meat products where the retention of added water can be effected by the use of polyphosphates.

The simple addition of water is relatively easy to detect by modern methods if a significant dilution results in a product whose composition is outside the established range of authentic values. In some cases, a product may contain a higher amount of solid than is typical. An example might be grapes intended for wine production in a particularly good growing year. In this case it might seem possible to add water to the grape must with the intention of diluting the must within the range considered authentic for the product. However, ground waters and plant juices have different isotopic compositions. This practice can therefore be defeated by measuring the hydrogen and oxygen isotopic composition of the water in the final product.

The substitution or extension of a product for economic gain will often require the addition of food solids that form one of the major components of the food. Proteins, carbohydrates and fats all represent bulk food components that might be misused as food adulterants. In some cases, there will be chemical differences between the authentic material and the adulterant, thus providing the basis for sound tests for the authentic product. The substitution of one fish species for another and its detection by electrophoretic examination of protein profiles is one example.

Food commodities will frequently contain the same food chemicals but perhaps in different amounts. Thus expensive food oils will often contain the same fatty acids as cheaper oils but in different proportions. In this case, the ratio of concentrations of fatty acids will enable the identification of the addition of oils of different composition.

The addition of a relatively pure chemical to a foodstuff that contains that chemical as a natural component is a particularly difficult problem to resolve. The principal components of orange juice are glucose, fructose and sucrose in a ratio of approximately 1:1:2, together with small quantities of fruit acids such as citric and malic. The extension of orange juice with sucrose is therefore a relatively straightforward adulteration provided that the chemical composition of the product is kept within the range established for authentic products.

The traditional approach to resolving this type of problem has been to try to identify small quantities of impurities in the potential adulterant and the presence of oligosaccharides has been used to indicate the addition of beet medium invert sugar as an adulterant. Those adding an adulterant will naturally respond to the development of any such tests by either trying to remove the marker chemical or using a different source of adulterant that does not contain the marker.

Tests that are able to measure some aspect of the major component of an adulterant are extremely desirable. Since the measurement is being undertaken directly on the adulterating component of the product, it will be present in relatively large quantity and it will be difficult to turn to a different source of adulterating material. It is for this reason that the use of isotopic measurements that depend on the species and region of origin of food components are now beginning to form a crucial weapon in the fight against adulteration. These measurements are now becoming an integral component of methods to ensure the authenticity of several products. Their use in the detection of the source of sugar is particularly well established.

A note of caution is required, however. Although the methods described may be expensive and highly technological, this is not sufficient to prevent attempts to subvert them. For flavours such as vanilla, the difference in price between the natural commodity and the synthetic chemical is very high because of a high demand for the natural product. It is therefore potentially very profitable for an adulteration with the synthetic chemical to be undertaken. The large price differential is sufficient for it to remain worthwhile for synthetic chemicals with a similar isotopic composition to the natural product to be synthesised. It therefore remains important for those laboratories who seek to guarantee the authenticity of natural products to remain vigilant to the ways in which their tests may be overcome.

The analysis of the major component of the food clearly has important advantages. Occasionally, there are minor components associated with the bulk adulterant that are not practicable to remove before addition to the higher value commodity. So long as methods are sufficiently sensitive and these analytes are specifically indicative of adulteration, they provide the basis for sound analytical techniques. The development of methods based on species-specific DNA fragments have been very important in this regard. DNA polymerase chain reactions permit the amplification of species-specific DNA fragments, which form the basis for several important authenticity tests (see section 4.3.7).

1.8 Interpreting authenticity data

The ideal tests for authenticity would comprise rapid, cheap measurement of an analyte that was present in the adulterant but not in the food being tested. Simply finding the analyte in the food would then be sufficient evidence that the food was not authentic. A good example of this kind of test was the measurement of D-malic acid in orange juice. Racemic malic acid was added to orange juice with sugars to ensure that the correct brix/acidity ratio was maintained but very little (if any) D-malic acid is naturally present in orange juice. A second example might be the presence of specific sterols in one vegetable oil that might not be expected in another.

A more difficult situation arises when the authentic food contains a low concentration of the analyte and the adulterating material a much higher concentration. It is then necessary to take into account the range of concentrations that might be expected from the adulterant and the range of concentrations that might be expected from the food. It may be useful to prepare a graph of expected analyte concentration against proportion of adulterant in the pure food. The mean concentration of adulterant in the two components will enable the calculation of amount of adulterant added. The ranges of the analyte in food and adulterant will enable an estimate of the accuracy of this determination.

The situation where the food contains none of the analyte therefore represents the optimum special case of this more general relationship. One of the great benefits that accrues from this approach is that it concentrates the mind on ensuring all the necessary information is available to perform an authenticity test. If you cannot prepare a graph of this kind, it usually means that some essential information is missing. Of course it is always necessary to ensure that other factors (e.g. processing conditions) will not lead to the possible presence of analyte in the authentic food.

It is more common for there to be more than one analyte of value when assessing authenticity. These analytes could be investigated in isolation, indeed this is how fatty acid profiles of vegetable oils are often interpreted. However, where links can be established between analytes, then it is often possible to obtain a more exacting test. One example of this might be the use of the citrate/isocitrate ratio when assessing the authenticity of citrus fruit juice.

One of the great difficulties in ensuring the authenticity of natural products is that, whatever the measurand, a range of values for the authentic product may be anticipated. This inevitably offers the opportunity to dilute those products having an extreme value within the authentic range with an adulterant having a value outside the authentic range. It is possible to obtain information from one component of the food which sets a limit upon the range of authentic values that might be expected on the component of interest. This 'internal reference' approach has been used for honey where the carbon isotope composition of the protein in the honey acts as a marker for the correct carbon isotope composition of the sugars that represent its main components.

This 'internal reference' approach has application outside isotopic measurements. The preferred technique for determining whether meat that is being sold as fresh has in fact been previously frozen and thawed uses the measurement of an enzyme (β-hydroxyacyl-CoA dehydrogenase HADH). This enzyme is released from disrupted mitochondria. In principle, it is possible to measure the amount of enzyme and compare this against a database value. However, a superior approach is to measure the amount of enzyme from one sample of tissue, then freeze a second sample of tissue, thaw it, and see if a change in enzyme activity is observed. Such a comparative approach to ensuring authenticity is usually stronger than comparisons against a database of authentic values.

In principle, if several separate pieces of information can be linked, a method can be developed that will be more precise than that afforded by the individual determinations on their own. In some cases, it may prove possible to develop a test for a particular authenticity issue where the individual pieces of information on their own were insufficient for this task. This kind of approach requires sophisticated statistical techniques. This represents an area of considerable development in recent years and it is certain that many new uses of these statistical approaches will occur in the future. One that has recently achieved considerable publicity concerns the use of an 'artificial nose', which uses changes in conductance of 'doped' silicon chips by the aroma components of a sample. It is perhaps too early to expect validated methods from such an approach and some analysts are reluctant to use methods in which the identity of the measurand is not defined. Nevertheless, this type of approach is quick and simple and may have important benefits in process control. There may be benefits in attempting to validate methods using this kind of rapid test against currently accepted analytical methods. If the methods were shown to be equivalent, it might be possible to use them as screening techniques so that more expensive tests were undertaken only on borderline samples.

Statistical approaches are particularly appropriate for studying the authenticity of a population of samples. As already noted, a particular commodity will have a range of authentic values for a particular determinand. This range may be expected to have a Gaussian distribution and hence can be defined by a mean and standard deviation. Typically about 95% of authentic samples will lie within 2 standard deviations of the mean. However, if a series of samples lie close to one edge of this range, it is highly likely that the composition of these samples is being manipulated. It is difficult for companies or regulatory authorities to take action in these circumstances but these conditions would provide a strong indication to governments or the EC that further research is required to develop methods in this area.

Determining the authenticity of a food product after processing can be a particularly challenging task. This is not such a problem for producers since they will usually have purchased commodities in a minimally processed condition. It is more often of concern to distributors and enforcement agencies. As yet there have been few attempts to address this problem but the successes achieved so far suggest that the problems are far from intractable. An excellent example is provided by the measurement of deuterium/proton ratios on the methyl site of ethanol to determine the plant source of the sugar used for fermentation of wine. A second example is provided by the determination of authenticity of canned tuna using DNA techniques (see section 5.7). The high temperature and pressures generated in the canning process materially alter the fish structure, which make techniques used for fresh fish inappropriate. These techniques are interesting in that they are both measuring major components of the food under test. As yet the problem of determining the authenticity of one component of a multicomponent processed food has not been addressed.

1.9 Due diligence

It is not our intention to produce detailed guidance on the legal aspects of food authenticity and its enforcement. Should the reader require such advice, they are strongly advised to take appropriate legal counsel. Nevertheless, as much of the detail described in the following chapters concerns the identification of fraudulently described or altered food, it seems appropriate to outline the legal framework within which these tests must operate. While being conscious that most of our information and experience is only relevant to the UK, it may well be that other countries adopt similar provisions or will decide to do so in the future. The UK Food Safety Act 1990 introduced significant changes pertinent to food authenticity. 'Guidelines on the Statutory Defence of Due Diligence' were published in February 1991 by the Institute of Environmental Health Officers, The Food and Drink Federation, the Local Authorities Co-ordinating Body on Trading Standards, the National Consumer Council, the National Farmers Union and the Retail Consortium. These guidelines provide only an indication of how the law operates since, ultimately, the courts must decide how the law is interpreted. Below is provided a brief summary.

In the above act, 'falsely describing or presenting food' was one of a series of 'strict liability' offences. Evidence that such an offence had been committed was sufficient for conviction to be inevitable. To introduce a balance of fairness, a statutory defence was included. If the defendant could show that the offence was caused by an event outside their control, that they had performed reasonable checks on the food and that they did not know an offence was being committed they could establish their innocence. For a such defence to be successful, it is essential that the defendant establish that they had taken all reasonable precautions and exercised all due diligence.

McLaughlin (1994) gives guidance on how to set up such a 'due diligence' defence. There is, as yet, little case law relating to the use of this defence in food law but similar legislation has been in place for consumer law for some time. It has become clear that more efforts are expected from a large retailer than, for instance, from a small shop. Nevertheless, the defendant will need to be able to show that they took some positive action. The Hazard Analysis and Critical Control Points (HACCP) system is designed to identify and control all hazards to the food and is suggested as a good basis for a due diligence defence.

1.10 Introduction to the series

In this book, we have brought together the latest opinion on the authenticity of the most important food commodities. However, pressure of space precludes us from attempting to include more than brief details of the analytical methods in use. It is our intention to remedy this deficiency by producing a second volume, which will deal in detail with such techniques. The more simple techniques will be presented as 'tools of the trade'. Such detail will be less appropriate for the more complex, expensive or high-technology techniques. These will be presented in sufficient detail for the reader to comprehend their usefulness, their drawbacks and how they may be extended to other products. It will become clear from the following chapters that food authenticity is a constantly changing subject. As methods for the detection of adulteration improve, so new methods of seeking an economic advantage will be sought. We are therefore confident that the need to provide an update on the important issues in food authenticity and the need for new methods to resolve them will mean that further volumes will be required in the not too distant future. For the present, however, we trust you will enjoy discovering the breadth of scientific problems and the many approaches to resolving them outlined in this volume. The study of food authenticity represents a truly inter-disciplinary science.

References

The Food Safety Act (1990), HMSO, London.

Guidelines on the Statutory Defence of Due Diligence (1991), The Food and Drink Federation.

McLaughlin (1994) Due diligence and how to set up a defence. *Frozen and Chilled Foods*, March, pp. 42–48.

2 Authenticity of fruit juices, jams and preserves

D.A. HAMMOND

2.1 Introduction

The adulteration of foods and beverages has been a problem ever since we stopped producing all our own food and had to rely on commercial supplies. There have always been some unscrupulous suppliers who were prepared to maximise their profits by the extension of their products with cheaper ingredients. In the latter part of the 19th century there was allegedly extensive dilution of milk and beer with water and this led to the introduction of some of the first laws laying down compositional guidelines for foods and beverages.

The driving force behind any adulteration is the substitution of an expensive component with a cheaper ingredient(s), hopefully without the consumer being able to perceive the extension/substitution, in order to make additional gain. To illustrate this point, the cost of orange juice concentrate is typically in the region of US$1300–$1700/tonne, and sugar within the Common Agriculture Policy (CAP) of the European Community (EC) is *ca.* US$900/tonne and outside the EC *ca.* US$600/tonne. Since orange juice is typically 85% water, 10% carbohydrate and 0.8% acid, there is a significant cost advantage to be made by substitution of even just 10% of the juice solids by sugar. This offers the possibility of making significant sums of money because of the large volumes of juice involved. It is estimated that the fruit juice market in the UK, which is by no means the largest in Europe, is worth something in the region of £800 million per annum.

Since sugars are a major constituent of most of the commercially important juices the majority of adulterations will involve the substitution of juice solids with sugar. In the case of citrus juices, where the addition of second extracts of the fruit, such as pulpwash, are prohibited by EC regulations, such adulterants may also be involved. With the more expensive red/black fruit juices and purées (e.g. raspberry, strawberry and blackberry), adulteration can involve extension with less expensive juices as well as addition of sugars and acids. Apple, pear and grape juices have been used to extend red/black fruit.

2.2 Legal investigations

In the USA there have been several legal actions against companies and their directors for producing and selling adulterated fruit juices.

Perhaps one of the most notable cases was that carried out in the early 1980s when a company and some of its directors were charged with the production and sale of adulterated apple juice and found guilty. The chief executive received a prison sentence. In fact it was found that the 'apple juice' they were selling did not actually contain any apple juice at all, as it was actually a blend of sugar syrup, malic acid, caramel colour and apple flavour. The adulteration was initially detected by analysis, which showed a complete lack of chlorogenic acid and a distorted carbon isotope ratio. This gave the legal authorities enough evidence to obtain permission from the courts to search the factory and seize records. These records proved what the company had been doing and gave blending details of the cocktail that went into the product. They also found that the production records did not agree with the quantities of apple juice concentrate purchased by the company.

Several other cases have occurred in the USA more recently. In one case, a company pleaded guilty to the sale of adulterated orange juice and the company and its executive officers were fined US$9 million. In another successful case, the Federal authorities were able to prove that a second company were selling adulterated orange juice that contained beet sugar and pulpwash. In this case, the authorities discovered a syrup room behind a false electrical panel. During the course of the trial the owner tried to justify the room, with a false door, on the basis of a 'lack of space' for a real doorway. However, the jury did not accept his argument and, with the help of evidence from some of the employees, the owner was not only convicted with defrauding the consumer but also of conspiracy because of the actions the company had taken in order to hide their illegal procedures. In this case, the owner/chief executive was given a 9-year prison sentence for his part in the conspiracy.

One aspect that has assisted the authorities in obtaining criminal prosecutions in the US, is that, generally, the adulteration of the fruit juice has taken place in the packing factory. This has meant that there have been available records to substantiate the evidence of adulteration obtained from the chemical analysis and has shown that the factory did not receive sufficient juice solids to meet the production records, thus confirming that the product must have been diluted. Often there has been other evidence to show that the shortfall in juice solids has been made up by the purchase of sugar and pulpwash etc. However, this type of information is frequently unavailable, particularly in Europe, since the juice is often purchased from a supplier after extension with pulpwash and sugars, so there are no records to prove fraud.

During the late 1980s, there was some press coverage (Jones, 1988; Butler, 1988) in the UK that highlighted concerns about the quality/authenticity of the orange juice on sale. Around this time, the industry had set up a scheme to control the quality of orange juice in the UK market. One of the first actions that this scheme took was to request its members to introduce a minimum value for the refractometric solids content for orange juice prepared from concentrate. This value was set at 11 Brix, uncorrected for acidity, which was equivalent to the minimum adopted in several other control schemes throughout Europe.

Although the scheme was very successful at controlling the minimum Brix level of its member's juices, it was less successful in identifying adulterated products. Concerns over the authenticity of orange juice were finally exposed to the public by the UK Ministry of Agriculture, Fisheries and Food (MAFF) in February 1991, when they published data from a survey that had been conducted in mid/late 1990 (Anon, 1991a). The report concluded that there was systematic and widespread adulteration of the orange juice on sale in the UK market.

In the MAFF study, 21 orange juices were examined. Using a range of techniques (RSK, oligosaccharides, isotopic and fluorescence procedures), all of which will be discussed later in this chapter, 16 of the products were said not to conform to their labelling description as 100% pure unsweetened orange juice. Just before the MAFF results were published, officers of the Trading Standards Authority, the body through which some of the UK's food laws are enforced, searched and took a number of 'official' samples from several factories that were preparing juice. These samples were intended to lead to several criminal prosecutions against the producers and retailers selling these products. However, owing to a number of procedural and legal difficulties, rather than technical or scientific problems, all of these actions were dropped.

As a follow-up exercise to the original MAFF survey, further samples were taken in July 1991 by several Trading Standards departments throughout England. In this survey, most of the samples were found to be authentic; however, two products were found to be adultered and a successful prosecution ensued.

2.3 Analytical approaches

Although there is a constant search for a single component or method that can prove that a sample is real or adulterated, so far none has been identified or developed and it is possible that there will never be one that is universally effective for an indefinite period.

2.3.1 Single components

Although there have been many conventional analytes that have been used to detect specific types of adulterations, their usefulness is often short-lived. These have included such parameters as the formol value, total nitrogen, phosphorus, isocitric acid and D-malic acid.

D-malic acid is not found at any significant levels in juices or purées (Beutler and Ara, 1992) and so is a good marker for some types of adulteration. The enzymic method for measuring this substance has a good specificity and sensitivity, and a detection limit of *circa* 10 ppm in juices (Beutler and Wurst, 1990a; 1990b). When it was used as part of the MAFF study (Anon, 1991a), D-malic acid was detected in many of the orange juices. If this material is detected in a juice it usually indicates that D,L-malic acid has been added to the product to hide other adulterations, or for other reasons. In orange, where the level of L-malic acid is relatively low, typically around 1600 ppm, we detected small amounts of D-malic acid (typically between 25–250 ppm) in some orange juices at the time of the MAFF survey. Conversely, in apple juices, where the level of L-malic acid is much higher (typically 4–5 g/l) we found much higher levels of D-malic acid, anything up to 1500 mg/l, which suggests that about half of the acid was added.

D-malic acid has not been detected in any samples of apple, grapefruit, orange or pineapple juices since mid-1992. This might indicate that it would be worthwhile to drop the analyte from a screening protocol. However, if this material was to be detected in a sample of pure juice, it would instantly *prove* that there was a problem with the product and the company selling the product would have no defence to counteract any criminal action brought against them.

With the advent of an inexpensive and reliable method for the detection of D-malic acid, unscrupulous suppliers have had to switch their approach; which is the normal reaction to a new test. Since L-malic acid is commercially available, although more expensive than the racemic mixture, it is now straightforward to replace the synthetic acid with the natural isomer.

Before the production of the enzymic assay method it was very difficult to detect low levels of added D,L-malic acid in citrus juices. However, in apple there was a well-developed method for determining the approximate level of D,L-malic acid. This procedure involved the assessment of the level of fumaric acid (Junge and Spandinger, 1982) and the total acids and L-malic acids (Zyren and Elkins, 1985). Fumaric acid, which is the dehydration product of malic acid, is often found as a low-level contaminant (up to *circa* 1%) in malic acid and it is possible to detect it in apple juice at leess than 1 ppm. Thus, if a sizeable concentration of fumaric acid is detected in an apple juice together with a difference of greater than

10% between the L- and total malic acid levels, this is considered indicative of adulteration.

Another single component that can be extremely useful for identifying adulteration in some red/black fruits is sorbitol. In fruits such as raspberry, strawberry, blackcurrants and redcurrants and blackberry, sorbitol is either absent or present at a very low level. The presence of sorbitol in these fruits is an indication that they have been extended with juices from apple, pear, plum or cherry.

In some red/black fruit pulps, where sorbitol should be undetectable, we have found up to 2.5 g/l. This indicates that up to *circa* 50% of the soluble acids could be coming from apple or *circa* 20% from pear, plum or cherry. The presence of apple and/or pear can also be confirmed by the analysis of the polyphenol profile as there are key markers that are indicative of apple or pear addition (phlorizin and arbutin, respectively). This will be discussed in more detail later in the chapter.

Another component that can be used to detect adulteration is tartaric acid. Grape juice is unique among the commonly used juices in containing tartaric acid so that if a juice/purée is extended with grape juice, tartaric acid should be readily detectable. However, since de-acidified ('rectified') grape juice is commercially available, the detection of an addition of this type of juice is more difficult because of the low level of tartaric acid remaining.

2.3.2 Detection of adulteration with sugars

As already mentioned, the extension of juices often involves the addition of sugars and water. The conventional approaches used to detect whether sugar has been added to a juice/purée are now discussed.

One method, used as part of the German RSK profile, is detection by inference. For instance, if there has been a general dilution of the non-sugar components and the 'sugar-free extract'[1] is low, or there is an elevated level of sugars relative to the soluble solids content (Brix), these features would all indicate sugar addition.

Another more direct approach relies on the distortion of the sugar profile, in other words distortion of the concentrations of sucrose, glucose and fructose. Most juices have characteristic proportions of these sugars, where distortions from these values can be used as an indication of either adulteration with a sugar or possibly the addition of another juice. A list of the typical proportions of the main sugars found in several juices are given

[1] The 'sugar-free extract' is the non-sugar component of the juice and is calculated from the density and the total level of sugars. The density is used to give the soluble solids content, the true Brix, and the total level of sugars (measured by HPLC, enzymic or other procedures) is subtracted from this value.

in Table 2.1. This table has been drawn up using several sources and gives the 'normal' values found in a range of juices. The values are not all encompassing and are intended only for illustration of the different patterns that occur.

For example, if a raspberry juice concentrate contained a high concentration of sucrose, more than 10 g/l at single strength equivalent, this would indicate that between 10 and 20% of the sugars had been added from another source. Another example would be that if an apple juice had a glucose:fructose ratio of 0.7 it would indicate that about 40% of the sugars in this sample came from an invert syrup.

By the judicious choice of the appropriate sugar mixture or invert syrup, it is possible to mask most of the distortions in the sugar profile. A clear example is the addition of beet medium invert syrup (BMIS) to orange juice. BMIS is prepared by the acid- or enzyme-catalysed hydrolysis of beet sucrose to give a mixture containing two parts of sucrose to one part of glucose and fructose, which happens to be roughly the proportions found in orange juice. As the sugars are now present in approximately the correct proportions, it makes it impossible to detect the addition of this type of syrup to an orange juice simply from the sugar profile. However, these types of syrups have other useful fingerprints that can be used to detect them.

2.3.3 *Oligosaccharide analysis*

In the MAFF survey of 1991, one of the new methods used in the study was a procedure to detect the addition of invert, high fructose corn syrups (HFCS) or glucose syrups (GS) by the presence of unusual 'oligosaccharides' (di, tri- and tetra-saccharides). This procedure was developed by Low and Swallow (1991) in Canada and relies on the formation of such oligosaccharides during the manufacture of the sugar syrup (inversion of

Table 2.1 Normal values for the major sugars found in a range of commercially important juices and purées (values given in g/l)[a,b]

Type of juice	Sucrose	Glucose	Fructose	G:F ratio
Orange	40	23	25	0.9–1.0
Apple	14	25	60	0.3–0.5
Grapefruit	25	23	23	0.9–1.0
Pineapple	45	25	25	0.9–1.2
Raspberry	Trace	23	29	0.7–0.9
Strawberry	10	23	24	0.7–0.9
Blackberry	Trace	33	34	0.8–1.0

[a]Trace = 0–5 g/l. [b]*Sources*: Wrolstad and Shallenberger (1981); Sourci *et al.* (1981); Anon (1987); Hammond (unpublished data).

the sucrose or the isomerisation of glucose to fructose) at a high solids level. These fingerprints oligosaccharides can be used as markers to identify if an orange juice contains an invert syrup. The oligosaccharides were analysed using high-pH anion-exchange chromatography coupled with a pulsed amperometric detector (HPAEC-PAD). Although the original procedure was time-consuming, various adaptations have since been made (White and Cancalon, 1992; Wiesenberger *et al.*, 1992; Hammond, 1993; Koziet and Bricout, 1994), which reduce the analysis time.

These procedures not only detect BMIS but also pick up the addition of any syrup where either inversion or isomerisation has been carried out at a high solids level. Under these conditions there is a limited amount of 'free', or unbound, water and so not only does hydrolysis occur but, in some reactions, instead of water acting as a nucleophile to displace one of the sugars, another sugar molecule reacts instead to form a di- or trisaccharide. Examples of the pattern of the oligosaccharides found in BMIS, high-fructose corn syrups (HFCS), a pure juice and one adulterated with 10% BMIS, are given in Figures 2.1–2.4, respectively. The peaks of interest are in the region of 15–25 min.

The analysis of juices for unusual oligosaccharides has three main limitations. Firstly, it will not detect the addition of pure sucrose, either from a cane or beet source. This is because these compounds have been highly purified and do not contain a significant level of oligosaccharides.

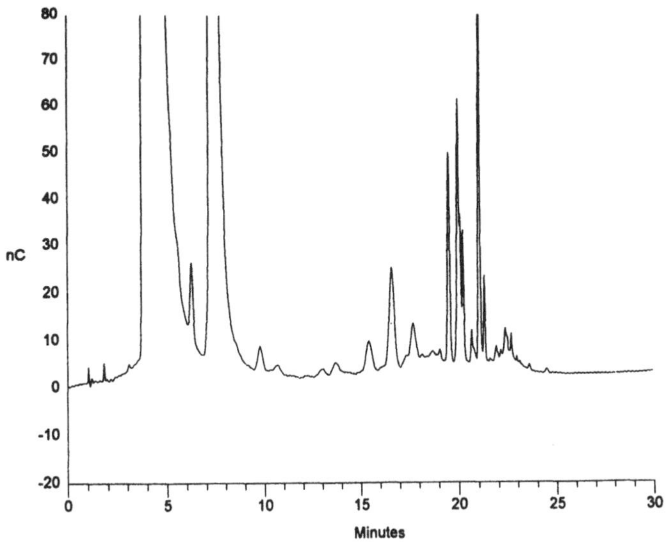

Figure 2.1 Oligosaccharide profile of a beet medium invert syrup.

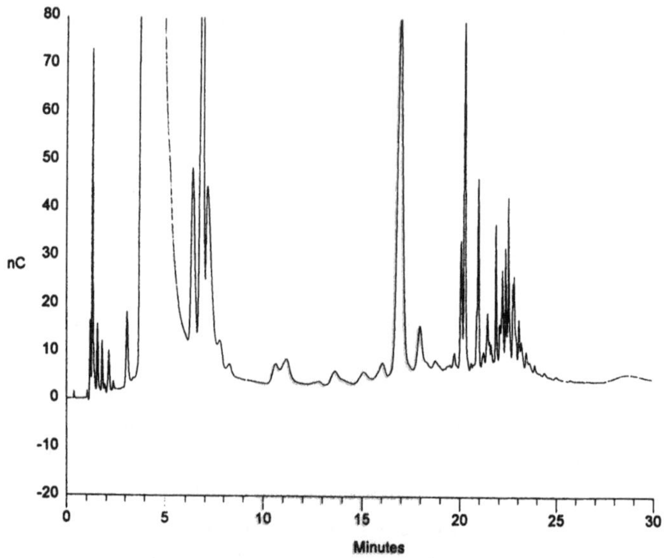

Figure 2.2 Oligosaccharide profile of a high-fructose corn syrup.

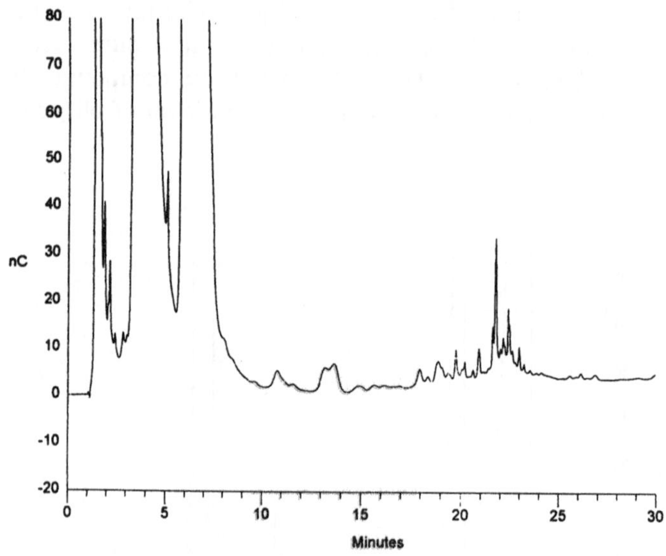

Figure 2.3 Oligosaccharide profile of a normal orange juice.

Secondly, if the syrup has been prepared at a low 'solids' content, it will contain very low levels of the fingerprint oligosaccharides. Although it is rumoured that syrups that have clean oligosaccharide profiles are available, the author has not seen any. The final and perhaps the most serious limitation of this type of approach is that, if a juice concentrate is

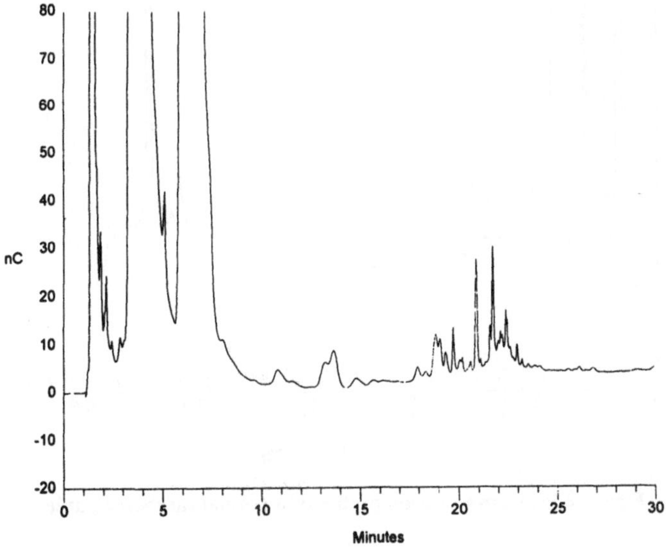

Figure 2.4 Oligosaccharide profile of an orange juice adulterated with 10% BMIS.

heated, a pattern of peaks similar to those found in sugar syrups is formed. This makes the interpretation of complex oligosaccharide profiles more difficult but does not wholly negate the technique. If a complex pattern is found in a juice, other markers can be checked to see if these are indicative of excessive thermal treatment. For instance, in studies on orange juice concentrate, the author has found that there is roughly a linear correlation between the total area of the peaks found in the oligosaccharide region and the level of hydroxymethylfurfural found in a heated juice.

Another situation where the oligosaccharide profile can be made more complex is when enzymes are used to assist in the clarification of a juice. This is the normal route where juices are destined to be made into clear concentrates (e.g. apple, pear, raspberry etc.). In this situation, oligosaccharide analysis is still of use but, as with heated concentrates, care is needed in the interpretation of a complex pattern. If a glucose syrup has been used to extend a juice, not only will the sample have a complex oligosaccharide pattern similar to that found in high fructose corn syrup (see Figure 2.2) but it will also contain maltose, which is not a normal component of juices (Figures 2.5 and 2.6).

There are several laboratories that are, at present, looking at the use of capillary gas chromatography to detect the presence of unusual oligosaccharides in juices. This approach is reported to have certain advantages over the HPAEC-PAD method and to be able to detect invert syrups at lower levels (Ooghe, 1994; Low, 1995) (see note at end of chapter).

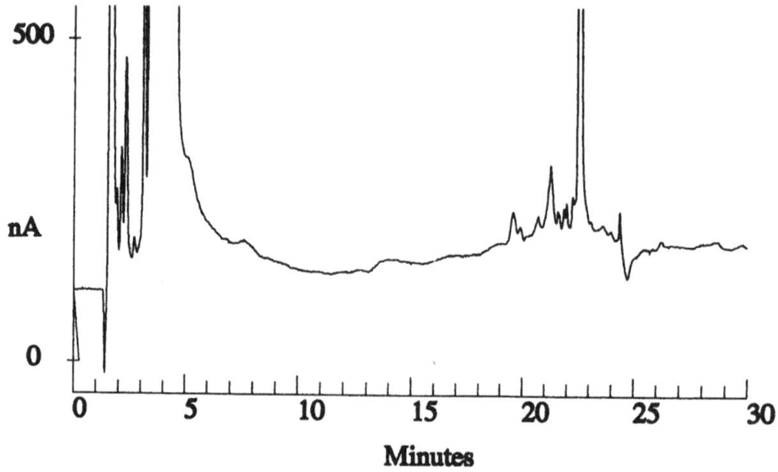

Figure 2.5 Oligosaccharide profile of a normal raspberry purée.

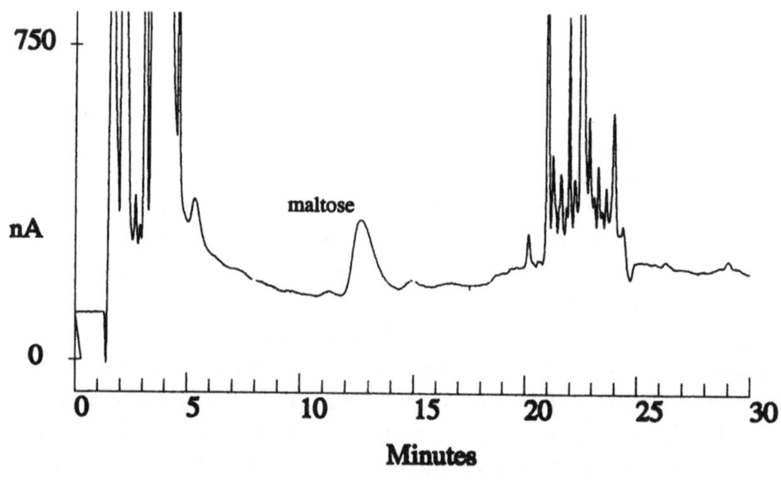

Figure 2.6 Abnormal oligosaccharide profile showing addition of glucose syrup to a raspberry purée.

2.3.4 Isotopic procedures

In the context of fruit juices, the elements of interest are carbon, hydrogen and oxygen. These three elements exist in several isotopic forms, which are present in known quantities, Table 2.2. These values express the normal isotopic balance of the various forms of the elements. However, it is possible for these proportions to be modified by isotopic fractionation during some biological and physical processes. This fractionation occurs

Table 2.2 Typical isotopic abundances for hydrogen, carbon and oxygen[a,c]

	Element								
	Hydrogen			Carbon			Oxygen		
Atom	^{1}H	^{2}D	^{3}T	^{12}C	^{13}C	^{14}C	^{16}O	^{17}O	^{18}O
Atomic mass	1	2	3	12	13	14	16	17	18
Proportions (%)	99.985	0.015	[b]	98.904	1.096	[b]	99.763	0.037	0.2

[a] *Source*: Heath (1974). [b] These isotopes are radioactive and are absent in nature or present at very low levels. [c] Atomic masses are given as whole numbers.

when there is competition between isotopes in the rate-determining step of a reaction/process. The heavier isotope will always react at a slower rate and this modifies the 'normal' values; this general behaviour is the guiding principle behind all of the isotopic procedures. There have been several general discussions of the application of these type of procedures in the detection of adulteration of juices (Bricout, 1982; Krueger, 1988; Doner, 1991). However, owing to the usefulness of these procedures in detecting adulteration, a brief background will be given here with some of their strengths and limitations.

2.3.5 *Stable carbon isotopic analysis*

Most scientists are probably familiar with radiocarbon dating of organic materials using the ^{14}C isotope. However, in adulteration studies, one is interested in the two stable, non-radioactive, carbon isotopes ^{13}C and ^{12}C. Craig (1953) discovered that different plants contained different amounts of ^{13}C, although at that time, it was not clear why this should happen. In the late 1960s Hatch and Slack (1970) identified a new pathway for the fixation of carbon dioxide from the atmosphere which rationalised these differences. Two more extensive surveys of the ^{13}C contents of plants (Bender, 1971; Smith and Epstein, 1971) found that all plants are depleted in ^{13}C relative to the atmospheric carbon dioxide that they used due to fractionation of the ^{13}C and ^{12}C isotopes during fixation of CO_2 by the plants.

As the differences in concentrations of the ^{13}C and ^{12}C atoms are small, in absolute terms, they are normally expressed using the delta notation (δ) in parts per thousand (‰ or permil) and are expressed relative to the carbon isotope ratio in a standard limestone, PeeDee Belemnite (PDB).

Several workers have shown that $\delta^{13}C$ values can be used to detect the source of the sugar in a given product. Owing to the availability of relatively cheap cane sugar and sweeteners derived from corn in the USA, it means that these materials are frequently used to extend juices and purées. However, because of the characteristic ^{13}C fingerprint of this sweetener, the use of carbon stable isotope ratio analysis (CSIRA) has been more widely used in the USA than within the EC. This can be illustrated clearly by the fact that there was a fully accepted method for juices published in 1984 by the Association of Official Analytical Chemists (AOAC) (Anon, 1990c; 1990d), yet it has taken authorities within the EC until 1995 to adopt and publish a similar procedure as a European standard. (A method to measure the ^{13}C content of fruit and vegetable juices will be published shortly as a temporary European standard.)

Another possible cause for the lack of interest in the isotopic procedure results from the EC common agricultural policy, which dictates that the price of sugar from non-beet sources is artificially higher than in the rest of the world. Thus the use of corn-derived sweeteners is comparatively rare within the EC.

Although the ^{13}C method can detect the addition of cane/corn sugars, it is not possible to distinguish between the sugars from a juice (e.g. orange or apple) and those from beet sucrose as sugars in these plants are synthesised by the same C_3 route. This means that, in sophisticated adulteration, the unscrupulous supplier is likely to use beet sugars in preference to any other, although it might be more expensive. However, since there is little or no difference in the price of cane, corn or beet-derived sugars in the EC, beet would normally be used.

It has been found that the organic acids in juices have similar ratios to the sugars. This has meant that the carbon isotope approach can also be applied to detect the addition of citric acid (prepared from cane or corn sugars via microbiological fermentation) to lemon juices (Doner, 1985). In studies in RSSL, the author has found that this approach has been very useful in confirming the adulteration of lemon juices that were believed to be adulterated based on other analytical data.

One way in which the sensitivity of the ^{13}C method can be improved is to use internal markers and compare the value of one component with another. This approach is discussed using several applications by Schmidt *et al.* (1993). It is used quite commonly in honey where the sugars and protein $\delta^{13}C$ values are compared to ensure that no HFCS has been added. This type of comparison has also been applied to juice where the soluble solids and the pulp are used (Bricout and Koziet, 1987; Parker, 1982).

It has also been found that stable carbon isotope ratio analysis can be useful when detecting the adulteration of certain flavour compounds. For instance a stable isotope ratio mass spectrometer has been coupled with a gas chromatograph (GC) (Schmidt *et al.*, 1993), enabling the measurement of the different proportions of the two isotopes in the flavour compounds as they elute from the capillary GC column.

2.3.6 Oxygen isotope analysis

Craig (1961a; 1961b) found that there were characteristic differences in the levels of ^{18}O relative to its ^{16}O isotope in waters taken from different parts of the world. The largest difference was detected between water taken from the equatorial region and from the poles. The level of these two isotopes is determined by the extent to which they are fractionated during evaporation and condensation. The light isotope evaporates more rapidly at the equator and when this water condenses and finally falls as rain at higher latitudes it is depleted of the heavy oxygen isotope.

The ^{18}O:^{16}O ratio in plants, is in general, dependent on where the plant grows, local latitude and rainfall patterns etc. and, in particular, by diffusion across membranes and by evaporation at the leaf/air boundary. In the latter case, the lighter isotope is lost preferentially and so the water becomes enriched in the heavy isotope relative to the groundwater. The extent of this enrichment is dependent on the plant physiology and local climatic conditions with values around 5‰ higher in several fruit types – oranges (Bricout *et al.*, 1972; Nissenbaum *et al.*, 1974) apples (Bricout *et al.*, 1973; Dunbar and Wilson, 1983) and grapes (Dunbar, 1982) – to that found in the local groundwater.

Significant illicit profits can be made if a supplier or retailer sells orange juice from concentrate as top quality 'not from concentrate' juice (NFC). Although this is unlikely in practice, owing to the significant differences in the taste of the two products, it is possible to differentiate NFC orange juice from reconstituted concentrate by depression of the ^{18}O content caused by the addition of groundwater.

The final process in which isotope fractionation occurs is in the commercial concentration of juices where there is again a preferential enrichment in the heavy isotope. Although orange juice would typically have a $\delta^{18}O$ value of *circa* 4‰, in a 65 Brix concentrate, this value will have increased to around 15‰ (Brause *et al.*, 1984; Brause, 1992). This gives the possibility of detecting the addition of groundwater to the concentrate at 65 Brix as well as at natural strength (as already discussed). A likely reason for groundwater appearing in a juice concentrate would be if sugar(s) or pulpwash had been added to 'extend it'. This type of addition could be detected if these solutions/suspensions were prepared in ground water. The use of $^{18}O/^{16}O$ in the water of a juice is another method that has received more attention in the USA than in the EC.

The use of ^{18}O:^{16}O ratio as a method of detecting the addition of groundwater/sugars to juices and concentrates has been applied successfully to orange and grapefruit but it is less effective in apple as the natural juice is normally slightly depleted relative to the reference standard (Standard Mean Ocean Water, SMOW), whereas the citrus juices are enriched.

This procedure gives the analyst another weapon in his armoury of tests to use; however, as with most if not all other approaches, there are ways around it. If water is recovered from a juice evaporator, or similar process, it is enriched in the heavy isotope, relative to groundwater. This water would then be used to mask the addition of sugar/pulpwash to a concentrate (Doner *et al.*, 1987).

2.3.7 Deuterium isotope analysis

As with the ^{18}O isotope, the deuterium content has been used as a marker for the addition of groundwater to juices (Bricout, 1973). Methods to differentiate between NFC and juice from concentrate were very important for the French market where the regulations stipulated that 'pure' orange juice should not be made from concentrate. Although this has now been changed, bringing the French regulations in line with the rest of the EC, it is perhaps one reason why there has been more interest in isotopic methods in France than in the rest of Europe.

In the sugar molecule there are two types of hydrogen/deuterium atoms. Those that are attached to oxygen (i.e. the hydroxyl protons) are readily exchangeable with the surrounding water, which make them unsuitable for this application. The other hydrogen/deuterium atoms are actually bound to carbon and therefore far less exchangeable, which means they can be of analytical use. Before any analysis of the 'fixed' hydrogens is possible, the exchangeable ones have to be removed so that they do not cause anomalous results. There have been two solutions to this problem, one converts the sugar into a derivative without any free hydroxyl groups. The most commonly used derivatisation is the formation of nitrate esters of the sugars, which are then analysed using isotope ratio mass spectrometry (IRMS) (Dunbar and Schmidt, 1984). This method has certain limitations as nitrate esters can be difficult to prepare cleanly and effectively, they are also not very stable and are liable to explode. Although there are limitations with this method, it has been used by some workers in the field to produce meaningful and reliable results. Distinguishing beet sugar from sugars obtained from other sources, such as juices and cane, has been demonstrated in several studies (Bricout and Koziet, 1985; Dunbar *et al.*, 1983; Dunbar and Schmidt, 1984; Doner and Phillips, 1987).

The second approach is to remove most of the exchangeable protons by converting the sugars into ethanol by yeast fermentation. This was first investigated by Bricout *et al.* (1975) and Rauschenback *et al.* (1979). In their studies, the overall deuterium:hydrogen (D:H) ratio in the resulting ethanol was measured by IRMS and they found that they were able to differentiate between different sources of sugars.

This approach was developed further by Martin *et al.* (1982; 1985; 1986) for wines and later fruit juices (Martin, 1993). However, instead of using IRMS to measure the overall isotopic ratio in the molecule, these workers used high-field deuterium magnetic resonance (DMR) spectroscopy, which enabled them to measure the ratios at the different sites in the ethanol molecule CH_3–CH_2–Oh. The hydroxyl hydrogen is exchangeable with the solvent and is therefore not significant. However, solvent hydrogen atoms are also incorporated into the ethanol molecule during its biosynthesis.

This does not mean that the technique is not a valid way to detect the addition of, for example, beet sugars to citrus juices but it is one reason why the information from the methyl site is more useful than the methylene site. This 'water' effect can be allowed for if the D/H value in the fermentation broth is measured.

In the DMR procedure, the juice is prepared at single strength and a yeast is added. In the original method for wine (Anon, 1990a), it was suggested that an amino-acid-free nitrogen base should also be added to the juice. However, in a recent ring test, conducted under the auspices of the EC technical committee for standardisation of methods for fruit and vegetable juices (CEN TC/174) and AOAC, it was indicated that no exogenous nitrogen source should be added to the fermentation medium, as this would limit hydrogen/deuterium exchange.

In this method, it is essential that a complete fermentation takes place, giving better than 95% conversion of the sugars into ethanol, to eliminate any isotopic fractionation in the process (Moussa *et al.*, 1990). The fermentation process can be monitored by high-powered liquid chromatography (HPLC) of the sugars, when the sugars are exhausted the ethanol is recovered by distillation. As in the fermentation step, it is critical that there is no isotopic fractionation in this process and so the majority of the ethanol (›95%) must be recovered. The purity of the ethanol is measured and a sample of pre-calibrated standard (tetramethyl urea – TMU) is added to the ethanol prior to analysis by DMR. The size of the peaks relative to the TMU are measured and, from this, the actual level of deuterium at the different sites can be calculated. A typical DMR spectrum of an ethanol derived from an orange juice is given in Figure 2.7. The origin of the peaks are, from right to left, the CH_2D, TMU, CHD and OD. From the deuterium level found at the different positions, it is possible to assess if sugars from another source have been added to the sample, since the addition of beet sugars depresses the deuterium content at the methyl site, whereas addition of cane sugars has the opposite effect.

Owing to the cost of the high-field NMR spectrometer, there are very few laboratories using this technique and very little data has been published on the values expected in juices and in sugars from other origins. A comparison of the isotope ratios found in different sources of ethanol is, however, given in Table 2.3. By using either the overall ratio, as determined by IRMS, or the level at the different sites in the ethanol, it is possible to differentiate between various sources of sugars.

Applying this technique in our own laboratory we have found that beet

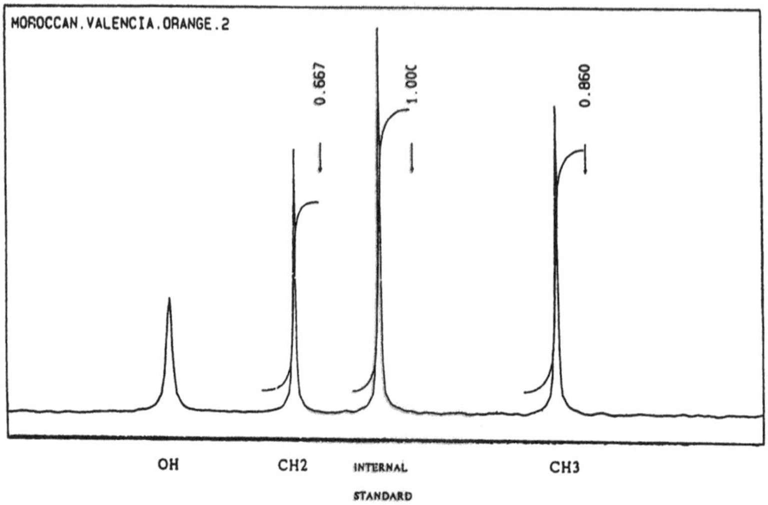

Figure 2.7 Typical DMR spectrum of ethanol derived from an orange juice.

Table 2.3 Typical carbon and hydrogen isotope values found in ethanols from various sources[a–d]

Source	D_1	D_2	Avg D(IRMS)	$\delta^{13}C$
Sugar cane	112.0	127.7	124.2	−12.2, −12.6, −11.6
Corn syrup	110.6	122.9	122.8	−10.2, −10.4, −11.9, −10.3
Sugar beet	92.7	124.1	125.3	−26.7, −27.2, −28.0
Tubers	97.2	127.1	113.2, 117.8	−26.9, −26.3
Synthetic ethanol	126.8	144.0	135.4	−31.0, −28.0, −27.4, −31.0
Fruits	97.2	125.2		−26.6

[a]*Source*: Modified from Martin *et al.* (1991).
[b]D_1 is the deuterium isotope value at the methyl site in the ethanol.
[c]D_2 is the deuterium isotope value at the methylene site in the ethanol.
[d]Avg D is the average deuterium content in the ethanol molecule.

sucrose typically has a deuterium content around 92 ppm. This is significantly lower than the values found in citrus juices from concentrate (typically around 103–105 ppm) and for cane sucrose (around 111 ppm). However, if the correct balance of beet and cane sugars is used, the DMR technique alone can not detect the adulteration. This is the reason why the technique should always be used in conjunction with carbon stable isotope analysis (CSIRA) to limit the addition of cane sugars.

Although we have found that the technique is very good at detecting beet sugar added to citrus juices, the sensitivity of the technique is much lower in juices prepared from fruit grown in cooler climates, for example apples. The use of sugar blends can also significantly affect the detection limit.

The traditional and isotopic procedures already discussed are equally appropriate for the detection of the addition of sugars to juices as well as to fruit purées, such as those used in jams and preserves. However, these procedures are ineffective in the finished product as these already contain added sugars that would distort the results.

2.3.8 Biological approaches

Other approaches to detect sugar addition have also been investigated over the last 5–10 years. Several groups have worked on enzyme-linked immunosorbent assay (ELISA) techniques, with one group actually patenting their method (Potter and Mansell, 1992). However, some of the methods are not always so robust as they might be. Another more recent biological approach, which is presently under investigation in several laboratories, is to determine specific DNA fragments that could be used to confirm the addition of beet sucrose. This work is still under investigation and it is not yet clear how successful it will be.

The application of DNA technology to the characterisation of fruit juices has been found to work well in one area. Workers at the Leatherhead Food Research Association (LFRA) in the UK demonstrated that DNA fingerprinting could be used to characterise Ambersweet juice (Patel, 1993). This would be of interest in Europe because this fruit is a hybrid cross between an orange, tangerine and sour orange, which would not be accepted as an 'orange' within the EC. However, the USDA and FDA in America have accepted Ambersweet as an appropriate fruit from which to make orange juice concentrate and, in a few years, when sufficient acreage has been planted in the US and other areas of the world, juice from this fruit may start to appear in 'orange' juice. Under the present EC regulations, the sale of orange juice that contains this hybrid is illegal and a method to detect its presence in a juice could be important.

The whole area of what constitutes 'orange' was the subject of a seminar organised by the International Federation of Fruit Juice Producers (IFU) in Parma, Italy, in September 1994. During this seminar, several papers were presented, giving the view of experts and interested parties in the area as to the precise definition of orange juice (Anon, 1994c). Although it is easy to define *Citrus sinensis*, it was much harder to decide what orange juice is and where some of the new fruit hybrids fit in with this definition. No firm resolution emerged from the meeting as to the status of the new hybrids and whether these should be allowed in orange juice without declaration. This whole area is under continued discussion, and at the time of writing, it is anticipated that an IFU expert committee will eventually produce an informed statement for the guidance of the fruit juice industry.

2.3.9 *Second extracts of fruit*

Citrus pulpwash and its detection remains a problem in several countries, especially those within the EC where the addition of both 'in-line' and 'off-line' pulpwash is prohibited. In other countries, such as the USA, the detection of the addition of 'off-line' pulpwash is the problem.

Pulpwash is prepared from the pulp screened out of the juice by the finishers. 'In-line' pulpwash is prepared by mixing the pulp with water in order to recover the small percentage of occluded fruit juice. After mixing for a while the water/pulp mixture is finished again to obtain the juice/water mixture. This material is then added back to the juice stream prior to concentration. If this process is carried out 'off-line' the soluble solids obtained from the pulp are often treated with enzymes to degrade some of the pectin, which allows the material to be concentrated to between 40 and 60 Brix. The normal and legitimate use of a pulpwash concentrate is in soft drinks production where it can be used to give 'cloud'. However, it can, in the hands of an unscrupulous supplier, be added to extend juice. Secondary processing can be taken a stage further, with the extraction of the peels as well as the pulp to make second extracts of fruit. Again these products are quite legitimate in the soft drinks industry as high-intensity natural cloudifiers but should not appear in 100% juice products.

Several methods have been proposed to detect the addition of pulpwash to juice; the major ones used in orange juice involve the detection of elevated levels of components that are found at higher levels in pulp/peel than in the juice, such as hesperidin, calcium and pectin. However, as there are large natural variations of these components, confirmation of pulpwash addition is often difficult.

Petrus and Attaway (1980; 1985) published a method to detect the addition of citrus pulpwash to Florida orange juice. This method involved the preparation of an ethanolic dilution of the juice and its evaluation by ultraviolet/visible absorbance and fluorescence spectroscopy. They found that the absorbance values at 443, 325 and 280 nm together with the sum of these values and the ratio of the values at 443 and 325 nm could be used to assess if a juice contained pulpwash. The test relies on the fact that the soluble solids extracted in the pulpwashing process are depleted in carotenoids and have enhanced levels of polymethoxyflavones and flavonoids. As pulpwash is added to juice the absorbance value at 443 nm falls and the values at 325 and 280 nm increase. The absorbance profile of the juice is also distorted as the inflection at 280 nm becomes more noticeable as a peak if pulpwash has been added. In the second part of the test, the fluorescence profile of the juice is altered by pulpwash addition.

This technique has been used to examine juice from sources other than Florida, and there has been considerable discussion as to the usefulness of the procedure and what the acceptable values for juices should be (Saguy and Cohen, 1984a; 1984b). Although juices from different origins have slightly different absorbance characteristics, the technique has some value. The fluorescence portion of the essay can, however, be difficult to replicate with some modern types of fluorescence spectrometers as they do not produce exactly similar profiles to those published in the original work.

One way that unscrupulous producers have tried to get round this method has been by adjusting the colour of the product to mask the suppression caused by the pulpwash addition. Several materials have been used to mask pulpwash addition for example, β-carotene, β-apocarotenal, annatto and turmeric extracts. The first two will be discussed later in the section on chromatographic techniques. Annatto, a natural carotenoid and turmeric can be detected by a HPLC procedure developed by Rouseff (1988).

Another technique that has been proposed to detect the addition of pulpwash in orange juices, involves determining the relative levels of hesperidin and narirutin (Wade *et al.*, 1992). Workers at Procter & Gamble in the USA have found that the addition of even small amounts of pulpwash significantly affected the narirutin:hesperidin ratio. The typical value that they quoted for orange juice was given as 0.4 and this ratio was elevated by the addition of pulpwash. A ratio of 0.721 was given for a juice that contains 20% pulpwash. This phenolic fingerprinting technique will be discussed in more detail later in this chapter. In our laboratories, the author has confirmed that this technique is very useful at detecting the addition of pulpwash to juices, although higher ratios were found in hard-pressed juices from certain types of fruit. Some fruit grown in the Mediterranean region, in particular the Shamouti orange, can also have higher ratios than those reported by Wade.

With all the techniques mentioned above it is often difficult to differentiate between juice that has been prepared from fruit where too high an extraction pressure has been used and one where pulpwash has been added. However, if the fruit has been over-pressed this will give a substandard juice and the product should be rejected on quality grounds as it may taste bitter or 'burnt' from the incorporation of components from the peel.

Another technique that has recently been proposed to detect the addition of enzyme modified pulpwash or peel extracts to juices is the use of high pH anion-exchange chromatography, coupled with a pulsed amperometric detector (HPAEC-PAD) (Farnell and Biesenbruch, 1994; Hammond, unpublished data). (This is similar to the procedure used by Low and others to detect trace levels of oligosaccharides in juices.) HPAEC-PAD is capable of detecting fragments of pectin at low levels, which are formed during the enzymic degradation used in production of pulpwash concentrates or peel extracts. Early work conducted by Professor Pilnik's group at Wageningen in Holland (Voragen, 1972) showed that, when pectin was degraded with pectolytic enzymes, oligogalacturonides with less than 10 units remained in solution during the ethanolic precipitation used in the first stage of the normal pectin assay procedure. If an enzyme-treated pulpwash is added to a juice, the small oligogalacturonide units are not precipitated by the ethanol used in the conventional analysis technique and so these units are discarded during the washing procedure. This loss suppresses the pectin level detected in the juice so that the sample appears normal.

Although the author's method takes considerably longer than that proposed by Farnell and Biesenbruch (1994) (80 min versus *circa* 30 min), it allows both the oligosaccharide and oligogalacturonide profiles to be monitored in one run. It is clear from our studies that different patterns of oligomers occur with different balances of enzyme (e.g. pectin depolymerase, esterase and lyase) found in commercial pectinase preparations.

With some enzyme preparations, a very regular pattern of oligogalacturonides is obtained as seen in Figure 2.8. With other commercial enzyme preparations the patterns are simpler, as seen in Figure 2.9. Similar patterns can also be detected in commercial enzyme-treated pulpwash and also in juices that have been adulterated with a low level of pulpwash (Figures 2.10 and 2.11, respectively).

Although this technique provides a further procedure to detect the addition of pulpwash to juices, it is only effective if the pulpwash/peel extract has been treated with pectolytic enzymes so that galacturonic acid and oligogalacturonide units are produced. This approach does not address the issue of 'in-line' pulpwash, which is an area of concern following

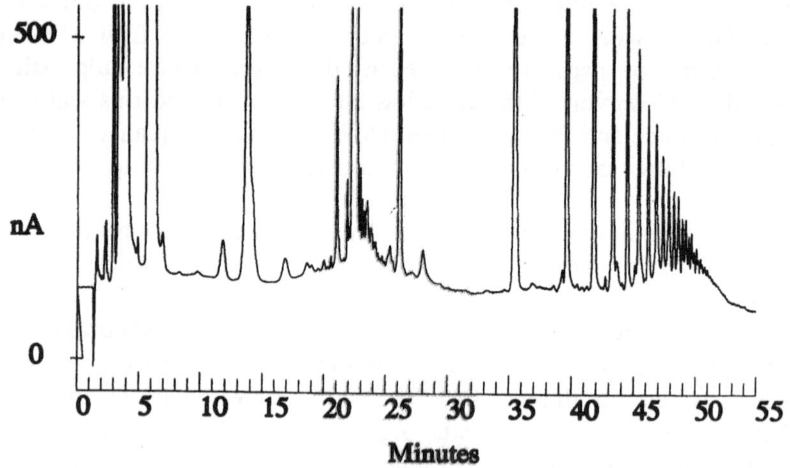

Figure 2.8 Pattern of oligogalacturonides found in a citrus pectin after treatment with Pectinol C.

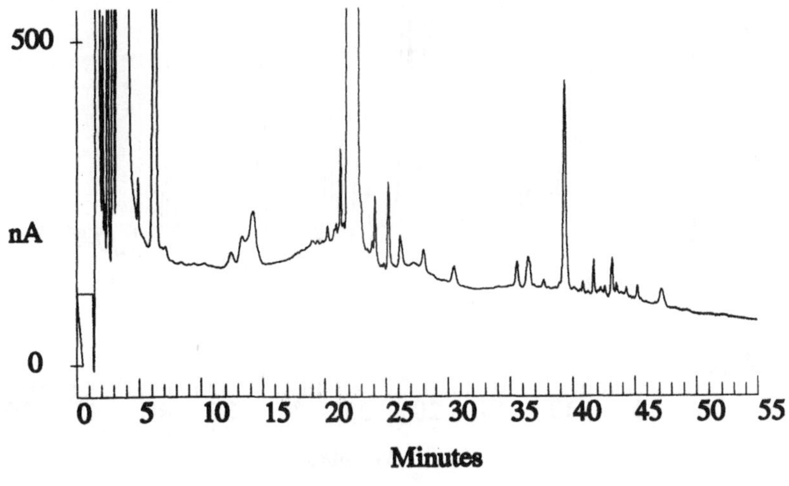

Figure 2.9 Pattern of oligogalacturonides found in a citrus pectin after treatment with pectinase.

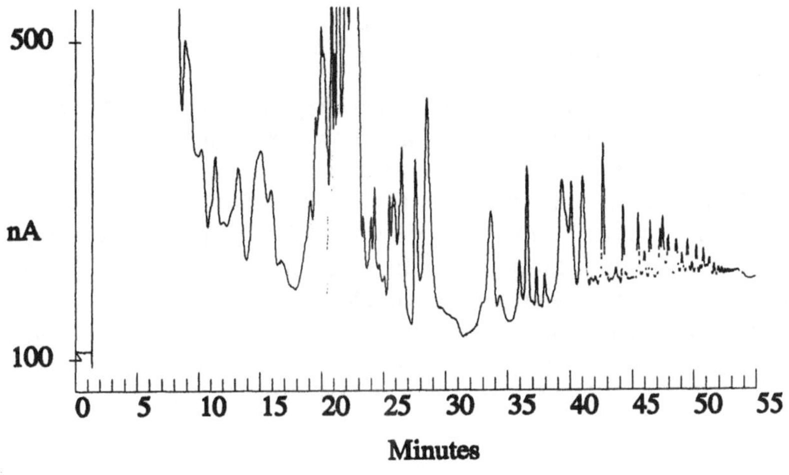

Figure 2.10 Pattern of oligogalacturonides detected in an enzyme-treated pulpwash concentrate.

changes in legislation relating to the addition of pulpwash to juices in Florida. This state had for many years prohibited the addition of any type of pulpwash to juice prepared within the state boundaries. However, owing to action by the Federal authorities in the US this regulation has now been countermanded and so it is now legal to add 'in-line' pulpwash to orange juice made in Florida. In a few years time it may be impossible for the EC to obtain orange juice without added pulpwash from any country.

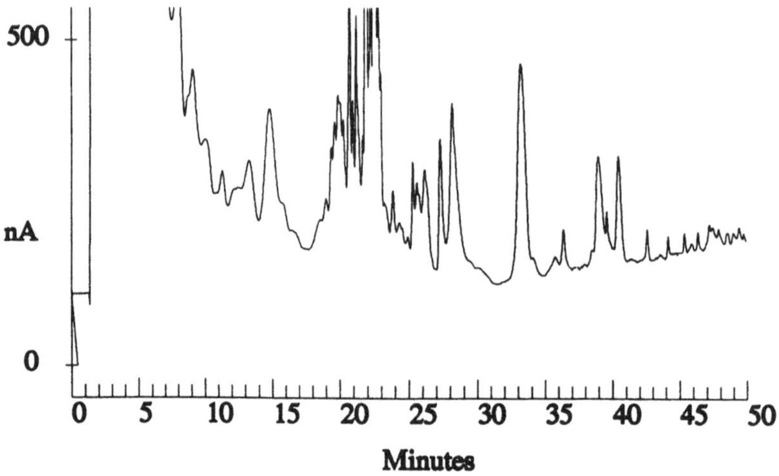

Figure 2.11 Pattern of oligogalacturonides detected in a juice containing an enzyme-treated pulpwash.

2.3.10 Multicomponent approaches

Most of the approaches discussed in this chapter so far have concentrated on specific techniques or analytes and how they can be used to detect a specific adulterant. However, most analysts involved in detecting adulteration believe that it is essential to use a multicomponent or matrix approach to have the best chance of success. A wide-ranging screen allows a range of different adulterants to be targeted, and gives the analyst the best chance to detect what an unscrupulous supplier is doing.

Two of the best known methods developed in the early/mid 1980s in Europe, which are still used today, are the French AFNOR system and the German RSK profile (Anon, 1988; Anon, 1987, respectively). Both of these procedures lay down guidelines and ranges for a wide range of components in a juice to allow the detection of various types of adulterations. These approaches use a range of components on which to base a judgement and this involves the analysis of sugars, acids, metals (e.g. K^+, Na^+, Mg^{2+}, Ca^{2+}), anions (e.g. Cl^-, SO_4^{2-}, PO_4^{3-}, NO_3^-), flavenoids, pectin fractions, carotenoid fractions, formol value, proline level and, in some cases, the whole spectrum of amino acids. This can involve the analysis of up to 54 compounds and is consequently time-consuming and costly. However, it provides much information on the overall composition of a juice.

One limitation of the multicomponent approaches is that the published ranges are often very broad. This leaves scope for manipulation by an unscrupulous supplier and, without expert knowledge of the values that are expected for a juice from a particular origin, an analyst could easily accept juices as authentic when in fact they were adulterated or *vice versa*. The RSK system stresses that the values are a guide for the competent analyst in making a judgement but such an approach is more in keeping with the German legal system than with the British or American ones. The AFNOR approach tries to cover this problem by promoting the use of multivariate statistical techniques to assist in the interpretation of the complex and interrelated data (Richard and Coursin, 1984).

The applicability of the RSK or ANFOR systems for the detection of highly sophisticated adulterations in the 1990s, rather than the less sophisticated approaches of the 1980s, was a point that MAFF questioned strongly in its survey of methods in 1991. However, it would be foolish to abandon all of the traditional approaches and rely solely on the 'newer' methods such as isotopic procedures. These procedures cannot answer all the questions that have to be asked when we assess if a juice is pure or not, and these 'new' procedures also have limitations of their own. Isotopic techniques should be incorporated into a broader multicomponent approach that will make it harder for an unscrupulous supplier to 'beat the system'.

Although the AFNOR and RSK approaches are still popular in Europe, they are increasingly becoming superseded by a more pan-European system organised under the auspices of the Association of the Industries of Juices and Nectars from Fruits and Vegetables of the European Economic Community (AIJN). This association has laid down normal ranges and guide values for the most commonly used juices in Europe and will, over the next few years, publish values for the broader range of juices.

AIJN codes of practice are presently published for orange, apple, grapefruit, grape and pineapple (Anon, 1994). These codes are based on RSK and AFNOR formats with ranges and guide values and have also adopted the RSK concept of a commentary to assist in the interpretation of the results. The actual ranges adopted within the respective codes have been decided by discussion and negotiation between experts from the various fruit juice industries of Europe. The continued development of this pan-European system over the last few years has meant that AIJN incorporated some of the isotopic procedures into its codes, which is a significant step forward. However, the isotopic data for some juices are not yet available and will require further research before they can be incorporated. Until the AIJN values are fully established for soft fruit, a useful reference for the levels of various analytes found in these types and other fruits is Herrman (1987).

Another recent development organised by the fruit juice industries of Europe is the formation of the European Quality Control System (EQCS). This system is designed to set up a network of quality control schemes throughout the EC to ensure that juices on sale across the community are pure and of an appropriate quality. This system will not define what should be done in each country but will define certain minimum standards that the schemes should conform to.

As there is more pressure on control schemes to detect increasingly more sophisticated adulteration and also continuing pressure to minimise costs and maximise profits, there will be a move towards broader screening procedures, which will often include the use of complex multivariate statistical procedures. This should allow costs to be kept down and money to be targeted towards suspect samples rather than a more random 'scatter gun' approach.

2.4 Fingerprinting techniques

These can be split into two main areas, chromatographic and spectroscopic procedures. These approaches have one thing in common, the fact that they involve the production of a pattern or 'fingerprint' that is characteristic of the material under investigation, as with human fingerprints. These fingerprints can then be used by an expert to make an assessment of the sample. Nowadays, owing to the complexity of the data obtained, they will often require multivariate statistical procedures.

2.4.1 Chromatographic procedures

2.4.1.1 Amino acids. Amino acid profiles are one of the chromatographic fingerprinting procedures that have been in use for the longest time. They have been used extensively in the characterisation of orange and grapefruit juices and, to a lesser extent, of other juices; however, they offer considerable possibilities. The most common way of obtaining amino acid profiles is by using a conventional amino acid analyser based on ion-exchange chromatography. However, this is expensive to set up unless there are large numbers of samples to run. Over the last 10 years or so there have been several HPLC methods developed; these do not require the dedicated system of the conventional analyser and therefore offer cost advantages to the authenticity laboratory. The methods often involve pre- or post-column derivatisation of the amino acids and use ultraviolet or fluorescence detection. However, some derivatising agents, such as ortho-phthaldialdehyde (OPA), do not derivatise secondary amino acids, such as proline, which is important in the characterisation of apple and other juices.

The application of amino acid screening was discussed by Wallrauch and Faethe (1988) who highlighted the types of adulteration that could be detected. For example, the addition of grape or pear juice to apple can be detected from an enhanced level of proline. Amino acid profiles can also be useful in some of the red/black fruits where proline is low, such as blueberry, blackcurrants, redcurrants, raspberry and strawberry.

Amino acid profiling has been used extensively in the citrus area where the juices are rich in amino acids. When simple tests, such as the formol value and chloramine-T index, were introduced to assess the total amino nitrogen content of juices to detect extension of the juice, the adulterators counteracted this by adding cheap sources of amino nitrogen such as ammonium salts, urea or glycine to adulterated samples. This led to the use of amino acid profiles to check for distortions in the pattern. The unscrupulous suppliers then switched to using 'tailor-made' blends of amino acids to give the correct balance. This was then counteracted by the analysts using sophisticated chiral chromatographic procedures to resolve the different optical isomers, D and L, thereby detecting the addition of synthetic (racemic) amino acids. The amino acids that are naturally found in juices all have the L configuration and Ooghe *et al.* (1984) used a gas–liquid chromatographic (GLC) method to resolve D and L-aspartic acids and so detect adulteration. Conversely, Duchateau *et al.* (1989) used high-powered liquid chromatography (HPLC) to resolve D and L-proline in adulterated juices. However, with the availability of L-amino acids in commercial quantities this approach has become less attractive.

2.4.1.2 Polyphenols. One chromatographic approach, mentioned above, is the polyphenol fingerprinting procedure developed at Proctor & Gamble in the USA (Wade *et al.*, 1992). Rouseff (1988) has also used a similar technique to characterise juices. This fingerprinting technique has found acceptance in some areas to detect the addition of pulpwash to juices and it has also proved very useful when differenting juice types. The procedure works best in juices where the polyphenols are not easily oxidisable because of their structure and/or where there is a high ascorbic acid level and/or a lack of polyphenoloxidase (PPO) activity. Its application to areas where there are active PPO systems (e.g. apple and grape juice) is a problem as the patterns can be very process-dependent.

As discussed previously, this fingerprinting procedure is good at detecting adulteration of orange juice with pulpwash (Figures 2.12 and 2.13). When pulpwash is added to a juice the level of soluble narirutin increases relative to the hesperidin level; the narirutin to hesperidin ratio also increases. This procedure is also able to detect very low levels of grapefruit addition to orange juice. When grapefruit juice is added to orange the bitter principal, naringin, which is absent in orange, is clearly visible in the polyphenol profile even at levels of 5% (Figure 2.14). The detection limit for the addition is in the region of *circa* 0.5%. The reverse adulteration, the addition of orange to grapefruit, is more difficult to detect as hesperidin in naturally present at low levels (10–20 ppm) in

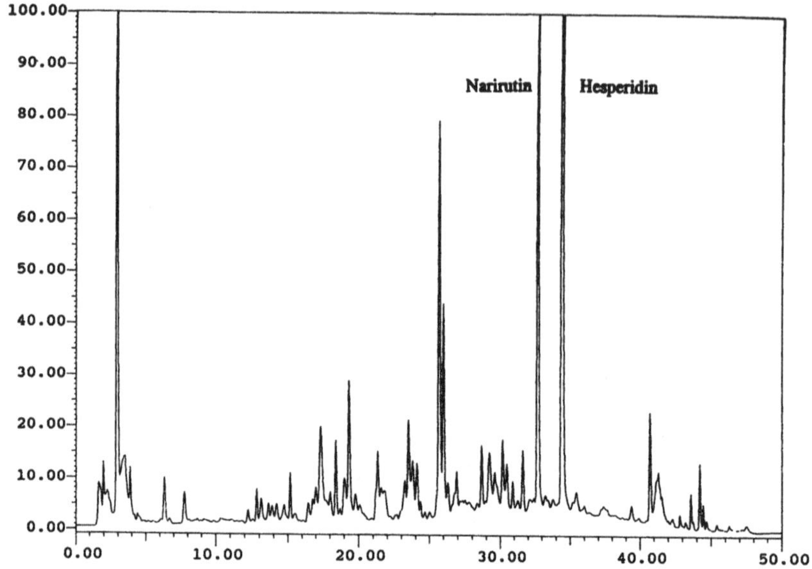

Figure 2.12 Typical polyphenol pattern found in orange juice.

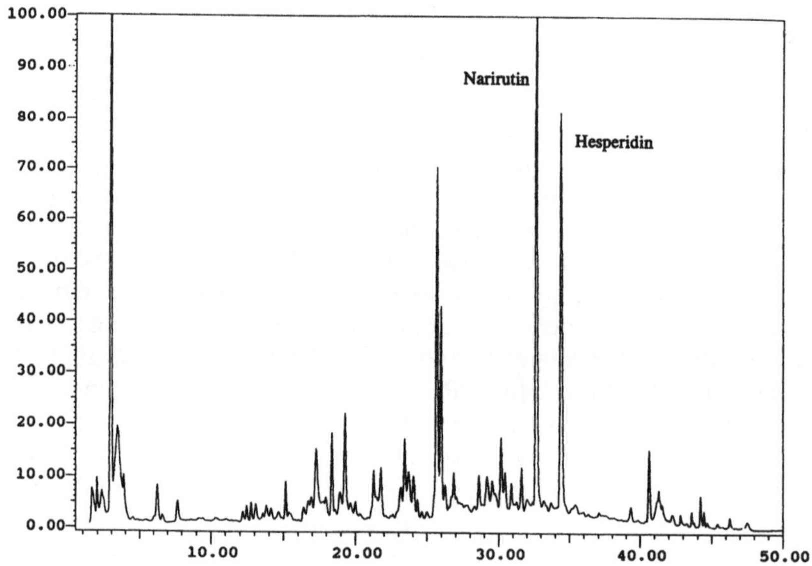

Figure 2.13 Pattern of polyphenols found in an orange juice containing pulpwash.

grapefruit juice. Polyphenols can also be used to detect the addition of sour orange (*Citrus aurantium*) to sweet orange (*Citrus sinensis*). A similar procedure has also been applied to the characterisation of exotic fruits such as passion fruit, Kiwi and mombin plum (Mareck-Engelke *et al.*, 1991).

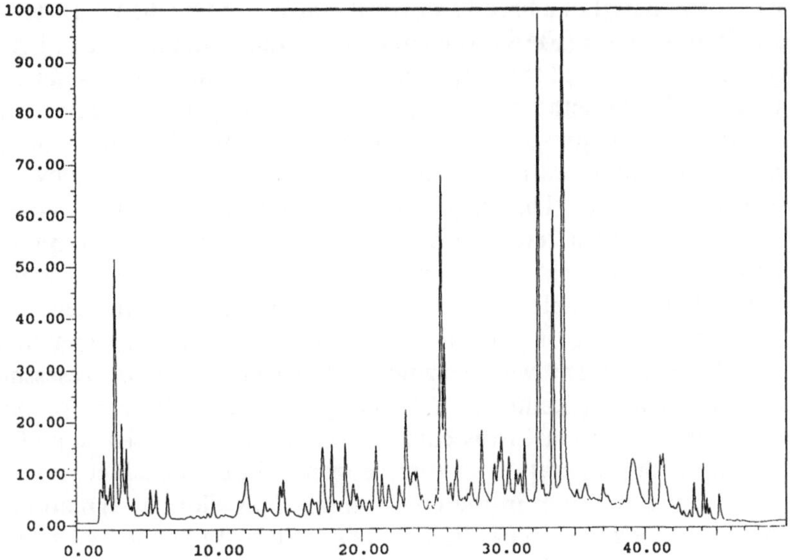

Figure 2.14 Pattern of polyphenols found in an orange juice containing 5% grapefruit juice.

Fingerprint data also offers the possibility of further interpretation by multivariate statistical techniques that have been applied by both Wade and Rouseff. Generally these procedures offer some improvement over the direct interpretation of the traces by eye. However, they are limited in so far as the human eye/brain is able to 'take in the whole picture' and make an overall judgement based on that pattern. Whereas small variations in the retention times can cause mis-assignment of the peaks, because of column ageing or day-to-day variations, which confuse the multivariate statistical interpretation.

Polyphenols can also be useful in the detection of the addition of apple and pear juice pomace to red/black fruits based on the analysis of two key markers. The presence of phlorizin in a red/black fruit such as raspberry or strawberry would be a confirmation of the addition of apple, which might already be suspected from an elevated level of sorbitol. Arbutin can be used in a similar way to confirm the addition of pear.

2.4.1.3 Polymethoxyflavones. A specific non-polar sub-class of polyphenols is the group of compounds generally referred to as the polymethoxyflavones (PMFs). As with the other polyphenols, they can be used to differentiate various fruits such as *Citrus sinensis* from *Citrus reticulata* and hybrid crosses (Ting *et al.*, 1979). They have also been used to detect the addition of pulpwash to orange juice and the addition of one type of fruit to another (Galensa, 1994).

PMFs formed part of a broad analytical matrix used by the US Federal & Drug Administration (FDA) to characterise orange juices (Perfetti *et al.*, 1988). The authors used the major PMFs found in orange (e.g. sinensetin, nobiletin, heptamethoxyflavone, tetra-O-methylscutellarein and tangeretin), the major flavonoid glucosides (hesperidin, narirutin and didymin) and 24 of the carotenoid peaks, most of which were unknowns, to assess orange juices. This data was then analysed using a range of multivariate statistical techniques to characterise the juices as pure or adulterated. In the best case, where they used the nearest neighbour technique, they were able to correctly assign all but two of the samples to their correct grouping of pure or adulterated. However, one component used in the assessment was benzoic acid, which made the assessment easier. The benzoate came from properly labelled pulpwash, prepared according to the Florida Citrus commission regulation, which stated that during pulpwash preparation sodium benzoate should be added at a level between 50 and 100 ppm in the concentrate. This Florida regulation has now been superseded by other Federal regulations. In this study Perfetti *et al.* (1900) used the assignment of the four nearest neighbours, in the multivariate space, from a training set of samples to classify the test samples as pure or adulterated.

2.5 Analysis of colours

Two types of colours are highly important to the quality of juices. The first class, the carotenoids, is very important in citrus, tomato and carrot juices. The other group, the anthocyanins, are responsible for the attractive colours of the red and black fruits.

2.5.1 Carotenoids

The analysis of the carotenoid pigments in orange juices has been used widely to detect many types of adulteration. As discussed earlier in the section on second extracts, one way of detecting the addition of pulpwash is from the depression in the colour of the juice. However, one way of circumventing such detection is by the addition of a red/orange colour, typically turmeric, bixin, β-carotene or β-apocarotenal (a synthetic carotenoid). There is a significant amount of information available as this area has been studied widely. A good, but somewhat old, review of this area was made by Gross (1977).

It is reported that there are approximately 40–50 carotenoids found in orange and this means that the chromatographic patterns are very complex. The carotenoids are also sensitive to normal and photochemical oxidations and so their analysis has to be conducted with care.

German experts have used the carotenoids for a while to assist in the assessment of juices to detect the addition of added colours and *Citrus reticulata* (tangerine) juice. In their procedure, not only is the total level of the carotenoids assessed by extraction and quantitation by visible spectroscopy but analysis of the various fractions within this class is undertaken by traditional column chromatographic techniques (Anon, 1991b). This method is similar to the AOAC procedure (Anon, 1990b), although the latter uses a different chromatographic phase. The AOAC column chromatographic 'clean up' procedure has been used as one of the initial stages in the analysis of carotenoids in vegetables such as carrots, green beans and in tomatoes (Siong and Lam, 1993) prior to analysis using HPLC. The fractionation of carotenoids using HPLC has also been studied recently as part of an EC funded FLAIR/BCR project on methods to detect the adulteration of juices (Hofsommer, 1994) and offers certain advantages over the conventional column separation. Although a range of procedures, both normal and reverse-phase, have been employed to separate the carotenoids, most authors have adopted a reverse-phase chromatography procedure since it is more robust.

One reason for the complex carotenoid pattern found in citrus juices results from the variety of esters present. Some authors have tried to simplify the analytical problem by using a hydrolysis step; however, this destroys some of the information and the general consensus seems to be that it is better to examine the carotenoids without an initial hydrolysis stage. The analysis of carotenoids can also be important in the examination of other fruits such as tomatoes (Wilberg and Rodriguez-Amaya, 1994), and the more highly pigmented pink and red grapefruit varieties (Rouseff *et al.*, 1992) to detect adulteration with other natural or synthetic colours.

2.5.2 Anthocyanins

HPLC is a critical element in the detection of adulteration of juices and purées of red and black berry fruits. In these fruits, the chromophores of some of the polyphenolic compounds, the anthocyanins, extend into the visible region of the spectrum, which gives these fruit their characteristic and striking colours. The anthocyanins also offer the analyst a very useful set of markers to detect the addition of one type of fruit to another. Generally red/black fruit species possess between one and up to four major anthocyanin pigments that are characteristic. The simplest pattern is that found in strawberries where there is only one major pigment and a few minor ones. At the other extreme is the pattern seen in blueberry or Concord grape (a North American variety) where there are *circa* 10–20 pigments, respectively. The very complex patterns seen in the last two fruits would make their adulteration with other fruits much more difficult to detect than, say, strawberry.

One advantage of this area, as with the carotenoids, is that there has been extensive work performed in the past so there is a good database on fruit pigments. Most of this work derives from more traditional procedures, such as paper and thin-layer chromatography. In fact one of the first applications of chromatographic procedures to the detection of fruit species was the differentiation of *Vitis vinifera* ('Old World' grapes) from inter-species crosses with varieties from the 'New World' (Ribereau-Gayon and Ribereau-Gayon, 1958). An excellent reference to the phenolic compounds in fruits is the book compiled by Macheix *et al.* (1990).

In the red/black fruit area the author has often found that unscrupulous suppliers will substitute part of the named fruit with, for example, elderberry, which has little commercial value but a very strong colour. This type of adulteration can easily be detected as elderberry has a very characteristic pigment pattern, Figure 2.15. This type of adulteration has been found in samples of all the commonly used fruits in the jams and preserves. The characteristic patterns of anthocyanins found in blackcurrant, blackberry, raspberry and strawberry juices and purées are shown in Figures 2.16 to 2.19. Although the cooking process used in jam production might be expected to destroy significant quantities of the anthocyanin pigments, we have found that this is not generally the case and the pigment pattern of the fruit persists into the final product. However, pelargonidin found in strawberries may be partly lost on cooking.

Another type of adulteration that is quite common in a range of red/black fruits is the addition of grapeskin extracts. Once again, this type of adulteration can be detected using anthocyanin profiles since red grapes

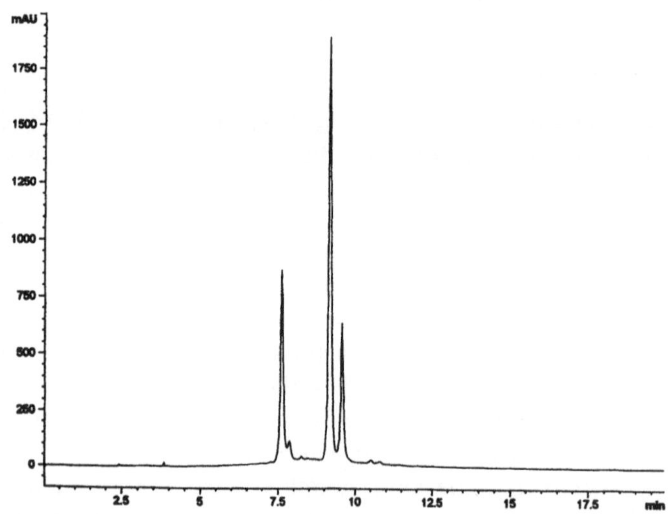

Figure 2.15 Pattern of anthocyanin pigments found in elderberry.

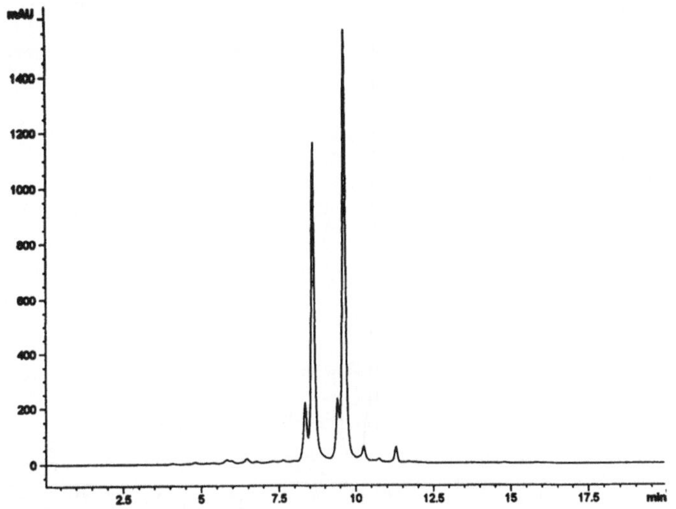

Figure 2.16 Pattern of antocyanin pigments found in blackcurrant.

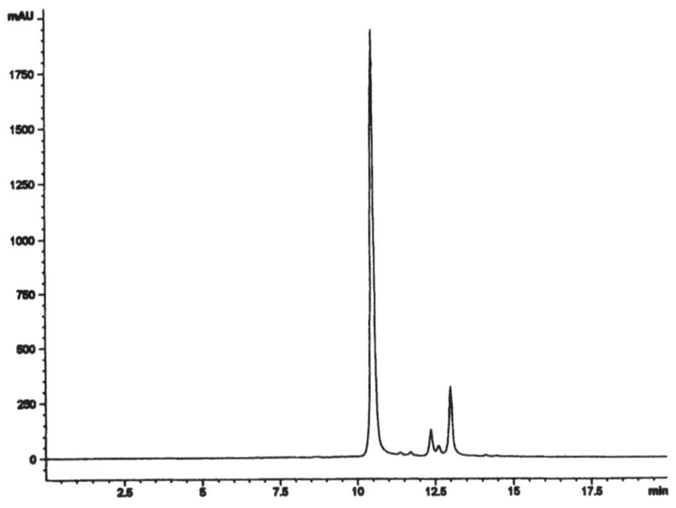

Figure 2.17 Pattern of anthocyanin pigments found in blackberry.

have significant levels of pigments based on the malvidin nucleus, which is not very common in the other commercially important fruits. These malvidin compounds elute late in the chromatogram (Figure 2.20) and also have diagnostic spectra with a longer wavelength maximum than most other pigments.

One type of adulteration that is not so common as might be expected, in

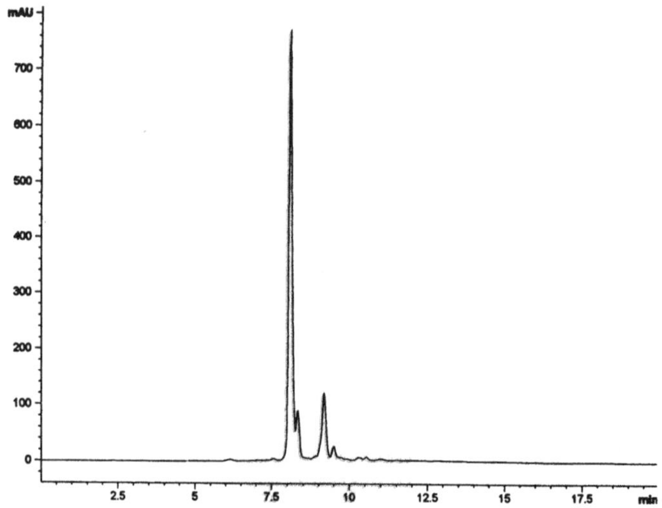

Figure 2.18 Pattern of anthocyanin pigments found in raspberry.

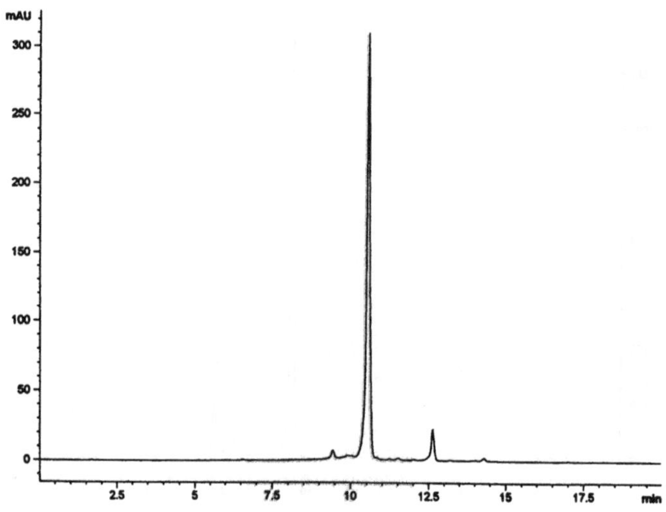

Figure 2.19 Pattern of pigments found in strawberry.

the red/black fruit area, is the addition of synthetic colours, such as Ponceau 4R. The author has only detected this type of adulteration in one or two cases, probably because this type of addition can easily be detected by conventional analytical techniques and is therefore too obvious to be worthwhile.

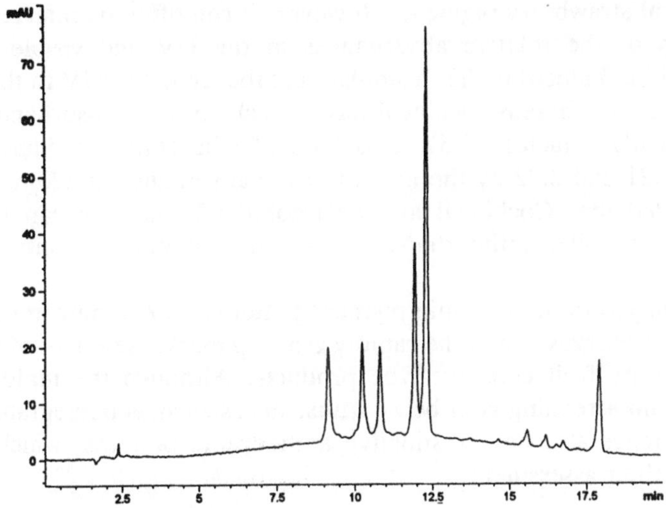

Figure 2.20 Pattern of pigments found in red grape.

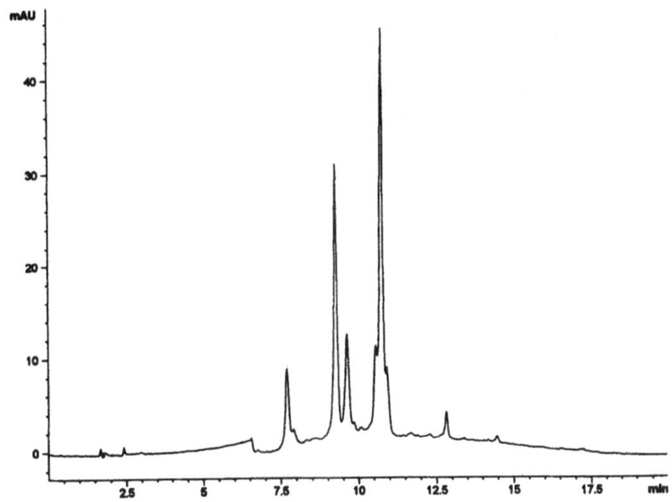

Figure 2.21 Chromatogram of strawberry juice adulterated with elderberry and cochineal using detection at 520 nm.

Two adulterants that seem to be quite common in strawberry are elderberry and cochineal. The former is readily detected using HPLC, by its characteristic triplet of early running peaks as seen in the adulterated strawberry juice (Figure 2.21). Cochineal can be difficult to detect because under the normal reverse-phase HPLC conditions, it elutes very close to the natural strawberry pigment. However, it can often be inferred by the distortion of the relative absorbances in the UV and visible regions; pelargonidin-3-glucoside has a similar absorbance in the UV to that in the visible region, whereas cochineal has a much stronger absorbance at 280 nm (typically a factor of 5). This type of adulteration is illustrated in Figures 2.21 and 2.22 by the much weaker absorbance at 520 nm to that seen at 280 nm. Cochineal and pelargonidin-3-glucoside can be easily separated on a silica rather than a reverse-phase medium, if confirmation is required.

The analysis of anthocyanin pigment patterns is extremely important in jams and preserves and in the rapidly growing market segment of yoghurt, with separate fruit conserve-type products. Although the major use of anthocyanin screening is in berry fruits, juices such as pomegranates and blood oranges also have distinctive anthocyanin patterns, which can be useful in their assessment.

2.6 Other methods

2.6.1 Trace element screening

One area that has been used by analysts in the US to characterise juices is trace metal profiling using such elements, as Al, Ba, Ca, Fe, K, Li, Mn,

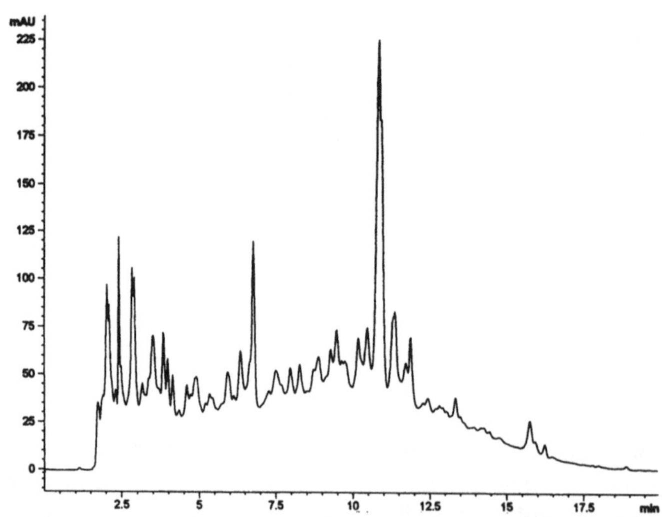

Figure 2.22 Chromatogram of strawberry juice adulterated with elderberry and cochineal using detection at 280 nm.

Na, P, Rb, Si and Zn. Trace metals have also assisted in typing the provenance of wines within the EC although when carried out by atomic absorption spectroscopy, this technique was time-consuming and hence expensive, the introduction of inductively coupled plasma atomic emission spectrometers (ICP-AES) or ICP-mass spectrometers (ICP-MS) has made multi-element mapping much cheaper and more rapid. Instruments of this type are expensive, which means that it is only available in relatively few laboratories; however, it seems to be particularly useful for characterising the geographical origin of samples – its application to the characterisation of orange juices has been reviewed by Nikdel *et al.* (1988).

2.6.2 *Multivariate data analysis*

Certain techniques such as near-infrared (NIR) and mid-range infrared spectroscopy and pyrolysis–mass spectroscopy (Py–MS) produce very complex data that requires the use of multivariate statistical procedures. Several statistical procedures may be used but they all attempt to reduce the very complex multivariate data to a much smaller number of independent variables that are easier to understand. Two types of data analysis that are used extensively, are principal components analysis (PCA) and canonical variates analysis (CVA). They produce similar results to other procedures such as statistical isolinear multi-category analysis (SIMCA), which is used by other groups such as the FDA in Washington DC (Johnson and Kaufmann, 1984; Page, 1986).

2.6.3 *Mid-range infrared spectroscopy*

The application of infrared (IR) spectroscopy to the detection of fruit juices is relatively recent and has really come about with the advent of Fourier transform (FT) spectrometers. Coupled with fast personal computers the FT process can now be performed very quickly, which allows multiple spectra to be processed rapidly, reducing the signal-to-noise ratio and improving the spectral information.

In the infrared region there are characteristic bands for all of the bonds found in a molecule. These not only involve the general stretching and bending of the functional groups (e.g. carbonyl, amino, hydroxyl groups etc.) but also how the overall molecule vibrates. In its conventional application, the IR spectrum is a characteristic fingerprint of the compound and it is this feature that offers the possibility to characterise juices and detect adulterations.

Work carried out at the Institute for Food Research (Norwich, UK) has shown that mid-range FT-IR, between *circa* 4000 and 800/cm, may offer a fast screening procedure to characterise and detect adulterations in fruit purées (Wilson, 1994). The authors have been able to identify several bands in the carbonyl region of the spectrum (1650–1750/cm) that are characteristic of fruit type (e.g. apple, raspberry, strawberry), which allows them to be separated from each other in multivariate space or just by visual inspection. These bands can also be used, for instance, to detect the addition of apple to the more expensive red/black fruit purées. Work is continuing in this area with trials of the IR equipment in a factory environment to assess if this could be adopted as a rapid screening procedure.

The use of mid-range IR spectroscopy can also be used to characterise pure and adulterated samples of orange juices (Hammond, unpublished results).

2.6.4 Near infrared spectroscopy

The use of near infrared (NIR) radiation is another spectroscopic method that has been used in the characterisation of foods. The NIR region of the electromagnetic spectrum covers the region 1100–2500 nm, which is above the IR region and below the visible region. Bands in this region result from overtones and combination bands of the absorbances found in the IR region and it is generally impossible to assign an origin to them, which is very different from the mid-range IR case where the bands are easily classifiable. NIR has some advantages over mid-range IR because of the ease with which radiation in this energy range passes into and through foodstuffs. One major application of NIR has been in the characterisation of the quality of flours, using a rapid screening technique. A good reference on the development of NIR and some of its application is the book edited by Osborne *et al.* (1993).

The application of NIR to the characterisation of fruit juices has recently been developed by Campden & Chorleywood Food and Drink Research Association (CCFDRA) in the UK, with funding assistance from the EC (Scotter and Legrand, 1994). The process involves drying single-strength juice onto a disc under controlled conditions, followed by analysis of the sample by reflectance NIR spectroscopy. The process is moderately fast and it has been found to give good results when presented with a range of blind samples of unknown origins and containing a range of adulterants. In a blind test as part of the final assessment of the methods examined in the EC-funded FLAIR/BCR project (Scotter, 1994), it correctly characterised the majority (80%) of the samples as either pure or adulterated.

Owing to the large amount of spectral information produced it has to be processed using multivariate procedures to allow any useful information to be obtained from it and, in this case, the authors used a discriminant analysis to differentiate between the various samples. Work in this area is continuing at CCFDRA.

2.6.5 *Pyrolysis–mass spectroscopy*

Pyrolysis–mass spectroscopy (Py–MS) is a process where a sample is placed in a high vacuum and heated under controlled conditions. The organic material is decomposed rapidly and the relatively low-molecular-weight products produced in the thermolysis process are led into a mass spectrometer for analysis. This approach has been used by our laboratory in several applications to characterise fruit juices (Aries, 1986; Evans and Hammond, 1989), micro-organisms (Shute *et al.*, 1984), whiskies (Reid, 1990) caramel colours and fruit juices (Hussain, 1993) and a range of other materials.

Its major advantage is the rapidity of the technique since it is possible to obtain a pyrolysis-mass spectrum of a sample, in triplicate, in less than 10 min after the sample has been dried onto a Curie-point foil. One problem in the past was with the reproducibility and stability of the mass spectrometer, which made it difficult to compare data for one set of samples with another set run at a later time. However, with the recent introduction of a new generation of pyrolysis-mass spectrometers, this problem should now be solved.

Py–MS was used to assess the authenticity of a range of commercial orange juices from the UK in the summer of 1990, as reported at the ACS meeting in Washington that year (Hammond, 1995). In this study it was found that most of the samples did not group with the Brazilian or Israeli control samples that were used in the study (Figure 2.23). This is perhaps

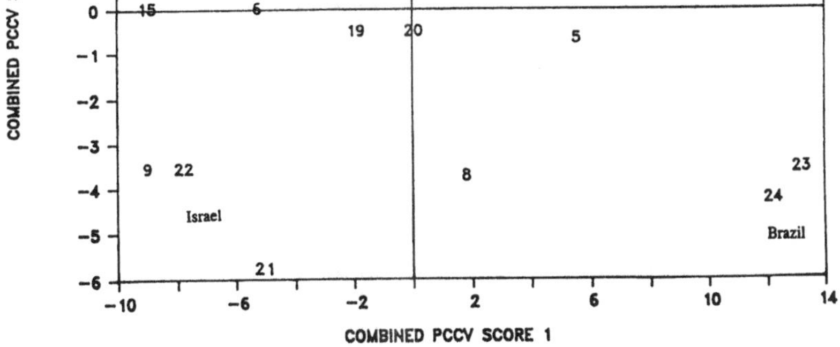

Figure 2.23 PCCV plot of Py–MS data for commercial orange juices taken from the UK market in 1990.

not surprising as commercial juices could easily contain juices from a range of origins. However, it was known from the product labels that several of the samples were supposed to be from Israel and yet they did not group with the controls. It was possible to determine why these samples were differentiated from each other by examination of the multivariate loadings plot, which shows how the measured variables (in this case the Py–Ms ions) are related to the variation observed between the samples. In this case, the loadings plot showed that some of the variation seen between the samples was caused by masses associated with sugars (97 and 126 m/z) or pectin (85 and 125 m/z). This was attributed to the addition of sugars and pulpwash to these samples, which was later confirmed by the MAFF study on samples collected at a similar time (Anon, 1991a). Work is continuing in this area to extend and investigate the limits of the technique.

2.7 Jams and preserves

The majority of the techniques discussed above are wholly appropriate for the examination of the fruit materials used in jam and preserve manufacture. However, most are completely useless if applied to the finished product. This is because during the jam manufacturing process the fruit material is effectively adulterated with sugars, acids and generally pectin to make the final product. This means that the isotopic procedures and many of the screening methods designed to detect sugar or acid additions would all give positive results but this would not tell you if the original material was authentic or not.

Some of the single analyte tests, such as sorbitol and tartaric acid can still be used to detect undeclared apple, pear or grape juices/pulps. Anthocyanins can also be used to detect the addition of other red/black fruit or added colours. The use of certain fingerprinting techniques, such as mid-range IR, may also be useful but work is still continuing in this area to assess its suitability. It is the author's opinion that the best approach to address the authenticity of jams and preserves is to assess the fruit material going into the product, as this gives the best chance of detecting any suspicious features and allows the whole range of methods, as discussed, to be used.

2.7.1 Fruit content

One topic of interest in jams and conserves is 'fruit content'. The traditional approach is to assess a range of components that appear in the fruit over a relatively narrow range, such as total nitrogen, phosphorus and potassium (NPK). However, it is relatively easy for these parameters to be adjusted in such a way that the product appears to contain more fruit than it actually does. This could be achieved simply by the addition of potassium and ammonium phosphates. The best way to overcome this problem, however, it to assess a range of other parameters as well, including amino acids, which allows some cross-checking between the various parameters to be made and detection of any distortion caused by the product being adjusted in such a way to pass the 'NPK' test.

2.7.2 *Light microscopy*

One area that is often overlooked in the assessment of fruit purées is light microscopy. This approach can be particularly useful at detecting admixtures of fruit from specific morphological featues. A very good example of this would be rhubarb as this has a very fibrous nature and the long filaments that make up the stem are easily detectable in a purée if added at a significant level (10–20%). The author has used this technique to detect the addition of pear to soft fruits by the presence of the hard 'stone cells', which are characteristic of pears.

Another area where microscopy can be of use is in the characterisation of the seeds in purées. It is possible to characterise the seeds of strawberries, which are small, smooth and tear-drop in shape, from those of raspberry, which are larger, sickle-shaped and ridged. Examination of the size of the fruit cells can also give clues as to the type of fruit that a purée contains. However, it can be difficult to be categorical from this evidence alone because of the variations seen in fruit cells.

Light microscopy is obviously only of use in jams/conserves that contain fruit pieces where the tell-tale features are carried through into the finished product. Products that are prepared from clarified juice concentrates will contain little if any useful information as particulate material will be lost during the filtration process.

2.8 Conclusions

The problem of the purchase of adulterated juices, purées and foodstuffs in general is one that is faced by all suppliers, importers and manufacturers who do not actually control the production of the raw materials. The approach adopted to examine the authenticity of samples has to remain flexible, because there are always some unscrupulous chemists and technologists who are working hard to find routes to circumvent the techniques that are used by the regulators/control laboratories so they can enhance their company's profits.

It is critically important for UK companies to do some testing of their own as under the 1990 Food Safety Act a 'warranty', supplied by a manufacturer outside the UK, stating that a product is pure will no longer be considered as an acceptable proof that reasonable steps have been taken ensure that a product is pure. This means that the onus has now shifted to the UK manufacturer or agent to provide evidence that the product that they are using/supplying is authentic.

Although a plethora of different methods have been used to assess the authenticity of samples, and new approaches are continuously being added, it is essential to keep funding research in this area to develop new techniques. Our aim should be to squeeze the unscrupulous suppliers into ever smaller corners, until they reach the stage where the additional profit they can make from adulteration is no longer worth the risk of being caught, with all the bad publicity that would ensure. However, with the price of sugar less than one-third of the cost of most juices, and considerably less than that in the case of the red/black fruit, this is a *very* difficult task.

A company that purchases juices and or purées should always use a range of different techniques to assess their supplies of fruit materials. This means that a multicomponent approach should be used so that a broad range of variables is assessed in a single sample. This does not mean that every sample has to be examined with *all* the established procedures and any new ones that happen to be 'flavour of the month'. What should be done is to establish a routine screening or 'due diligence' programme that uses a broad range of techniques including isotopic, compositional and fingerprinting procedures. If a problem with a particular supplier or origin occurs, it should be identified by this screening regime. If a problem is identified, it is essential that more-thorough testing is undertaken, maybe batch-by-batch, until it can be proved that the problem has been solved.

If a manufacturer is using juices from well-established contacts it is possible, by routine testing and factory visits, to assess how reliable these supplies are and if they have a record of proven delivery of 'good' material. This type of approach, control and assessment is a critical feature of the concept of 'due diligence' as envisaged in the UK Food Safety Act, 1990. If, on the other hand, a juice/pulp/purée is bought on the open/spot market, one cautionary note should *always* be considered: *some good deals are just too good to be true* and, in this situation, '*the buyer should always beware*!'.

What should always be borne in mind is the consequences of not identifying a problem with your own raw material. If a problem is detected by the authorities this could involve a sizeable fine and/or prison term and at least, bad publicity, especially if you are in the USA. Conversely, if it is detected by a consumers interest group there could be much bad publicity, which could destroy a brand name.

Recent developments

The adulteration of apple juice in Europe and America with high fructose syrups derived from inulin (a polyfructose) has recently been discovered. This procedure is based on the method developed by Low (1995) but involves new marker components which are specific for the inulin syrups (Low and Hammond, 1996). Testing using this procedure has caused some products to be withdrawn from the market in USA and may involve a number of companies in criminal and civil law suits (Butler, 1996).

References

Anon (1987) *RSK Values – The Complete Manual*. Flussiges Obst, Schonborn, Germany.

Anon (1988) *Trilingual Fruit Juice Specifications*. AFNOR, Paris.

Anon (1990a) Wine method to detect sugar addition to wines. Detecting enrichment to grape musts, concentrated grape musts, rectified concentrated grape musts and wines by application of nuclear magnetic resonance of deuterium (SNIF-NMR/RMN-FINS). *The Official Journal of the European Communities*, L272, **33**, 61–64.

Anon (1990b) Carotenoids in fresh plant materials and silages. Spectrophotometric method. *AOAC Official Methods of Analysis,* **941(15)**, 1048.

Anon (1990c) Corn syrup in apple juice by carbon ratio mass spectrometry method. *AOAC Official Methods of Analysis,* **981(09)**, 926.

Anon (1990d) Corn syrup in orange juice by carbon ratio mass spectrometry method. *AOAC Official Methods of Analysis,* **982(21)**, 926–27.

Anon (1991a) Comparison of methods for the detection of added substances in orange juice. UK Ministry of Agriculture Fisheries and Food. HMSO, London.

Anon (1991b) Determination of total and individual carotenoids groups. In: *The International Federation of Fruit Juice Producers Handbook of Analytical Methods. Method Number 58.* Swiss Fruit Juice Union, Zug, Switzerland.

Anon (1994) *AIJN Code of Practice*. Association of the Industries of Juices and Nectars from Fruits and Vegetables of the European Economic Community (AIJN), Brussels, Belgium.

Anon (1994c) Definition of orange and orange juice. In: *Conference Proceedings of the IFU Seminar, Parma, Italy, September 1994*. Brief summaries available from the Swiss Fruit Juice Union, Zug, Switzerland.

Aries, R., Gutteridge, C.S. and Evans, R.E. (1986) Rapid characterisation of orange juice by pyrolysis–mass spectrometry. *J. Food Sci.* **51**, 1183–86.

Bender, M.M. (1971) Variations in the $^{13}C/^{12}C$ ratios of plants in relation to the pathway of photosynthetic carbon dioxide fixation. *Phytochem.* **10**, 1239–44.

Beutler, H.O. and Wurst, B.A. (1990a) A new method for the enzymatic determination of D-malic acid in foodstuffs, Part I: Principles of the enzymatic reaction. *Deutsche Lebensm. Rdsch.* **86**, 341–44.

Beutler, H.O. and Wurst, B.A. (1990b) A new method for the enzymatic determination of D-malic acid in foodstuffs, Part II: The influence of other substances and investigations of foodstuffs. *Deutsche Lebensm. Rdsch.* **86**, 386–89.

Beutler, H.O. and Ara, V. (1992) Enzymatic determination of D-malic acid in fruit juices. *Fluss. Obst* **9**, 140–41.

Brause, A.R. (1992) Detection of juice adulteration 1991 perspective. *Internatl Food Ingred.* **1**, 4–11.

Brause, A.R., Raterman, J.M., Petrus, D.R. *et al.* (1984) Verification of authenticity of orange juice. *J. Assoc. Anal. Chem.* **67(3)**, 535–538.

Bricout, J. (1982) Possibilities of stable isotope analysis in the control of food products. In: *Stable Isotopes*, (Eds H.L. Schmidt, H. Foerstel and K. Heinzinger). Elsevier, Amsterdam, pp. 483–93.

Bricout, J. and Koziet, J. (1985) Detection de l'addition de sucre dans les jus d'orange par analyses isotopique. *Science des Aliments* **4**, 197–204.
Bricout, J. and Koziet, J. (1987) Control of the authenticity of orange juice by isotopic analysis. *J. Agricol. Food Chem.* **35**, 758–60.
Bricout, J., Fontes, J.C. and Merlivat, L. (1972) Sur la composition en isotopes stables de l'eau des d'oranges. *C. R. Acad. Sci. Paris Ser. D.* **274**, 1803–6.
Bricout, J., Merlivat, L. and Koziet, J. (1973) Fractionnement des isotopes de l'hydrogene au cours de la biosynthese de certains constituants aromatiques des vegetaux. *C. R. Acad. Sci. Paris Ser. D.* **277**, 855–88.
Bricout, J., Fontes, J.C., Merlivat, L. and Pussett, M. (1975) Stable isotope composition of ethanol. *Ind. Aliment. Agric.* **92**, 375–78.
Butler, D. (1988) Action on juice falsification. *Food News* **16(14)**, 1.
Butler, D. (1996) Apple juice alert goes out. *Food News* **23**(7), 1.
Craig, H. (1953) The geochemistry of stable carbon isotopes. *Geochim. Cosochim. Acta* **3**, 53–92.
Craig, H. (1961a) Standard for reporting concentrations of deuterium and ^{18}O in natural waters. *Science* **133**, 1833–34.
Craig, H. (1961b) Isotopic variations in meteoric waters. *Science* **133**, 1702–04.
Doner, L.W. (1985) Carbon isotope ratios in natural and synthetic citric acid as indicators of lemon juice adulteration. *J. Agric. Food Chem.* **33**, 770–72.
Doner, L.W. (1991) Verifying the authenticity of plant-derived materials by stable ratio and chromatographic methodologies. *J. Assoc. Off. Anal. Chem.* **74(1)**, 14–19.
Doner, L.W. and Cavender, P.J. (1988) Chiral liquid chromatography for resolving malic acid enantiomers in adulterated apple juice. *J. Food Science* **53(6)**, 1898–99.
Doner, L.W. and Phillips, J.P. (1987) Correlation of growth location of sugar beets and ratios of the $^{18}O/^{16}O$ and carbon bound D/H in derived sucrose. In: Abstracts of the 193rd National Meeting of the American Chemical Society, Denver, April 5–10.
Doner, L.W., Ajie, H.O., Sternberg, L. *et al.* (1987) Detecting sugar beet syrups in orange juice by D/H and $^{18}O/^{16}O$ analysis of sucrose. *J. Agric. Food Chem.* **35**, 610–12.
Dunbar, J. (1982) A study of the factors affecting the $^{18}O/^{16}O$ ratio in the water of wine. *Z. Lebensm. Unters. Forsch.* **174**, 355–59.
Dunbar, J. and Wilson, A.T. (1983) Oxygen and hydrogen isotopes in fruit and vegetable juices. *Plant Physiol.* **72**, 725–27.
Dunbar, J. and Schmidt, H.L. (1984) Measurement of the $^{2}H/^{1}H$ isotope ratios of the carbon-bound hydrogen atoms in sugars. *Z. Anal. Chem.* **317**, 853–57.
Dunbar, J., Schmidt, H.L. and Woller, R. (1983) Moglichkeiten des Nachweises der Zuckerung von Wein uer die Bestimmung von Wasserstoff-Isotopenverhaltnissen. *Vitis* **22**, 375–86.
Duchateau, A., Crombach, M., Munsters, B. *et al.* (1989) Selective analysis of proline enantiomers by high performance liquid chromatography. Paper given at 13th International Symposium on Column Liquid Chromatography, Stockholm, Sweden.
Elkins, E.R. and Heuser, J.R. (1994), Detection of adulteration in apple juice by L-malic/total malic acid ratio: a collaborative study. *J. Assoc. Off. Anal. Chem.* **77(2)**, 411–15.
Evans, R.E. and Hammond, D.A. (1989) Orange juice authenticity testing. *Eur. Food Drinks Rev.* February, 41–44.
Farnell, P. and Biesenbruch, S. (1994) Detection of peel extracts and pulpwash in orange juice. Paper presented as part of a Comett seminar held in London, November 1994.
Francis, F.J. (1982) Analysis of anthocyanins. In: *Anthocyanins as Food Colours*. (Ed. P. Markakis). Academic Press, London, pp. 181–207.
Galensa, R. (1994) Authenticity and detection of adulteration of orange juices by HPLC-determination of flavonoids. Paper presented as part of the SGF symposium '*Progress in the authenticity-assurance for fruit juices*', Parma, Italy, September 1994.
Gross, J. (1977) Carotenoid pigments in Citrus juices. In: *Citrus Science and Technology* Vol. 1: Nutrition, Anatomy, Chemical, Composition and Bioregulation. (Eds. S. Nagy, P.E. Shaw and M.K. Veldhuis). AVI, Westport, Connecticut, USA, pp. 302–54.
Hammond, D.A. (1993) Going sweet on ion chromatogrophy. *Lab. Products Technol.* **April**, 26.
Hammond, D.A. (1995) Authentication of fruit juices using traditional multicomponent

methods and a novel fingerprinting technique Pyrolysis–mass spectrometry. In: *Methods to Detect Adulteration of Fruit Juice Beverages*. (Eds S. Nagy and B. Wade). AgScience, Auburndale, Florida, USA.

Hatch, M.D. and Slack, C.R. (1970) Photosynthetic CO_2-fixation pathways. *Ann. Rev. Plant Physiol.* **21**, 141–62.

Heath, R.L. (1974) Isotopic data for the elements. In: *Handbook of Chemistry and Physics*, 55th edn. (Ed. R.C. Weast). CRC Press, Boca Raton, Florida, USA, pp. B247 and B250.

Herman, K. (1987) Chemische Zusummensetzung von Obst und Fruchstaften enschließlich Wichtiger Gemusesafte sowie deren ernahurungs-physiologische Bedentung. In: *Hanbuch der Lebensmittel-Technology, Frucht und Gemusesafte; Technologie, Chimie, Mikrobioloige, Analytik, Bedeutung*. (Ed. U. Schobinger). Ulmer, Stuttgart, FGR, pp. 39–88.

Hussain, S.F. (1993) Application of pyrolysis–mass spectrometry as a rapid analytical method for food science. PhD Thesis, Reading University, Reading, UK.

Johnson, M.R. and Kauffman, F.L. (1984) Orange juice adulteration detection and then actions of the FDA. *J. Food Qual.* **82(2/3)**, 81–89.

Jones, D. (1988) $700M orange juice scandal. *Today* (UK newspaper) 16 November, 1–4.

Junge, C. and Spandinger, C. (1982) Detection of the addition of L- and D,L-malic acid in apple and pear juices by quantitative determination of fumaric acid. *Fluss. Obst* **49(2)**, 57–62.

Koziet, J. and Bricout, J. (1994) Determination of minor sugars in fruit juices. Paper presented as part of the SGF symposium '*Progress in the authenticity-assurance for fruit juices*', Parma, Italy, September 1994.

Krueger, D.A. (1988) Application of stable isotope ratio analysis to problems of fruit juice adulteration. In: *Adulteration of Fruit Juice Beverages*. (Eds S. Nagy, J.A. Attaway and M.E. Rhodes). Marcel Dekker, New York, pp. 109–124.

Low, N.R. (1995) Personal communication.

Low, N.R. and Hammond, D.A. (1996) Detection of high fructose syrup from inulin in apple juice by capillary gas chromatography with flame ionization detection. *Fruit Processing* April (in press).

Low, N.R. and Swallow, K.W. (1991) Detection of beet medium invert sugar addition to orange juice by high-performance liquid chromatography. *Fluss. Obst.* **1**, 2–4.

Macheix, J-J. Fleuriet, A. and Billot, J. (Eds) (1990) *Fruit Phenolic*. CRC Press, Boca Raton, Florida, USA.

Mareck-Engelke, U., Galensa, R. and Herrmann, K. (1991) Identifizierung von exotischen fruchten in fruchtproduckten mittels HPLC. *Flussig. Obst.* **58(12)**, 659–63.

Martin, G.J. (1993) SNIF–NMR, a new method to detect added beet sugar to fruit juices and characterise their authenticity. *Fluss. Obst.* **2**, 123–28.

Martin, G.J., Martin, M.L., Mabon, F. *et al.* (1982) Identification of the origin of natural ethanols by natural abundance hydrogen-2 nuclear magnetic resonance. *Anal. Chem.* **54**, 2380–82.

Martin, G.J., Sun, X.Y., Guillou, C. *et al.* (1985) NMR determination of absolute site-specific natural isotope ratios of hydrogen in organic molecules. Analytical and mechanistic applications. *Tetrahedron* **41**, 3285–96.

Martin, G.J., Zhang, B.L., Naulet, N. *et al.* (1986) Deuterium transfer in bioconversions of glucose to ethanol studied by isotope labelling at the natural abundance level. *J. Am. Chem. Soc.* **108**, 5116–22.

Martin, G.J., Danho, D. and Vallet, C. (1991) Natural isotope fractionation in the discrimination of sugar origins. *J. Sci. Food Agric.* **56**, 419–34.

Moussa, I., Naulet, N., Martin, M.I. *et al.* (1990) A site-specific and multielement approach to the determination of liquid-vapour isotope fractionation parameter. The case of alcohols. *J. Phys Chem.* **94**, 8303–09.

Nikdel, S., Nagy, S. and Attaway, J.A. (1988) Trace metals: defining geographical origin and detecting adulteration or orange juice. In: *Adulteration of Fruit Juice Beverages*. (Eds S. Nagy, J.A. Attaway and M.E. Rhodes). Marcel Dekker, New York, pp. 81–105.

Nissenbaum, A., Lifshitz, A. and Stepak, Y. (1974) Detection of citrus juice adulteration using distribution of natural stable isotopes. *Lebensm. Wiss. Technol.* **7**, 152–54.

Ooghe, W. (1994). HPTLC and Capillary GC in the detection of sugar addition to orange juices. Paper presented as part of the SGF symposium '*Progess in the Authenticity-Assurance for Fruit Juices*', Parma, Italy, September 1994.

Ooghe, W., Kasteleyn, H. and Sandra, P. (1984) Amino acid enantiometer separation for the detection of adulteration in fruit juices. *J. High Res. Chromatogr. Commum.* Short communication No. 10488, **7(5)**, 284–85.

Osborne, B.G., Fearn, T. and Hindle, P.E. (Eds) (1993) *Practical NIR Spectroscopy With Application in Food and Beverage Analysis.* 2nd edn, Longman Scientific & Technical, Harlow, UK.

Page, S.W. (1986) Pattern recognition methods for the determination of food composition. *Food Technol.* **xx**, 104–09.

Parker, P.L. (1982) Uber die Anwendung des $^{13}C/^{12}C$ verhaltnisses zum erkennen einer sussung von fruchsaftkonzentraten. *Fluss. Obst.* **49(12)**, 672, 692–94.

Patel, P. (1993) Molecular techniques for fruit juice authenticity. Paper presented at the *2nd European Symposium on Food Authenticity*, 20–22 October, Nantes, France.

Perfetti, G.A., Joe, F.L. and Page, S.W. (1988) Liquid chromatographic methodology for the characterisation of orange juice. *J. Assoc. Off. Anal. Chem.* **71(3)**, 469–73.

Petrus, D.R. and Attaway, J.A. (1980) Visible and ultraviolet absorption and emission characteristic of Florida orange juice and orange pulpwash: detection of adulteration. *J. Assoc. Off. Anal. Chem.* **63(6)**, 1317–31.

Petrus, D.R. and Attaway, J.A. (1985) Spectral characteristics of Florida orange juice and orange pulpwash. A collaborative study. *J. Off. Anal. Chem.* **68(6)**, 1202–06.

Potter, R.L. and Mansell, R.L.L. (1992) Assay for the detection of beet sugar adulteration of food products. US Patent No. US 5 128 243 (US5128243).

Rauschenbach, P., Simon, H., Stichler, W. and Moser, H.Z. (1979) Verleich der Deuterium- und Kohlenstoff-13-gehalte in Fermentation- und Synthese Ethanol. *Z. Naturforsch.* **34c**, 1–4.

Reid, K. (1990) The application of pyrolysis–mass spectroscopy to the role of wood in the maturation of scotch whiskies. PhD Thesis, Herriot Watt University, Edinburgh, UK.

Ribereau-Gayon, J. and Ribereau-Gayon, P. (1958) The anthocyans and leucoanthocyans of grape and wines. *Am. J. Enol. Viticol.* **9**, 1–9.

Richard, J.P. and Coursin, D. (1984) Essais de characterisation de jus d'agrumes d'Israel application de la method FRANCINE. In: *Advances in the Fruit and Vegetable Juice Industry.* (Ed. International Federation of Fruit Juice Producers). Juris Druck and Verlag, Zurich, Switzerland.

Rouseff, R.L. (1988) High-performance liquid chromatographic separation and spectral characterisation of the pigments of tumeric and annatto. *J. Food Sci.* **53(3)**, 1823–26.

Rouseff, R.L., Sadler, G.D., Putman, T.J. *et al.* (1992) Determination of β-carotene and other hydrocarbon carotenoids in red grapefruit cultivars. *J. Agric. Food Chem.* **40(1)**, 47–51.

Saguy, S. and Cohen, E. (1984a) Evaluation of spectral characteristics of Israeli citrus products for authenticity determination. In: *Advances in the Fruit and Vegetable Juice Industry.* (Ed. International Federation of Fruit Juices Producers), Juris Druck and Verlag, Zurich, Switzerland.

Saguy, S. and Cohen, E. (1984b) Spectral characteristics of Israeli orange products. *Z. Lebensm. Unters. Forsh.* **178**, 386–88.

Schmidt, H.L., Butzenlechner, M., Rossmann, A., *et al.* (1993) Inter- and intramolecular isotope correlations in organic compounds as a criterion for authenticity identification and origin assessment. *Zeits. fur Lebens. Unters. und Forsh.* **196(2)**, 105–110.

Scotter, C.N. (1994) Near infrared spectroscopy in fruit juice discrimination. Paper presented as part of the SGF symposium '*Progress in the Authenticity-Assurance for Fruit Juices*', Parma, Italy, September 1994.

Scotter, C.N. and Legrand, A. (1994) NIR qualitative analysis – a new philosophy with special reference to rapid NIR screening of fruit juice authenticity. *Food Sci. Technol. Today* **8(3)**, 167–171.

Shute, L.A., Gutteridge, C.S., Norris, J.R. *et al.* (1984) Curie point pyrolysis–mass spectrometry applied to characterisation of selected *Bacillus* species. *J. Gen. Microbiol.* **130**, 343.

Siong, T.E. and Lam, T.C. (1993) Analysis of carotenoids in vegetables by HPLC. *Food J.* **7(2)**, 91–99.

Smith, B.N. and Epstein, S. (1971) Two categories of $^{13}C/^{12}C$ ratios in higher plants. *Plant Physiol.* **74**, 104–107.

Souci, S.W., Fachmann, W. and Kraub, H. (1981). In: *Die Zusammersetzung de Lebensmittel-Nahruert Tabellen 1981/82*. (Eds H. Von Scharze und G. Kloss). Wiss Verlagesges, Stuttgart.

Ting, S.V., Rouseff, R.L., Dougherty, M.H. *et al.* (1979) Determination of some methoxylated flavones in citrus juices by high-performance liquid chromatography. *J. Food Sci.* **44(1)**, 69–71.

Voragen, A.G.J. (1972) Characterization of pectin lyases on pectins and methyl oligo-galacturonates. Thesis, Wageningen, Netherlands.

Wade, R.L., Kirksey, S.T. and Lee, J.A. (1992) New analytical methods in the USA for detecting fruit juice adulteration. *Flussig. Obst.* **59(5)**, 62–72.

Wallrauch, S. and Faethe, W. (1988) Amino acids: criteria for the evaluation of fruit juices. In: *Adulteration of Fruit Juice Beverages*. (Eds S. Nagy, J.A. Attaway and M.E. Rhodes). Marcel Dekker, New York, USA, pp. 21–48.

Wisenberger, A., Kolb, E., Haug, M. *et al.* (1992) Detection of the addition of beet medium invert syrup to orange juice by HPLC. Application of the method to grapefruit juice concentrate and experience with quantitative evaluation. *Fluss. Obst.* **9**, 137–39.

White, D.R. and Cancalon, P.A. (1992) Detection of beet sugar adulteration of orange juice by liquid chromatography/pulsed amperometric detection with column switching. *J Assoc. Off. Anal. Chem.* **75(3)**, 584–87.

Wilberg, V.C. and Rodriguez-Amaya, D.B. (1994) Quantification of *β*-carotene and lycopene in tomatoes and some tomato products by HPLC. *Ciencia e Technologia de Alimentos.* **13(2)**, 132–41.

Wilson, R. (1994) Rapid authentication of fruit purées by mid range-infrared spectroscopy. Paper presented as part of a Comett seminar held in London, November 1994.

Wrolstad, R.E. and Shallenberger, R.S. (1981) *J. Assoc. Off. Anal. Chem.* **64(1)**, 91–103.

Zyren, J. and Elkins, E.R. (1985) Interlaboratory variability of methods used for detection of economic adulteration in apple juice. *J. Assoc. Off. Anal. Chem.* **68(4)**, 672–76.

3 Wine authenticity

B. MÉDINA[1]

3.1 Introduction

Wine is a very easy product to adulterate because of its strong chemical basis (high alcoholic content, low pH) and its availability throughout the world; it is also probably one of the food products most often analysed and for which legislation is the most complete (not to say complex). The imagination of potential deceivers seems to be without bounds and the extent of such fraud remains unknown. Even if they are scarce, fraudulent products cause considerable damage to the wine trade and, in addition to a potential loss of prestige, the damage to the image of the product in terms of consumer confidence can be immense, especially in the current context of surplus, where products of the 'new world' (new compared with the millenial European wine tradition) are increasingly competitive. Until recently, tasting by experts was the only way of verifying authenticity but, because experts are not readily available and because of the subjectivity of the technique, methods were developed that sometimes mixed advanced mathematical tools and the latest analytical technology to guarantee the authenticity of wine. In this chapter the problems posed by the geographical origin of wines, their typicity and year of production will be discussed. Other cases of authenticity, such as blending, and special examples, such as Champagne, Port wines and botrytised wines (e.g. Sauternes) are also discussed.

The search for authenticity can be carried out using several groups of substances, which will be discussed, but first a short overview of the mathematical tools generally used in the course of such studies will be given.

In general, volatile substances are used to characterise varieties, whereas minerals are used for geographical differentiation. Amino acids as well as phenolic compounds are used for both. The bulk of research uses a variety of substances and is not restricted to one particular group of substances; for academic purposes, the author will make the distinction whenever possible.

[1] The opinion expressed in this chapter reflects only the view of the author and does not engage the responsibility of the DGCCRF in any way.

3.2 Multidimensional analysis applied to wine

Since the pionnering work of Kowalski (Kwan and Kowalski, 1978; 1980a, 1980b; Kwan *et al.*, 1979) using the well-known software ARTHUR (Informetrix Corporation, 1981) and the invention of chemometrics, many studies of the subject have been performed. Pattern recognition or fingerprint techniques are among the most popular methods for ensuring the distinction of 'individuals' (the term 'individual' here refers to any sample on which numerous determinations, called variables, are performed). These methods became more appealing as it became evident that the absolute content of an element is not so important as the relationship between several elements (e.g. correlation, inverse correlation or independence). Usually pattern recognition follows four steps according to the collection of data coming from different groups:

1. Preprocessing to convert all the data to the same unit to avoid an effect of scale (a common technique is subtracting every number by the mean of all the individuals for a variable and dividing it by the standard deviation).
2. Selection of the most efficient variables for classification (number of individuals/number of variables ⩾3). It is not usually essential to use numerous variables; a classification with few variables works just as well and is more robust and efficient.
3. Group classification (K nearest neighbours, discrimination analysis, SIMCA (ARTHUR, Infometrix, 1981), Nuées dynamiques (ADDAD, 1985)).
4. Classification evaluation with test samples (that do not participate in the construction of the model but that are well known). Several refinements can be added such as the 'bootstrap' method (Diaconis and Efron, 1983).
5. Identification of unknown samples.

3.2.1 Database

Usually a database for wine consists of a wide range of data coming from different techniques, such as HPLC, GC, GC.MS, AA, ICP and ICP/MS, which produce huge amounts of data. Generally the lack of reference samples and the specificity of the techniques limit the validity of the data to one laboratory. The other remaining problem is the representative nature of the database and the authenticity of each of its members. If the sample is authentic and was chosen randomly, commonsense suggests that it retains a fraction of the representativity of its origin, every one adding bits of the characters that make a dominant picture (Diaconis and Efron, 1983).

The 'security' of the database can be evaluated by having grapes picked by authorised people (e.g. official agencies etc.) and making the sample wine. That strategy was chosen for the detection of 'chaptalisation' (sugar addition) where ratios are important; however, real samples are quite different because very limited (a few kilograms) quantities of grapes can be picked and made into wine. In addition, the white wine-making process usually performed on those samples includes immediate separation of the pomace, seeds and so on, from the juice by pressing the whole grapes. It is well known to experimenters that the quality and the characteristics of wine produced in the wineries and chateaux is never reached with a microvinification. At the pilot level, similar results only begin to be obtained with 1 or 2 hectolitres (hl).

3.2.2 Multidimensional versus reduced dimensions

A second case of misunderstanding when using multidimensional methods arises when they are the only method used. Very simple solutions often exist. These are frequently more meaningful than the multidimensional approach as the following example shows (Baldi *et al.*, 1984).

An accurate separation of three regions (Piemontese, six wines; Oltrepo Pavese, six wines; Veneto, 1 wine) is accomplished using discriminant analysis with seven determinations (out of a total of 20) – volatile acidity, alcoholic content, boron, silicon and magnesium concentrations. Volatile acidity and alcohol cannot be taken into consideration because they are the components that can vary the most from wine to wine according to the degree of maturity, the fermentation process etc. Moreover, the number of variables is excessive for only 13 samples and three regions.

Using Principal Component Analysis we have shown that the area of production is not the dominant factor in the natural variation of the samples (Figure 3.1a). A careful selection of the best variables for separating (using the procedure SELECT in ARTHUR (Infometrix Corporation, 1981) indicates that only boron and phosphorus are necessary for a perfect separation (Figure 3.1b). This fact can be attributed to the different types of fertilization according to the zone. Using ratios and the same process produces an even better result (Figure 3.1c).

This is a simple example and the successful separation of groups of wines sometimes requires many more factors; nevertheless, numerous well-developed approaches to pattern recognition exist and can be tested. For further interesting examples for wines and food see Fiorina *et al.* (1987). One promising technique is the 'artificial neural networks', which has been applied successfully in the classification of other products and to detect adulterations (for olive oil see Goodacre *et al.*, 1993).

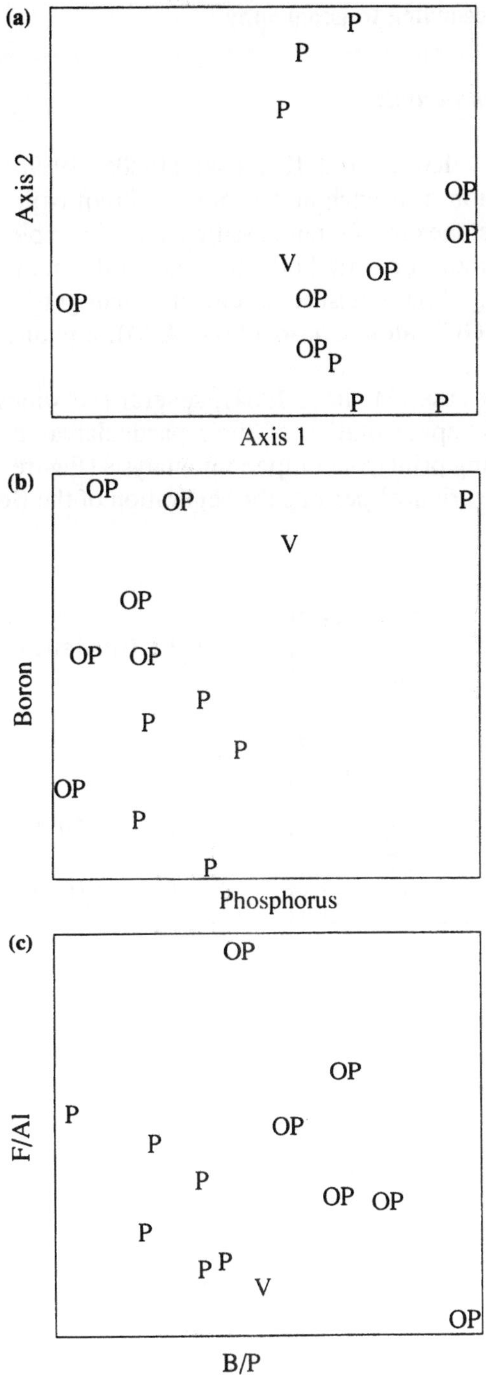

Figure 3.1 Principal component analysis: (a) plot of axis 2 *versus* axis 1. Separation of regions of production; (b) using two features; (c) using the ratio of two features. Source: Baldi *et al.* 1984. P = Piedmont; OP = Oltrepo Pavese; V = Venetia.

3.3 Distinction according to geography

3.3.1 Organic compounds

3.3.1.1 Volatiles. Kwam and Kowalski (1980a; 1989b) carried out a very good separation of French and American Pinot Noir wines based on 1-hexanol and cyclohexane (1 misclassified on 30 samples). Using data-processing and laboratory methods that generate numerous relatively accurate data (e.g. high-pressure liquid chromatography (HPLC), gas chromatography (GC), atomic absorption (AA)), a pilot database can be developed.

In the following case (Médina, 1984), several red wines from all over France, chosen as representative of their particular area of production, were separated using principal component analysis (Figure 3.2a). The first axis (16% of total variance) permits the separation of the Bourgogne wines

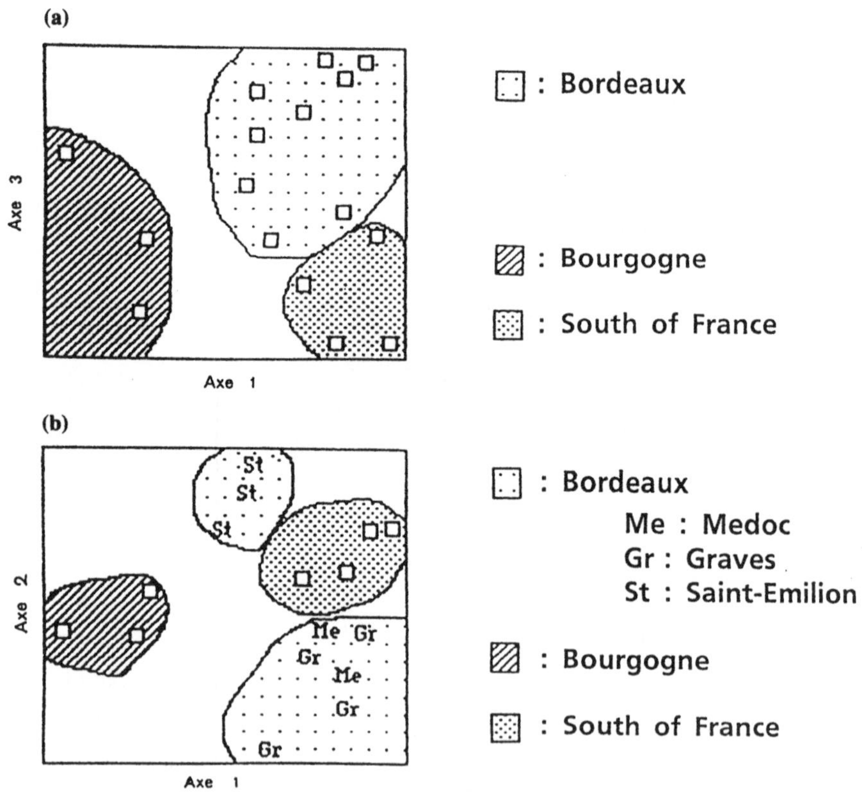

Figure 3.2 Principal component analysis (PCA) on wines from Bordeaux (Médoc, Graves and St Emilion), Bourgogne and the South of France. Projections: (a) axes 1 and 3; (b) axes 1 and 2. Source: Médina and Van Zeller (1984).

from the rest and is a linear combination of fatty acids, ethylic esters, 2-phenylacetate, isoamyl esters and diethyl succinate. The second axis (13%) constructed essentially with procyanidins, flavanes, and hexanol, enables the differentiation of products from different Bordeaux areas (e.g. Médoc and St Emilion) and the wines from the South of France (Figure 3.2b).

More recently, using purge and trap followed by capillary gas chromatography and mass spectrometry, Garcia-Jares *et al.* (1995) performed the differentiation of some white wines of Galicean origin (Rias Baixas) using 19 highly volatile compounds.

3.3.1.2 Amino acids. Amino acids are a useful group of substances to differentiate wines. Production of wines such as sparkling wines where the second fermentation producing the overpressure of CO_2 is performed in the bottle, leads to an increase in amino acids.

Tusseau *et al.* (1994) used this technique to separate authentic Champagnes from sparkling wines; unfortunately wines made from the Pinot Noir grape from the same place but different vintages show wide variations that do not assist characterisation. However, Champagnes are richer than sparkling wines in all amino acids, except arginine, because of the second fermentation in the bottle and the long contact with the lees. Using a linear discriminant function and amino acid analysis alone on a total of 110 wines, 4% of the Champagnes and 15% of the sparkling wines are misclassified. This performance – certainly far better than any achieved by tasting – is not judged to be sufficient (Tusseau *et al.*, 1992) and more parameters have to be introduced. Using linear discriminant analysis of 20 amino acids, Derde *et al.* (1993) correctly separated wines from Bourgogne, Bordeaux and Beaujolais.

3.3.2 Minerals

Minerals seem to be primary candidates for a 'fingerprint', as they are stable. The use of trace elements for the differentiation of wines closely followed the advance of analytical techniques. Strangely, 20 years ago neutron activation analysis (NAA) was capable of determinations smaller than parts per billion (ppb) but was costly and time consuming.

Atomic absorption, although adequate for parts per million – ppm (flame) and ppb determination (furnace), lacked the ability to carry out simultaneous determinations that are mandatory in this kind of research; the ICP, although simultaneous, lacked adequate sensitivity. It was not until the appearance of ICP/MS that analysis of most of the elements in wines was readily achievable.

3.3.2.1 Trace elements. The first work to appear in the literature concerning the differentiation of wines based on trace element analysis was by Siegmund and Bächmann (1977; 1978). They used the powerful neutron activation method; today very few authors have been able to match the level achieved for 15 trace and ultra-trace elements in wines. Their method was very cumbersome and required a nuclear reactor. The analysis itself, after irradiation for more than 2 days, took more than 3 months to perform. Nevertheless, 15 elements were correctly determined using this technique:

- elements currently determined: K, Na, Fe, Zn and Rb at ppm and sub-ppm levels
- trace elements: Cr, Co, Sb, Ag, Cs and Br at ppb levels
- ultra-trace elements: Sc, Eu, Hf and Ta at sub-ppb levels.

This performance was, at the time, remarkable. The authors realised that monodimensional statistics, simple ratios or graphical representations were unable to perform a good identification, and therefore used taxonomic analysis (cluster analysis) with Euclidian distances to group wines. With the elements: Na, K, Rb, Cs, Cr, Fe, Co, Zn and Ag, they showed that clusters were formed according to the area of production as well as the grape variety (Figure 3.3). They assumed that the transfer of elements

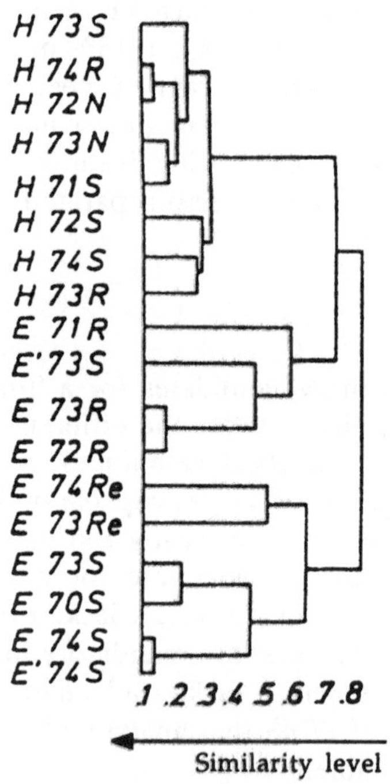

Figure 3.3 Cluster analysis for 18 German wines with nine minerals: Na, K, Rb, Cs, Cr, Fe, Co, Zn, Ag. Source: Siegmund and Bächman (1977). E = Alzeyer Galgenberg; H = Oppenheimer Zurcherbuerg; R = Riesling; S = Sylvaner; N = Nohling; Re = Regner.

from the soil to the plant had the greatest influence on their content in the wine. When analysing the soil, however, the only correlation found with the content in the wine was related to the alkaline metals (Na, K, Rb, Cs) and this was not true for all the sites. The year of production introduced some variations in the levels of the 15 elements in the same wine (place and grape). One important conclusion they drew was that, if a control wine exists, the profile of the mineral content proves the similarity of wines for a given producer. This is obviously not the case with classical determinations such as alcoholic content, total acidity, volatile acidity, sulphur dioxide etc.

Partly because of the fact that this technique was not convenient for the normal or advanced wine laboratory, no other determinations were conducted on a sufficient scale at this level for several years.

Using ARTHUR (Informetrix Corporation, 1981), Kowalski and his team (Kwan, W.O. *et al.*, 1979; Kwan and Kowalski, 1980) used only two mineral elements, calcium and barium, to classify correctly 14 French (Bourgogne) and 26 American Pinot Noirs (there was a single misclassified sample). They used a simultaneous emission technique (DCP) and 17 elements for this study. Following this pioneering work, more extensive use of minerals was applied to French and other wines.

Lacasta (1982) separated wines from French regions of Médoc and Saint-Emilion, where 64 wines (85%) were classified correctly (Figure 3.4).

Van-Zeller and colleagues (Médina and Van-Zeller, 1984) determined 14 mineral elements by atomic absorption and emission spectrometry (with flame and furnace) and with methods developed in our laboratory (Médina, 1978; Lacasta, 1982).

A correct classification of a reduced set of wines from three French regions was obtained using Li, K, Ca, Cu, Ni and linear discriminant analysis as well as other classification techniques (SIMCA) (Informetrix Corporation, 1981).

Gonzales-Larraina *et al.* (1987), found that Mn, Li, K and Fe were the most important elements for a classification of the three regions of Spanish Rioja. Classification reached 93% whatever method was used (e.g. multilinear regression, K nearest neighbours, SIMCA). Using the first two elements, a good distribution was obtained (Figure 3.5), the distinction even occurring between two types of production from big wineries and a family style operation (Rioja Industrial and Rioja Cosechero).

Maarse (1987) used those elements together with Na and Fe to distinguish German wines from the Rhein-Pfalz and Mosel regions.

The difference in mineral elements was shown as the basis of the typicity for Spanish wines from Galicia (northern Spain) and the rest of Spain (Herrero-Latorre and Médina, 1991).

At the regional level in Galicia, Latorre *et al.* (1994) showed that using only rubidium and lithium, a very good differentiation is found for Rìas

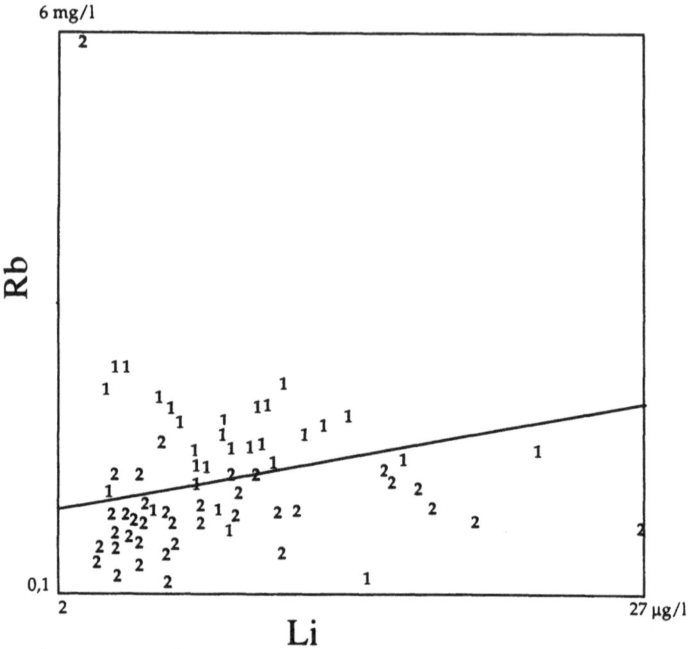

Figure 3.4 Classification of Médoc and Saint Emilion wines (Bordeaux) using rubidium and lithium. 1 = Médoc; 2 = Saint Emilion.

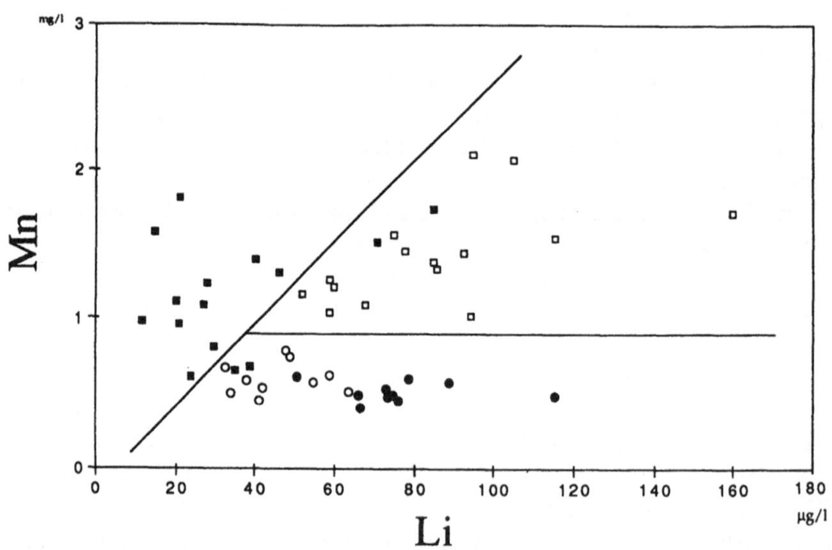

Figure 3.5 Classification of wines according to the three Rioja regions using manganese and lithium: ■ = Rioja Baja; □ = Rioja Alta; ○ = Industrial Rioja Alavesa; ● = Artisanal Rioja Alavesa.

Baixas compared with other white wines from the two other regions: Ribeiro and Valdeorras (Figure 3.6).

Using Ca, Na, K and Fe, in the same data set, they pointed out that these elements can be influenced by fertilisation and the wine-making process and are not relevant in a classification as shown in pattern recognition techniques (Fisher weights in ARTHUR, Infometrix, 1981).

3.3.2.2 Discussion. It is important to know to what extent the mineral content of the must reflects that of the soil and also how much is transferred to wine. However, it is not the total content of the minerals in the soil that is important but its availability, which, in turn, depends on several factors such as pH, humidity, porosity, clay and humic acid complex etc.

The composition of vegetation reflects to some extent that of the soil. This fact has given rise to geobotany, a geological science devoted to the search of species characteristic of certain ores and deposits (Brooks, 1972). It has been proven (Kim and Thornton, 1992) that vegetation (e.g. lettuce and rice) growing on soils such as uraniferous black shales that are rich in elements such as Cd, Mo and Se, accumulate these elements up to levels that may be of concern for health. Grape varieties seem to selectively take up certain elements, as discussed in the following paragraphs.

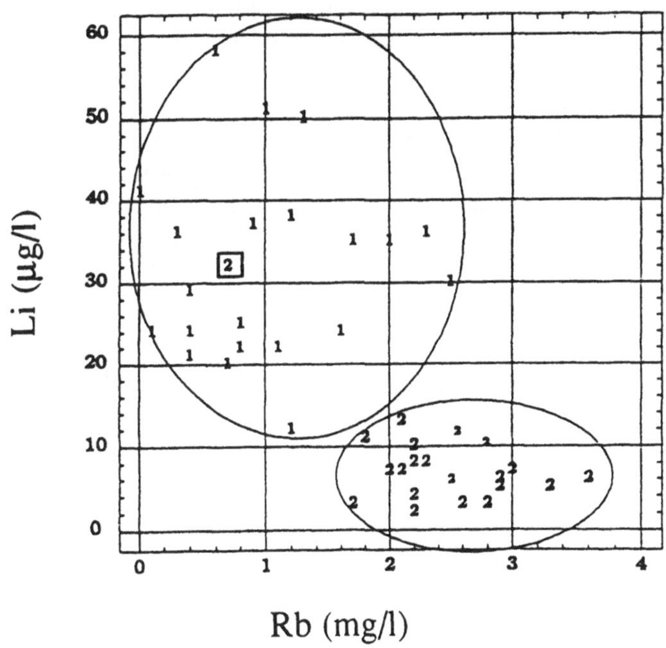

Figure 3.6 Plot of first selected feature, lithium, *versus* second selected feature, rubidium. 1 = non-Rìas Baixas; 2 = Rìas Baixas.

It must be kept in mind that grape cultivation and harvest is by no means a sterile process and that pollution by dust and soil is unavoidable. This pollution varies from year to year (because of rain and wind) and must account for a significant part of the mineral content in wine. The bunching of grapes that facilitates the retention of dust is also a factor not found in larger fruits. It is also clear that the wine-making process can alter considerably the mineral content of wine. For example, the use of technological additives or processes such as cold treatment can lower the content of potassium. In the future, more-advanced technology, such as reverse-osmosis, will probably introduce more factors of variation.

Wine-making equipment is also of great importance for metallic elements such as Fe, Cu, Zn, Pb (Legrand *et al.*, 1991).

When using mathematical analysis, the important thing to consider is the selection of elements that can vary during the process in the same way in all of the wines; it is not so much the absolute concentration of one element that is important but its relationship with the others.

Taking all the previous arguments into consideration, the choice of the mineral elements based on experience can be summarized as follows: potassium, calcium and sodium are elements that can vary widely, both with the process, and from wine to wine. Potassium and calcium can combine with tartaric acid and form potassium bitartrate and calcium tartrate respectively; these molecules can precipate the potassium influenced by temperature. The deacidification of wine can be carried out with calcium carbonate or potassium bicarbonate, thus producing an increase in the content of those elements. Bentonite fining adds sodium and aluminium to the wine (Etchats, 1986).

Metallic elements likely to migrate into wine from poor equipment are not the ideal choice because even if the trend toward stainless steel is making progress, traditional apparatus will be used for some time to come. That is why iron, copper, zinc, coming from bronze or brass, are inappropriate for our purposes. High lead content (above 150 μg/l) shows that the wine has been in contact with material not suitable for food manufacture. This may also lead to other elements, previously mentioned, also having an elevated value, sometimes going over the OIV recommendations of copper:1 mg/l, zinc:5 mg/l and lead:0.25 mg/l.

One problem also arises with the oenological treatments such as potassium ferrocyanide. This practice, authorised in wines by the EC regulation (JOCE, 1987) and performed only under the supervision of an oenologist, removes excess iron and other metals, namely zinc, manganese and copper (Etchats, 1986). French legislation is more restrictive and this treatment is forbidden in red wines.

As pointed out by Siegmund and Bächmann (1977), the only correlation found between soil and wine was shown for alkaline elements, even more so for lithium, which with rubidium appears to play a key role in geographical classification. Both elements are, however, relatively easy to determine in wines using flame-emission spectrometry (Lacasta, 1982). Lithium is found at a very low level in wine (1–200 ppb) but flame-emission spectrometry using acetylene fuel is very sensitive and permits an accurate evaluation of this element.

Rubidium is also easily determined in wines and has values in the range 0.5–5 ppm. Consistent work has been performed using these metals.

Table 3.1 (Sudraud and Médina, 1993) summarises the elements selected for a correct classification by several authors although it is not intended to be exhaustive. The elements most often cited are, by order of weight: rubidium, lithium, barium, strontium, magnesium, aluminium, manganese. The last two elements have to be considered carefully as already indicated.

These consistent results must be considered with caution because elements such as nickel, chromium and lithium are components of the glass of the bottle and can with time migrate into the wine (Médina and Sudraud, 1979; 1980). During ageing, the relative proportion of these latter elements may change according to the glass composition leading to inaccurate results in terms of geography. This fact is partially confirmed by a careful study by Siegmund and Bächmann (1977) where high chromium levels have also been found in wines (up to 2.5 ppm).

Apart from these considerations, alkaline metals, being easily mobilised in the soil and easily transported in the plant, are good indicators of geographical identity.

3.3.2.3 Ultra-trace elements: the lanthanides. In the light of the previous discussion, it was considered interesting to investigate one group of elements having the same chemical properties, namely the lanthanides or rare earth elements (REE). These elements have a mass of 140–180 and their patterns give insight into the history of a particular rock or soil as pointed out by the abundant bibliography on the subject in the geological sciences.

Extensive research on these elements has lead to efficient determination

Table 3.1 Mineral elements used for classification

Reference		Country or region	Elements used
Kwan *et al.*	1979	USA/France	Al, Ba, Ca, K, Mg, Sr
Médina and Van-Zeller	1984	France	Ca, Cu, K, Li, Ni
Maarse *et al.*	1987	Germany	Fe, Na, Rb
Gonzales-Larraina *et al.*	1987	Spain (Rioja)	Li, Mn, Rb
Etievant *et al.*	1988	France	Li, Mg, Rb
Herrero-Latorre and Médina	1992	Spain	Ba, Li, Mn, Rb
Day	1993	France	Al, Ba, Fe, K, Rb, Sr, Zn
Latorre *et al.*	1994	Spain (Galicia)	Li, Rb

techniques, even at the extremely low levels at which they are found in wine. ICP/MS is frequently the best technique for such determinations, largely because of the simple, easily interpreted spectra and the exceptionally low limits of detection possible, even with standard ICP/MS instruments. This technique has evolved rapidly and the development of new instruments allows accurate determination of ng/l or ppt in wines without preconcentration.

The authors used this technique to investigate the case of rare earth elements in relation to the geographical determination of the origin of wines. Figure 3.7 gives the spectra of rare earth elements obtained directly on a wine. Table 3.2 presents the quantitative results of all 14 REE in four wine samples from France (Augagneur *et al.*, 1996).

In geology, it is a widely accepted practice to normalise REE concentrations using chondrite values (Masuda *et al.*, 1973). The chondrites are meteorites whose values represent the natural pattern of REE levels, which were unaffected by geological/geochemical processes on earth (they act as control for the state of matter when the solar system was formed). For rocks of volcanic origin, chondrite is normally used, whereas marine sedimentary rock REE patterns are usually corrected using values from a reference marine shale.

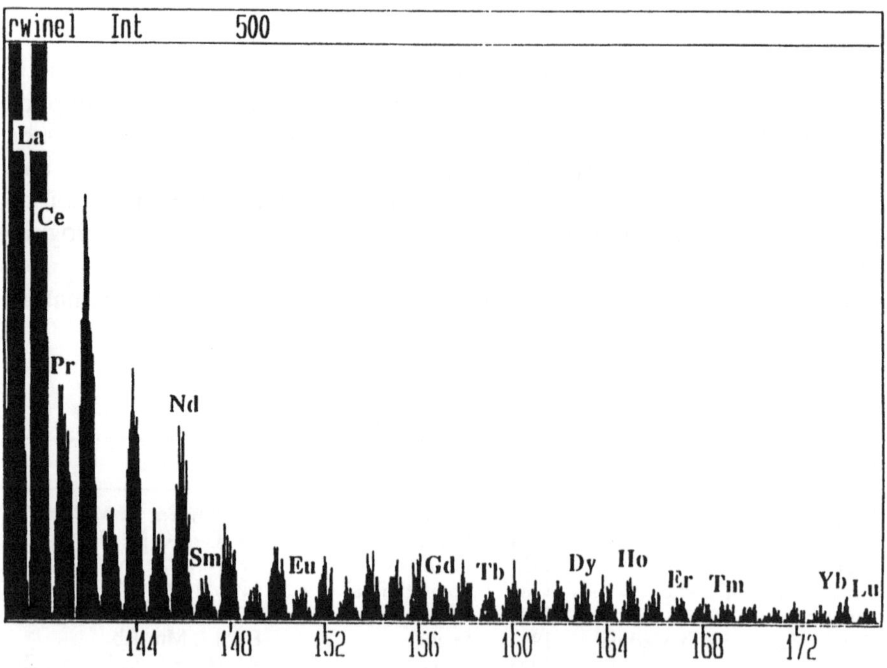

Figure 3.7 Spectrum of the REE in the pommard sample.

Table 3.2 Concentrations (pg/ml) of rare earth elements in four wines[a]

Element	Sancerre	Pommard	Pomerol	St Emilion
La	553.0 (11.5)	1370.0 (41.7)	1420.0 (4.82)	1030.0 (16.5)
Ce	1130.0 (22.5)	2140.0 (44.6)	117.0 (9.15)	63.2 (1.97)
Pr	136.0 (2.42)	369.0 (6.49)	51.0 (2.95)	7.20 (1.75)
^{146}Nd	575.0 (13.8)	1540.0 (6.98)	201.0 (9.78)	34.9 (5.69)
^{147}Sm	146.0 (10.9)	294.0 (23.9)	55.7 (2.99)	24.6 (2.62)
Eu	48.6 (3.00)	99.9 (5.32)	37.9 (5.32)	28.6 (0.86)
Gd	171.0 (4.38)	324.0 (17.1)	74.8 (3.98)	19.6 (7.53)
Tb	29.7 (3.21)	41.5 (2.26)	10.7 (0.32)	5.44 (0.22)
Dy	232.0 (6.21)	239.0 (1.44)	58.6 (3.84)	22.3 (8.00)
Ho	54.2 (2.87)	49.1 (2.46)	12.4 (0.31)	5.87 (1.81)
Er	196.0 (4.71)	141.0 (6.44)	40.2 (1.45)	18.9 (2.89)
Tm	42.7 (2.91)	21.7 (2.09)	7.22 (0.91)	4.45 (0.68)
^{174}Yb	326.0 (5.36)	138.0 (7.06)	35.4 (2.27)	18.2 (2.22)
Lu	67.1 (3.01)	23.7 (2.62)	12.0 (2.98)	8.16 (1.19)

[a]Concentrations in pg/ml; Standard deviations (SD) shown in brackets.

The corrected values when plotted against the atomic number of the element should give a smooth curve or straight line if the measured set of REE values is internally consistent. Chondrite plots are frequently used (Jarvis, 1988; Jarvis and Jarvis, 1988), as they are much easier to interpret and compare than raw REE data, in particular, anomalies for cesium and/or europium are more noticeable on an otherwise smooth plot.

The soils are the compositional reflection of the underlaying rocks that gave birth to them. Plants growing on those soils take up trace elements according to the composition of the soil as well as their chemical properties, which the REE composition also reflects (Brooks, 1972). Wine will probably retain the REE patterns of the soils and the possibility exists to identify the region of origin as shown in Figure 3.8 (McCurdy *et al.*, 1992). Clear anomalies exist for Ce and Eu for wines of the Bordeaux area.

Other representations using the whole set of REE such as principal component analysis (PCA) can be used. For red wines, a clear distinction between regions is made based on those elements (Figure 3.9).

3.3.2.4 Stable isotopes. A very appealing and sophisticated technique using 'stable' isotopes such as mass spectrometry of isotope ratios (MSIR) and nuclear magnetic resonance (NMR) has been used for geographical determination of wines. However, the results are difficult to interpret in terms of variations in the year, climate, grape varieties, fermentation process, etc. and the reference to a physical or biochemical process even remains somewhat empirical. The use of lead isotopes is related to atmospheric pollution and seems much more practical to use because this phenomenon is fairly well understood and is less dependent on small local variations.

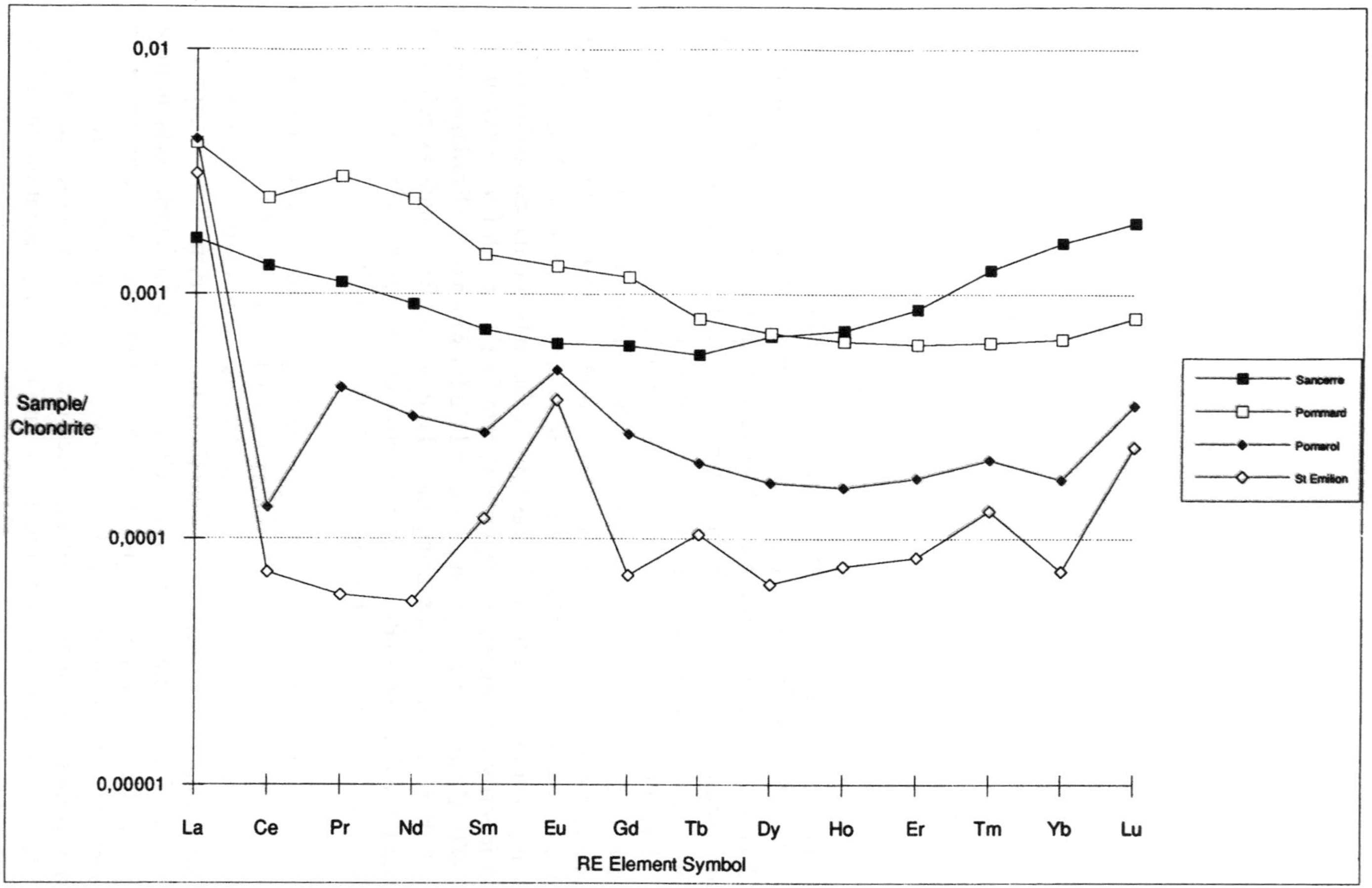

Figure 3.8 Normalised concentrations of REE in wines (chondrites).

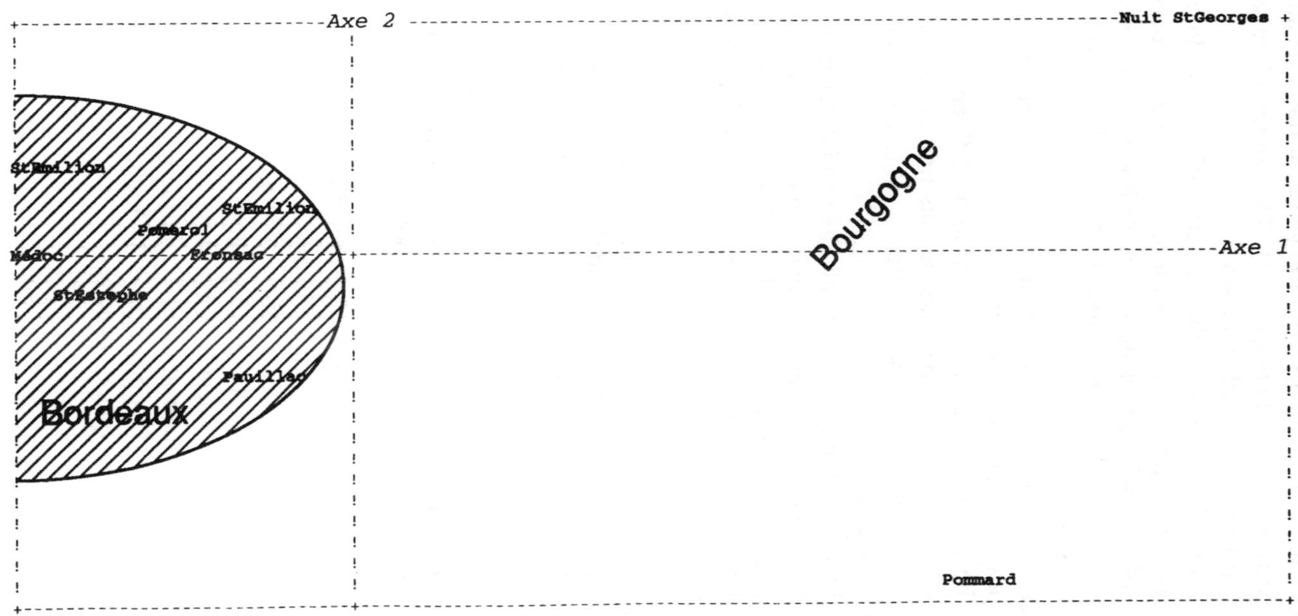

Figure 3.9 Principal component analysis: REE in red wines.

3.3.2.4.1 Carbon (^{13}C), hydrogen (^{2}H) and oxygen (^{18}O) isotopes. Several isotopes can be used: those from the water of the wine and those from the ethanol (main organic component). The source of variation of these products can be summarised as follows.

Water absorbed by the roots is combined with carbon dioxide from the atmosphere to produce sugars by photosynthesis. During this process there is fractionation of the different isotopes. A well-known process is the isotopic fractionation of ^{13}C during photosynthesis, which divides plants into three main groups.

The evapotranspiration of water in a plant has a concentrating effect on the heavy isotopes of oxygen and hydrogen.

Rain during the ripening stage can also have an influence by dilution, especially just before the harvest because, at this time, it contains fewer 'heavy' isotopes. The addition of concentrated must (rectified or not) will complicate the matter further.

Climate, continent and topography can all have an influence on isotopic fractionation: as little as a 150 m elevation is significant (Förstel, 1995).

Owing to all these sources of variation and relativity of the measurements, a well-conceived database is absolutely essential for these methods to be useful and accurate.

3.3.2.4.2 Deuterium. Deuterium is specifically distributed in the ethanol molecule produced in fermentation and is related to the deuterium content of the water and the sugar molecules. The deuterium of the methyl group (D/H)I of ethanol comes from a non-exchangeable site of the sugars. On the methylene (D/H)II site it contains a contribution from the fermentation water, and the hydroxilic group is in equilibrium with the deuterium of the wine water (Martin *et al.*, 1988). The site-specific isotope fingerprint has been used for detecting exogenous sugars in wines (Martin and Martin, 1983).

The use of these parameters permits the differentiation of wines produced in countries with substantial climatic or latitudinal differences. Using the (D/H)I (D/H)II and (D/H) of the water from wine and must, Martin *et al.* (1988) accomplished a good separation even when the wine produced from different vine varieties gave different values in the same region, and even when the maturation effect influenced the values (non-matured grapes lead to a 0.3–0.8 ppm increase). A good distinction is made between the three different types of climates:

- cold and humid
- cold and dry or hot and humid
- hot and dry.

Even for the same country (France) the regions (Alsace, Anjou, Bordeaux) were correctly separated for the same year of wine production.

As it is so dependent on the climate and its variation, the use of these isotopes, although useful, cannot be extrapolated from the year of production. Without this information, the sample can be attributed to several regions of origin in France (Day, 1993). These variations will be more severe for 'Atlantic' climates than for Mediterranean ones. As discussed later in this chapter, determining the year of production is not an easy task.

3.3.2.5 A useful metal: lead. Lead is a wine contaminant with an average of 60 μg/l in wines (OIV, 1995). The current limit set by the OIV is 250 μg/l.

This metal is a special case in the determination of authenticity. It is a very old toxicological problem known since the Roman empire, whose decline was supposed to be caused partly by excess lead intake from food and especially wine because lead acetate was added as a sweetener. A famous theory says that this high level of lead hampered the intellectual capacity of the ruling class at the time.

Currently, wines with a content under 100 μg/l are considered not to be polluted by the material during the wine-making process or storage. This very old problem has recently found a new development with the application of sophisticated analytical techniques until now confined to geological and environmental studies.

3.3.2.5.1 Organo lead. Using techniques developed for use in the Arctic circle icefield for the determination of organo-lead compounds from gasoline additives, these compounds were found in wines.

The determinations were set up at Anvers University by Lobinski and coworkers (1994), who used the special hyphenated technique (GC/MIP/AES (gas chromatography/microwave induced helium plasma/atomic emission spectrometry). The authors found ethyl lead and methyl lead but the latter was predominant. The compounds derive from tetraethyl and tetramethyl lead respectively, which were added as anti-knock agents in gasoline. Partially burned in the motor (40–80%), they are decomposed in the atmosphere under the action of ultraviolet light, ozone etc.

Extremely low levels were found – in the range of 10–500 ng/l for trimethyl lead and 0–50 ng/l for triethyl lead. The maxima were found in wines produced from grapes from vineyards close to the junction of two major highways in France in the Rhone valley. This study compared levels in 'vertical' series of wines covering over 40 years in the same area. The concentration of trimetyl lead is no higher than 25 ng/l in rural areas but increases with the proximity of highways and cities. With the increased use of unleaded fuel in the EU in the last 10 years the level of those compounds in wine has diminished sharply. For a chronological series of wines between 1950 and 1991, there is an increase of trimethyl lead after the introduction of the tetramethyl compound in 1960, a maximum *circa* 1978 and then a sharp decrease from 1980 to the present (Figure 3.10). Triethyl lead is present in 1960 but levels decrease after 1980.

Until recently, there was no European wine that did not contain alkyl lead compounds. In the USA and Australia, which have used unleaded fuel for a longer period of time, there is no organic lead in wine at present, however, before 1987, levels similar to those of European wines are found. The use of this time shift and the relative proportion of methyl and ethyl lead allows one to plot concentration curves to give an evaluation of the age of the wine from the 1960s to the present.

The use of these curves permits several deductions with reference to the year of production and the origin of the wine:

- no organo lead: the wine originates from countries having banned those compounds many years ago (USA and Australia) or is a very recent European wine
- according to the vintage, the relative proportion of trimethyl and triethyl indicate the year of production (i.e. pre-1960 there is no methyl lead present)

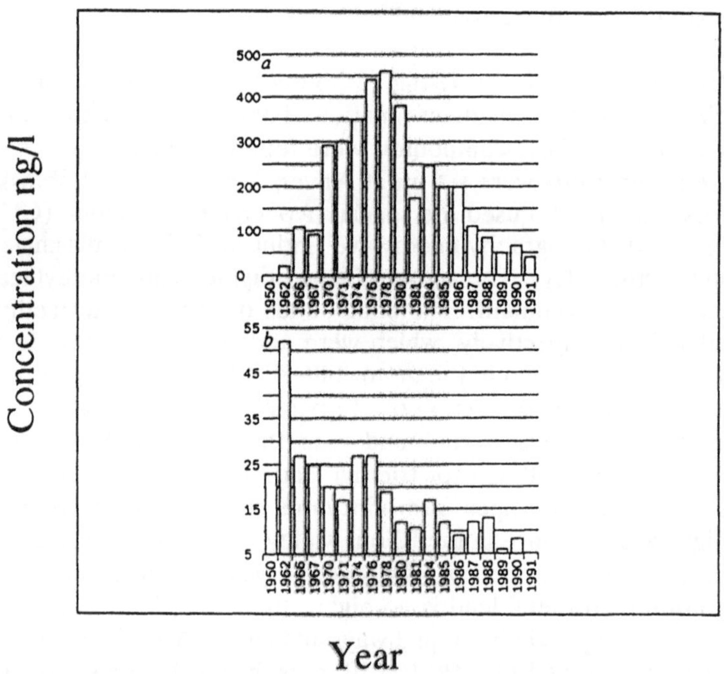

Figure 3.10 Changes in the concentration of organolead as a function of vintage: (a) trimethyl lead; (b) triethyl lead. Source: Lobinski *et al.* (1994).

- high levels of organoleads implies proximity to a heavily used road or a city (trimethyl lead > 100 ng/l and triethyl lead > 15 ng/l). For example, Rhône wines could be distinguished from other wines. The opposite is true for a rural wine (trimethyl lead < 25 ng/l and no triethyl lead present).

3.3.2.5.2 Lead isotopes. Lead has four isotopes, three of which come from the decay of uranium and thorium. Consequently, lead ores have a different proportion of lead isotopes according to their original content in these radioactive minerals. Figure 3.11 shows an ICP–MS spectrum of lead.

One major reason to use lead isotope variation as an indicator lays in the fact that their variation is much greater than that of lighter elements such as carbon or oxygen, which are in the part per million range. The variation between the two major lead mines (Missouri, USA, and Broken Hill, Australia) is close to 30%. Indeed, lead isotope ratios have been used for 20 years to trace lead pollutants (Chow *et al.*, 1975). Isotopic studies have shown correlation between lead in gasoline and soil and plants (Chow, 1970). In addition, heavily polluted soil has been found for metropolitan areas and is proportional to the proximity and the volume of traffic, ranging from 100–1000 ppm; these values should be compared with the level of lead in the Earth's crust, which is 16 ppm (Krauskopf, 1956).

Geographical origin. A graphical representation (Figure 3.12) of two of the isotopic ratios shows a clear distinction of wines by 'continent' (MacKay, 1994). These values have been obtained with ICP–MS and confirmed in our laboratory. This pattern is certainly caused by differences in the pollution of lead in the air spread by automobile exhaust. Organolead is volatile and can be found as far away as the poles but most of it is burned and remains in the same geographical area. The variation therefore reflects the origin of the lead used as anti-knock additives: from Missouri in the USA and Broken Hill in Australia. The situation in Europe is intermediate since lead comes from different sources (Morocco, Australia etc.). Even if the total content of lead emitted in the atmosphere is reduced, current levels of pollution will remain for some time as will the lead ratios.

Age using lead isotope ratios. The increased use of unleaded gasoline in Europe also leads to variation in specific ratios. As seen in Figure 3.13, the variation of $^{206}Pb/^{204}Pb$ for French wines (determined by ICP–MS) is closely related to their year of production (correlation coefficient of 0.94). To confirm this trend, two data points taken from the bibliography (Elbaz-Poulichet *et al.*, 1984 – referenced in Grousset *et al.*, 1994) in a study of aerosol samples, match those for wines. Wine is therefore strongly marked

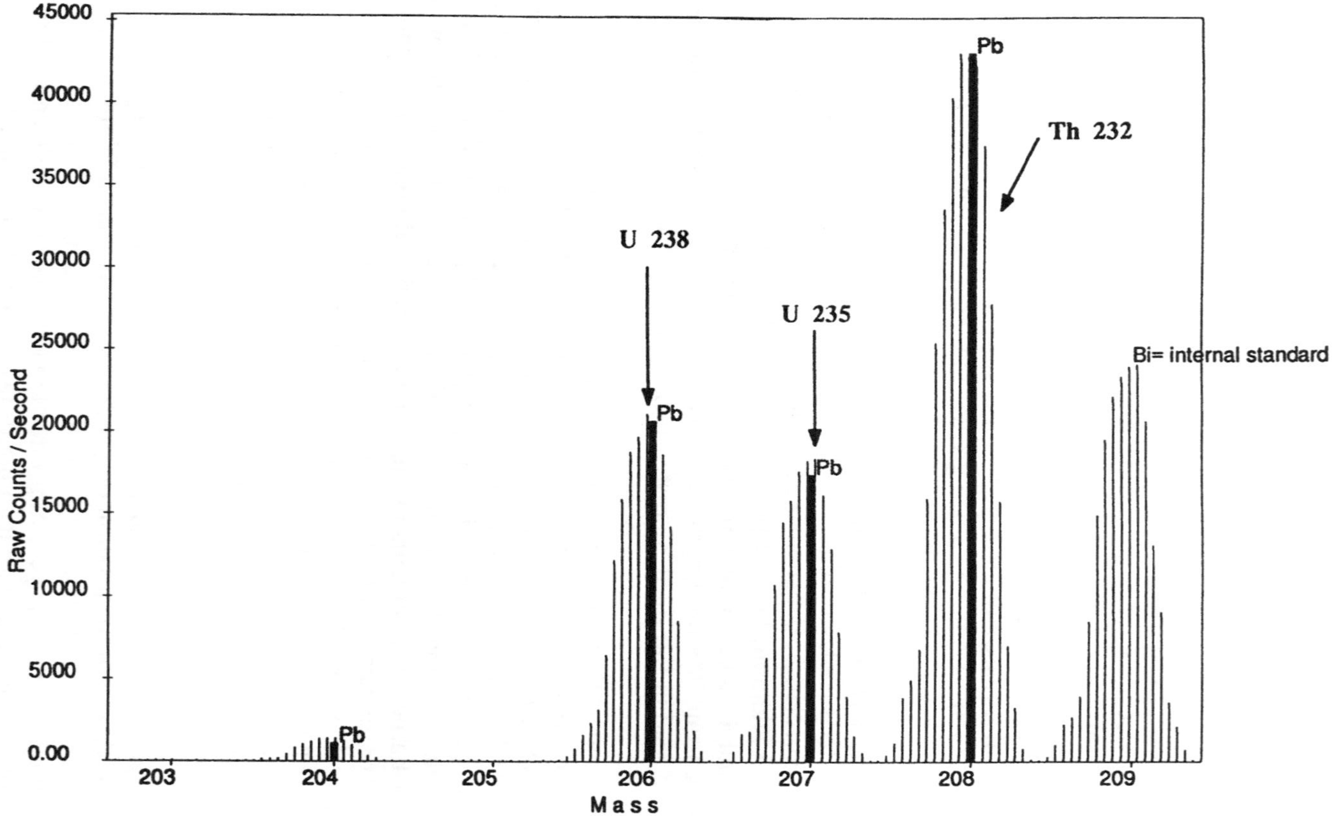

Figure 3.11 ICP/MS spectra of lead in wine.

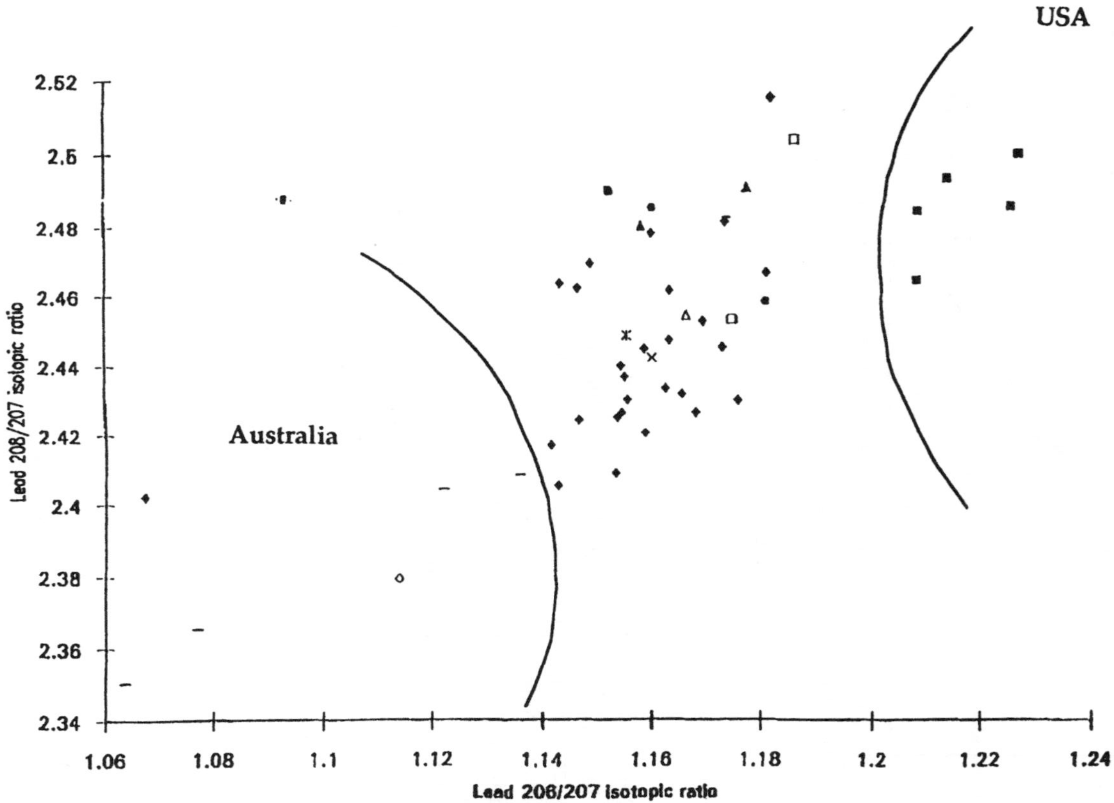

Figure 3.12 Separation of wines according to their 'continent' of origin using isotopic ratios. Source: MacKay (1994).

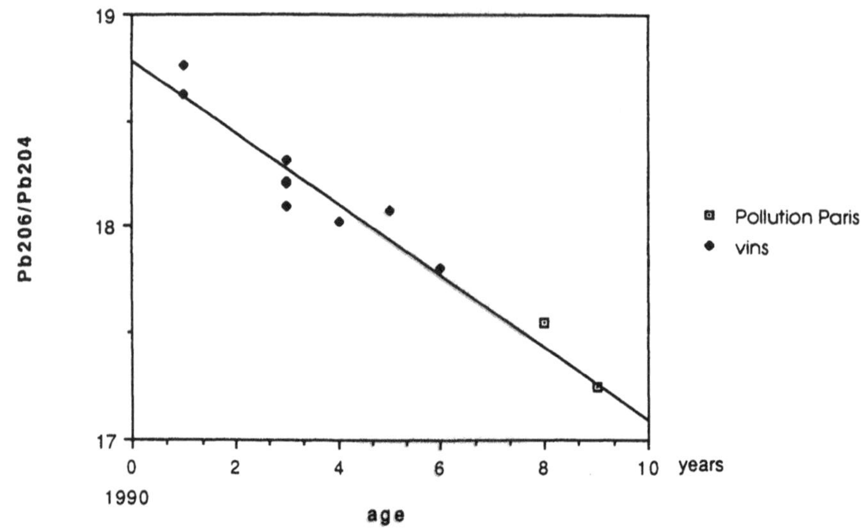

Figure 3.13 Isotope ratio and wine vintage ICP/MS.

in its composition by the environment and this provides valuable data. The concentration trend for lead is from a situation of lead pollution toward natural earth crust ratios. Indeed, this trend was found for aerosols in a study in San Diego, California, from 1964–1974 (Chow *et al.*, 1975) and confirmed by Grousset *et al.* (1994) working on European aerosols. Another factor in favour of reduced pollution is the fact that very low levels of lead are now found in wine, regardless of the source of the constituents. It is interesting to note that lead concentrations in aerosols are decreasing in a similar fashion (Grousset *et al.*, 1994).

Conclusions on lead. Differences existing in wines from Australia and the USA are mainly caused by atmospheric pollution by lead from the exhaust of automobiles burning leaded gasoline. Organic lead can be found at extremely low levels (0.5 μg/l) testifying to the influence of this phenomena. Further study shows that the extent of the atmospheric pollution can be traced more extensively in wine. Only a small fraction of organic lead introduced into gasoline exhaust remains unaffected by combustion (1–10%); in addition natural reactions occur that decompose the remaining compounds (ultraviolet light etc.). That is why only very sophisticated techniques (such as GC–AED) are able to detect alkyl lead in wine. However, isotopic ratio determination of $^{206}Pb/^{204}Pb$ shows a definite variation according to the year of production, following the increased use of unleaded fuel throughout Europe.

3.3.3 Organic and mineral features

It is appealing to construct a database where minerals and organic substances are mixed in the hope of encompassing several factors of variation such as grape variety, soil, climate, and so on, all in one step. Several authors have used this approach, following the pioneering work of Kowalski and coworkers (Kwan and Kowalski, 1980a; 1980b; Kwan *et al.*, 1979). Using the database in their work and the partial least squares (PLS) method, Frank and Kowalski (1984) showed that elemental analysis performed much better than organic analysis when differentiating French, Californian and Pacific Northwest wines. A combination of the two chemical blocks gave the best prediction.

In our laboratory several studies on the subject were also conducted. Médina and Van Zeller (1984) used minerals, volatile, polyphenolic and organic acids and classical determinations (62 variables) in a study on French red wines. Wines representative of their area of production were selected. In this feasibility study, the principal component analysis (PCA) technique was able to distinguish between wines of the three regions examined (Figure 3.14). The second axis allowed the differentiation of the Bordeaux regions (Figure 3.15). Searching for the fewest substances that could successfully characterise groups lead to the selection of lithium, propanol, procyanidin B2 and decanoic acid (see Figure 3.14). Each group of substances taken individually (e.g. phenols, minerals etc.) could be used for an accurate separation using group classification methods (KNN, LDA, SMICA in ARTHUR, Infometrix, 1981).

Following this work and having examined the most useful elements, Manterola (1985) used three mineral elements (lithium, manganese and rubidium) and one all-purpose HPLC technique for 13 neutral polyphenolic substances (Salagoïty-Auguste and Bertrand, 1984). The aim of this study was to separate the wines of the main Bordeaux areas using variables meaningful in terms of soil and grape variety.

The separation of 36 wines from Chateaux from the right bank of the Dordogne (Pomerol, St Emilion) compared with those from the left bank of the Garonne (from North to South: St Estèphe, Pauillac, St Julien, Graves) fell along the second principal component axis when using a principal component analysis of all the variables (see Figure 3.15a). The construction of this axis is dependent on:

- manganese and lithum levels being characteristic of the soil
- epicatechin and procyanidin B2 and B4 being representative of the variety; Cabernet Sauvignon is used in Médoc and Merlot in St Emilion and Pomerol
- tyrosol being dependent on the fermentation process.

The group classification was sufficiently accurate with two misclassifications

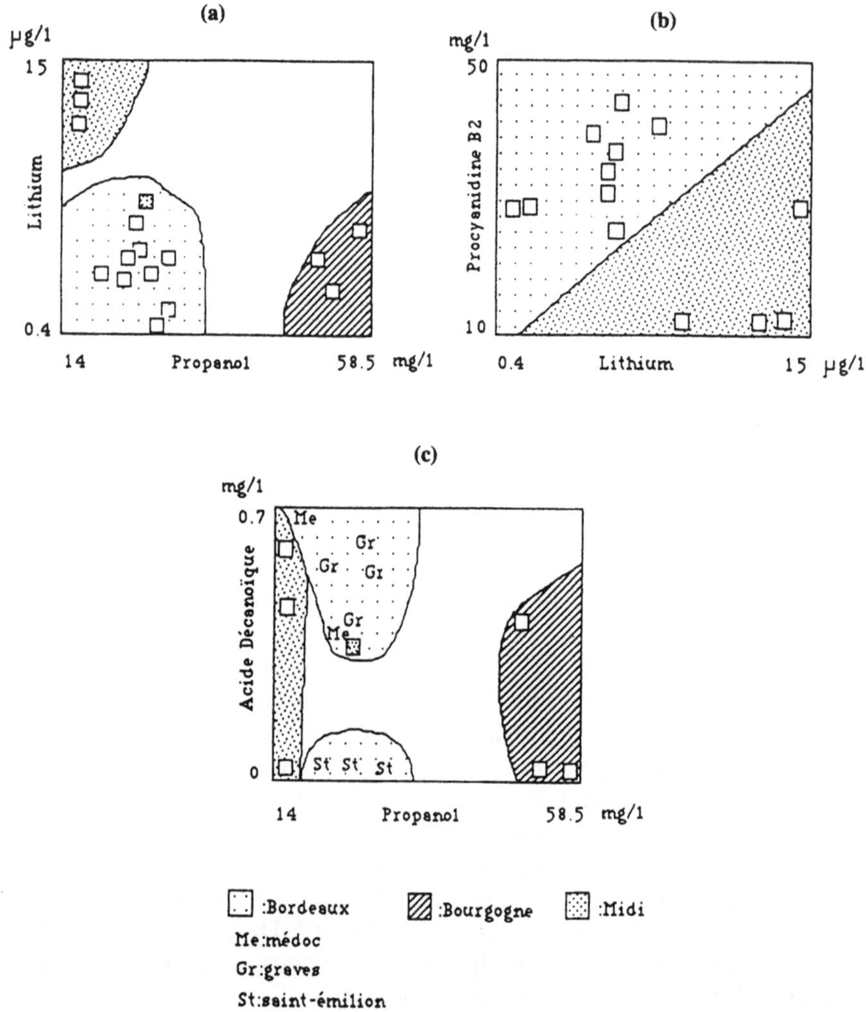

Figure 3.14 Plots of the best features for a regional classification of French red wines: (a) lithium *versus* propanol; (b) procyanide B2 *versus* lithium; (c) decanoic acid *versus* propanol.

using linear discriminant analysis and five substances. Using the first two substances, a satisfactory picture is drawn (see Figure 3.15b). For wines from the Médoc area, discrimination is achieved by two-dimensional analysis of myricetin and epicatechin levels (see Figure 3.15c).

St Emilion and Pomerol are two regions lying close together but when sophisticated ratios using tryptophol (alcohol from the benzene series), catechin and an unknown, but constant, HPLC peak (B) are used, a perfect separation is achieved (Figure 3.15d).

3.15(a)

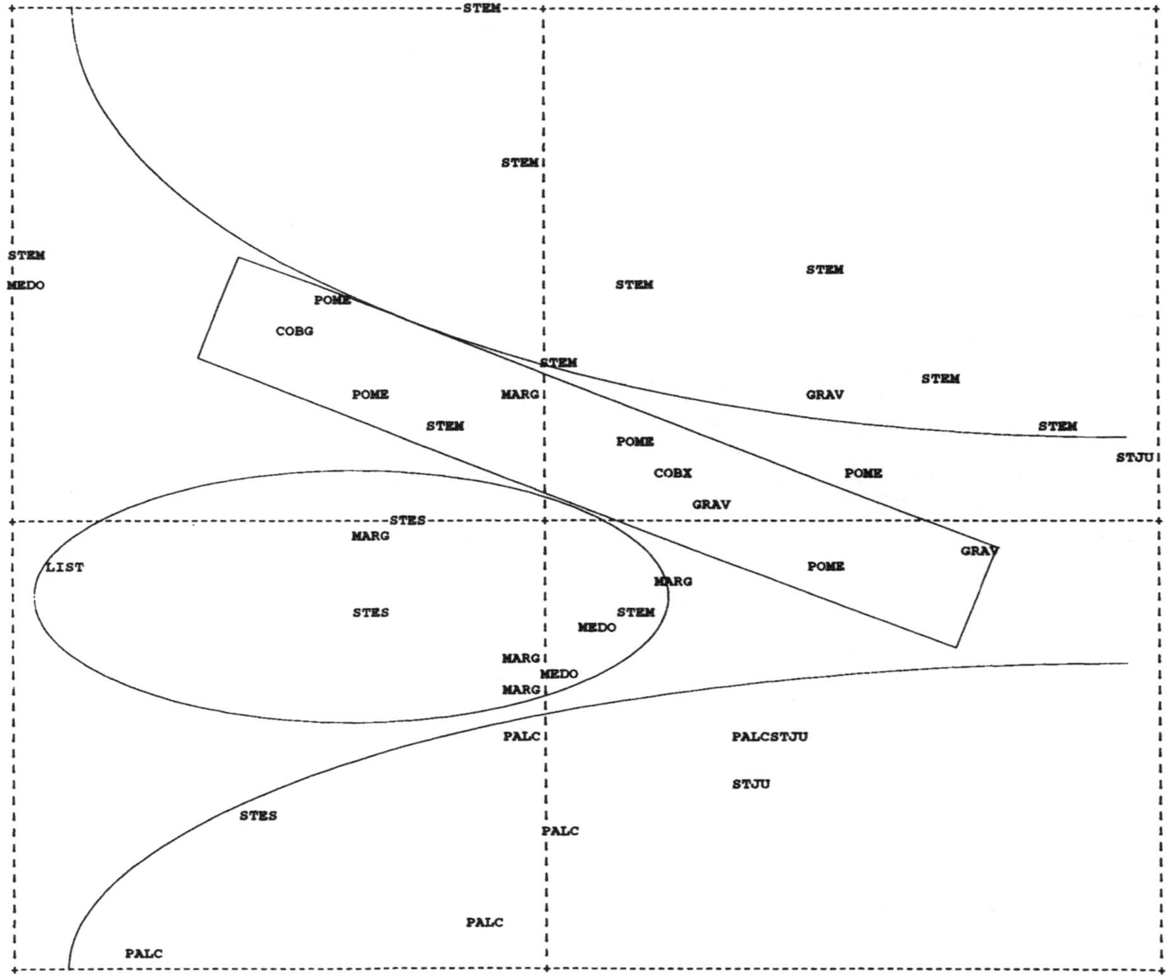
STEM
STEM
STEM
MEDO
POME
COBG
STEM
STEM
STEM
POME
MARG
STEM
GRAV
STEM
STEM
POME
COBX
POME
STJU
GRAV
STES
MARG
LIST
GRAV
POME
MARG
STES
STEM
MEDO
MARG
MEDO
MARG
PALC
PALCSTJU
STJU
STES
PALC
PALC
PALC

3.15(b)

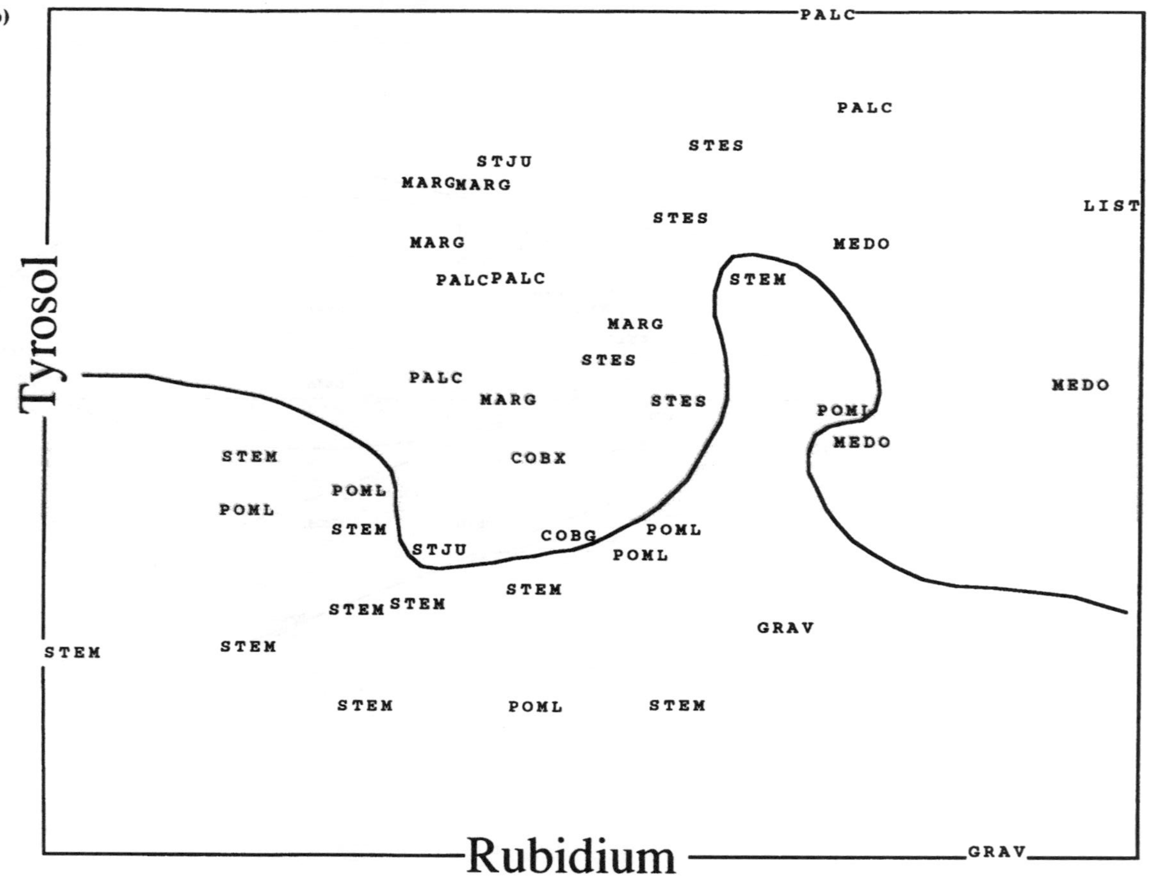
PALC
PALC
STES
STJU
MARGMARG
STES
LIST
MARG
MEDO
PALCPALC
STEM
MARG
STES
PALC
MARG
STES
MEDO
POML
MEDO
STEM
COBX
POML
POML
STEM
STJU
COBG
POML
POML
STEM
STEMSTEM
GRAV
STEM
STEM
STEM
POML
STEM
Tyrosol
Rubidium
GRAV

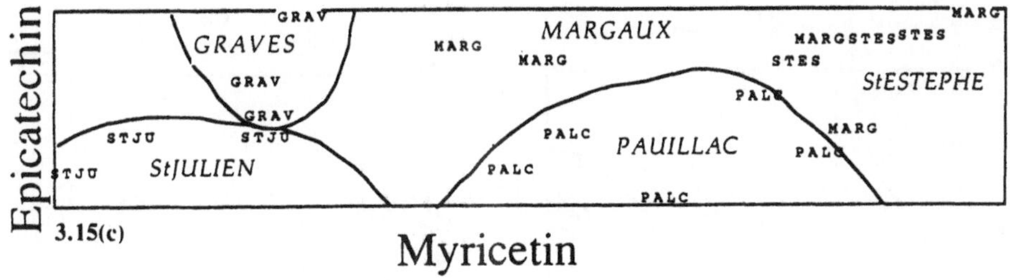

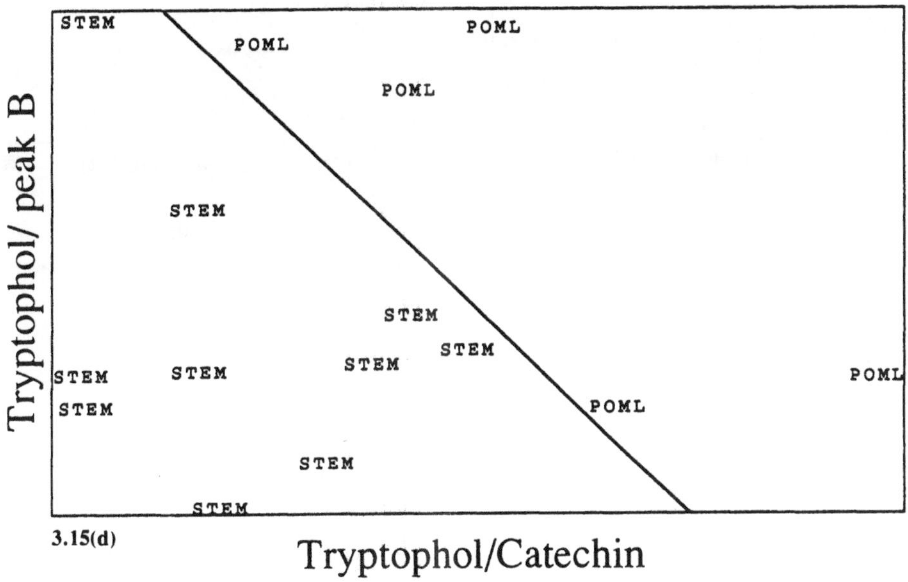

Figure 3.15 (a) Bordeaux wines characterisation using PCA: 36 wines, minerals and neutral polyphenolic compounds were tested. (b) Bordeaux wine separation using tyrosol and rubidium. (c) Médoc wine separation using myricetin and epicatechin. (d) St Emilion *versus* Pomerol separation using ratios of organic substances. STEM: St Emilion; POM: Pomerol; GRAV: Graves; MARG: Margaux; STES: St Estèphe; LIST: Listrac; PALC: Pauillac; STJU: St Julien; MEDO: Médoc (generic); COBG: Côtes de Bourg; COBX: Première Côtes de Bordeaux.

In our laboratory in 1984, research into the classification of French white wines was conducted (Sudraud, 1985). Figure 3.16 shows the separation obtained for a limited set of French white wines, two regions at a time and with only two variables. It is somewhat difficult to obtain a good differentiation for all the regions at once using the metals selected. This tends to demonstrate that the contact of the solid parts of the grapes (seeds and skin) has an influence on the content of the metals selected for separation.

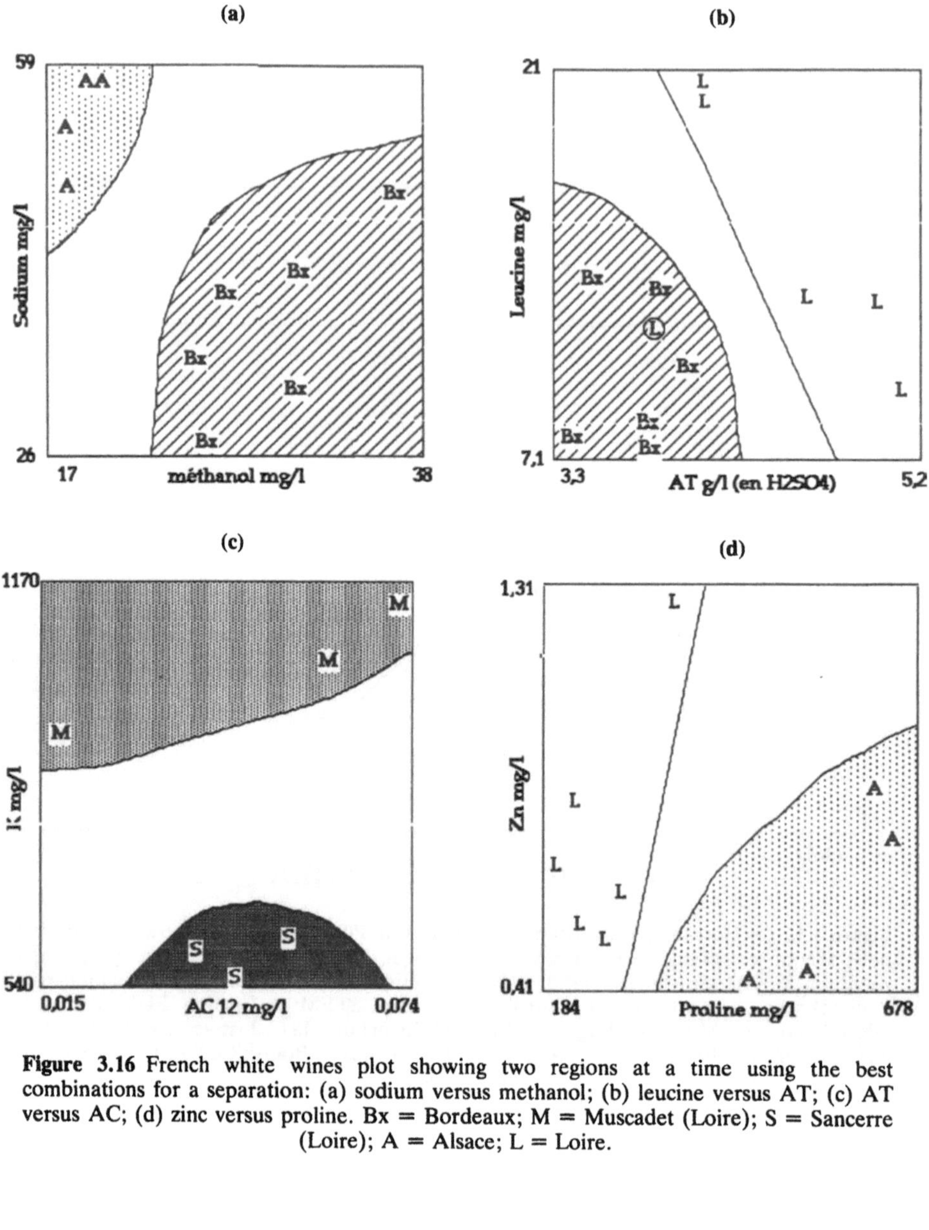

Figure 3.16 French white wines plot showing two regions at a time using the best combinations for a separation: (a) sodium versus methanol; (b) leucine versus AT; (c) AT versus AC; (d) zinc versus proline. Bx = Bordeaux; M = Muscadet (Loire); S = Sancerre (Loire); A = Alsace; L = Loire.

Etiévant *et al.* (1988) used minerals and amino acids to achieve a 94% accuracy when classifying 68 French red wines provided by three agronomical stations and made from pure grape varieties typical of the Loire (Angers), Southern France (Narbonne) and Bordeaux regions.

Fisher's test showed that only rubidium, lithium and magnesium could discriminate between the regions, although six amino acids showed potential. Stepped discriminant analysis selected rubidium, ethanolamine, arginine and magnesium.

It is well-known, however, that the amino acids are dependent upon fertilisation, climatic conditions, and the duration of skin contact in the must; those indigenous to the grape are metabolised by the yeast, excreted and released during the autolysis of the dead yeasts, more complex phenomena also exist.

In contrast, the minerals selected do not vary after the time needed to obtain red wine – 48 h of skin contact (Rizzon, 1985).

3.3.4 Tasting

Tasting was, until recently, the only way to classify wines according to vintage and origin; however, no extensive and objective research has been performed on the subject. Is tasting accurate enough for a correct classification? An expert taster will answer positively and every year there is an amazing worldwide contest of 'sommeliers' where even the name of the chateau is guessed (not to mention the vintage).

Sensory evaluation can be easily rationalised and the use of wine-tasting cards is now widespread. Frank and Kowalski (1984) showed that sensory data do not provide enough information to separate wines from France and the USA. A spin-off of this study demonstrates that elemental analysis contains more information relevant to the organoleptic characteristics than the organic composition analysed at this time.

In a very interesting study (Thurin, 1984), where the expert tasters were asked to guess the name of the pure variety (Cabernet Sauvignon, Merlot, Cabernet Franc) the wine was made of, a surprisingly low percentage (20%) of correct answers was achieved; indeed less than achieved from random guessing. However, when using all the 28 criteria of the wine-tasting card and the classification method SIMCA in ARTHUR, (Infometrix, 1981), which performs a reduction of the variables, the classification reached 90% (using 110 tasting cards). The typicity of the grapes are characteristic enough to be inscribed in a wine-score card, even when grown on different soils (Figure 3.17). The wine-score cards are designed to use the tasters as sensitive machinery without a cognitive function, that in this case, interfered with the result.

3.4 Varieties

Correct classification according to pure variety has been attempted by several authors using amino acid profiles (Vasconcelos *et al.*, 1989; Etiévant *et al.*, 1988), aroma (Rapp *et al.*, 1976; Maarse *et al.*, 1987; Marais *et al.*, 1981), and anthocyanins (Larice *et al.*, 1989).

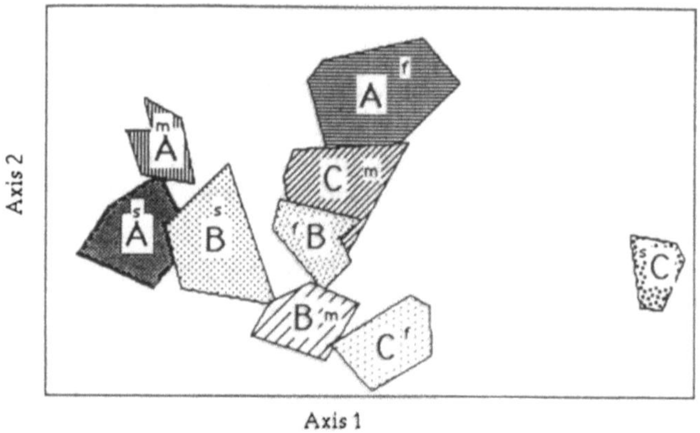

Figure 3.17 Classification of pure varieties of wines growing on different soils using wine-tasting score cards (PCA double stage, 110 tastings). Soils: A = argilo-calcaire; B = graves; S = Sancerre (Loire); C = pallus. Grape varieties: m = Merlot; s = Cabernet-Sauvignon; f = Cabernet-Franc.

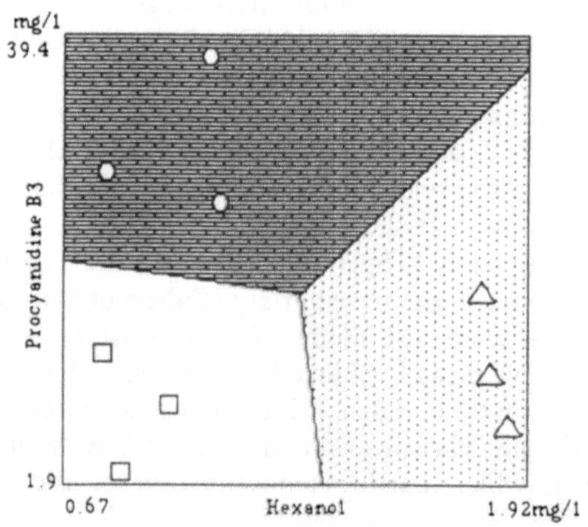

Figure 3.18 Classification of pure varieties of wines growing on different soils using two characteristic features. Grape varieties: O = Merlot; Δ = Cabernet-Sauvignon; È = Cabernet-Franc.

A grape variety seems to have the ability to concentrate selectively some metals, as pointed out by Etiévant *et al.* (1988), allowing discrimination between the typical varieties cultivated in three different regions (this was also performed with amino acids). Latorre *et al.* (1992) also performed a separation into three groups of the main grapes from Galicia: Mencía and Jerez from Godello, Dona Branca, Treixadura and Torrontés. Linear discriminant analysis performed the classification using lithium, nickel, calcium and rubidium. The grapes came from an experimental vineyard and grew on the same soil.

Thurin (1984), on a reduced set of carefully selected data, achieved a perfect separation of the main grape varieties used for Bordeaux wine (Merlot, Cabernet Sauvignon and Cabernet Franc) with hexanol and procyanidin B3, even when the grapes were grown on three different soils (Figure 3.18).

Polyphenolic content is so specific to a grape variety that, more recently, Vercauteren *et al.* (1995) used multidimensional NMR to distinguish between clones of the same species cultivated (Cabernet Sauvignon, Merlot Noir, Cabernet Franc) in the Bordeaux region. The control of the label of a varietal wine would be possible using this technique and with the development of a data bank. The technique requires a simple preparation of extracts and is of short duration (30 min to 1 h).

3.5 Age

3.5.1 Carbon isotope ^{14}C

One of the oldest methods for determining the age of an alcohol takes advantage of nuclear fallout by determining the ^{14}C content of the ethanol (Martinière *et al.*, 1979).

Cosmic rays colliding with nitrogen produce ^{14}C. The radioactivity of carbon from ^{14}C remained constant for two centuries (1752–1952) at a level of 13.6 dpm/g (disintegration per minute per gram). Atomic bombs and, especially, the hydrogen bombs tested in the atmosphere, produced neutrons that reacted with nitrogen and produced radioactive carbon. This carbon is rapidly oxidized to form CO_2 and is incorporated in plants. Unlike deuterium (^2H) and ^{18}O, the diffusion of the radioactive $^{14}CO_2$ was worldwide and homogeneous. There is little variation (±5%) between plants or with latitude and longitude.

The maximum value was obtained in 1963 with a value of 28 dpm; because most atmospheric nuclear testings stopped in 1964, a decrease has been observed since then but currently remains above the natural level (Mongereau and Evin, 1993). Figure 3.19 shows the shape of the curve.

This technique can also be applied to wines and spirits after a distillation process that must reach at least 95% by volume ethanol (double distillation for the wines).

Nevertheless, several factors can affect the results, such as chaptalisation if it is carried out with sugar of a different age than the grape. Given the shape of the curve, it is not possible to distinguish between the age laying on each side of the curve along the same horizontal line. Given the curve's tendency to flatten out and the uncertainty associated with the

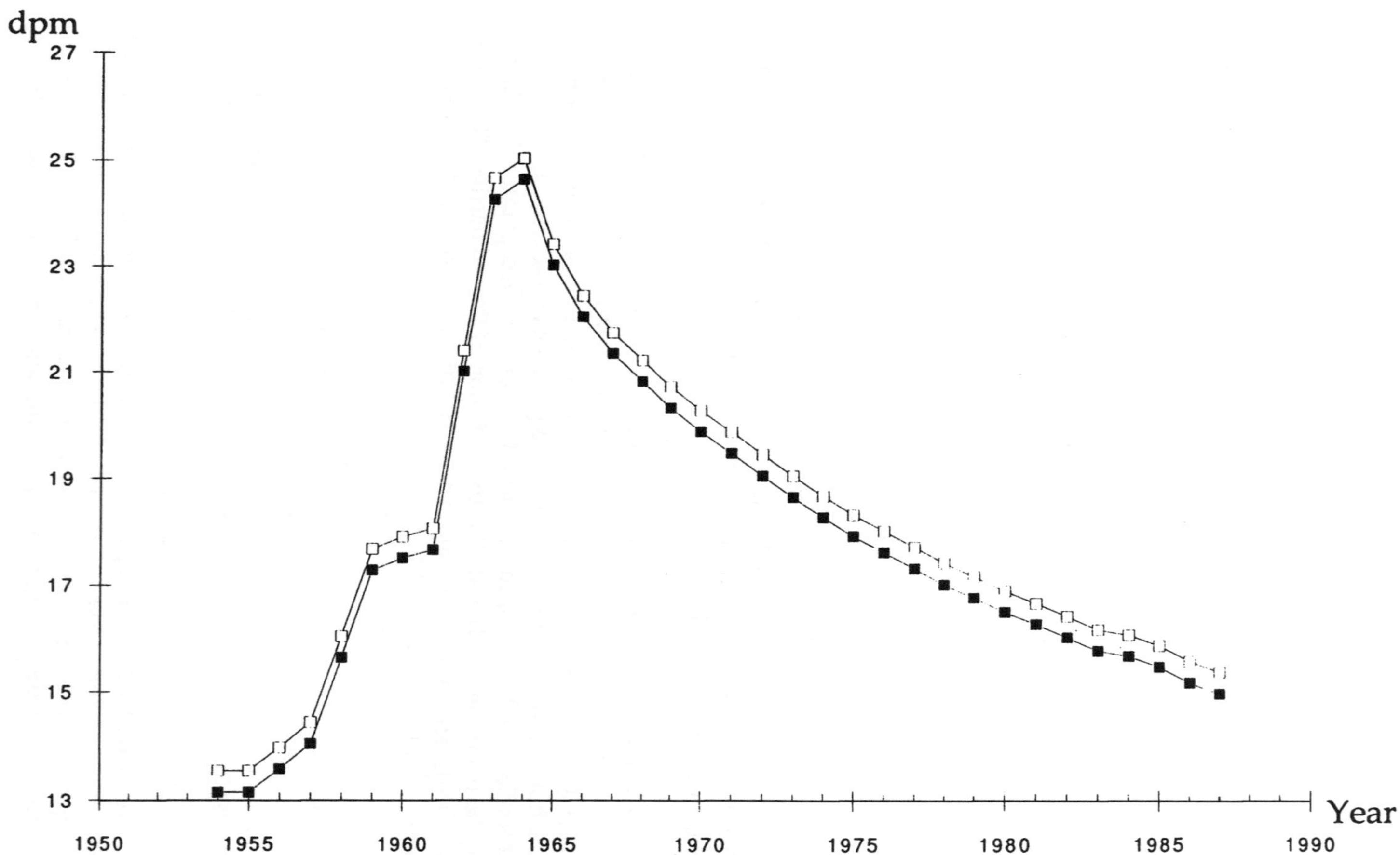

Figure 3.19 Evolution of the ^{14}C activity with the year of production of alcoholic beverages (lower and upper limits of uncertainty).

technique, the method is not sufficiently accurate at present and no improvement is forseeable.

This technique is, however, useful in some circumstances, such as for very old spirits or products of special vintages (±1 or 2 years). For red wines, this technique, combined with the analysis of polyphenolic compounds (polymerisation, tanin indicators (Glories, 1978) or colour indexes (the tile colour increasing with age), will narrow the margin. The addition of sugar to the wine complicates the process (because it is sugar of a different age) and widens the uncertainty to over 1 year (Mongereau and Evin, 1993).

Remaining above the natural level, ^{14}C levels can be used to date older wines. Recently, however, two cases that involved a 1472 Alsace wine in barrel and another *circa* 1777 were submitted to the laboratory. Unfortunately for those two cases, an anomaly in the ^{14}C from the solar activity prevented accurate dating; an error of plus or minus three centuries affects this period of time. Wines from the Greeks or Romans would be dated more accurately.

3.5.2 *Other elements for dating*

Other techniques such as tritium determination would probably not lead to better results because of the ultra low quantities, the enormous concentration of water involves and the complex disturbances induced by hydrogen bombs (Evin, personal communication).

It is reasonable to think that the nuclear fallout of atmospheric testing has left its mark in the wine other than for the ^{14}C. A technique used for dating the sediments (Shirahata *et al.*, 1980) that relies on the radioactive determination using beta and alpha spectrometry of ^{210}Pb, ^{210}Po, ^{239}Pu, ^{240}Pu and ^{137}Cs could be used. Dating has been performed by these authors for the period from 1850 to 1968 using 6-year intervals. It remains to be seen whether current apparatus is sensitive enough for such determinations in wines.

3.6 Special examples

3.6.1 *Rosé wines*

3.6.1.1 French and Spanish rosé wines. Even with unsophisticated techniques it is possible to distinguish wines from different countries. In one study (Garcia-Jares and Médina, 1993a), 43 French rosés and 35 Spanish rosés were analysed for alcohol content, volatile acidity, sugar content and colour (optical density at 280 nm, anthocyans, ionisation index and tanins) according to the methodology of Sudraud (1958) and Glories (1978, 1984). Tri-stimulus colorimetric analysis was also performed using a

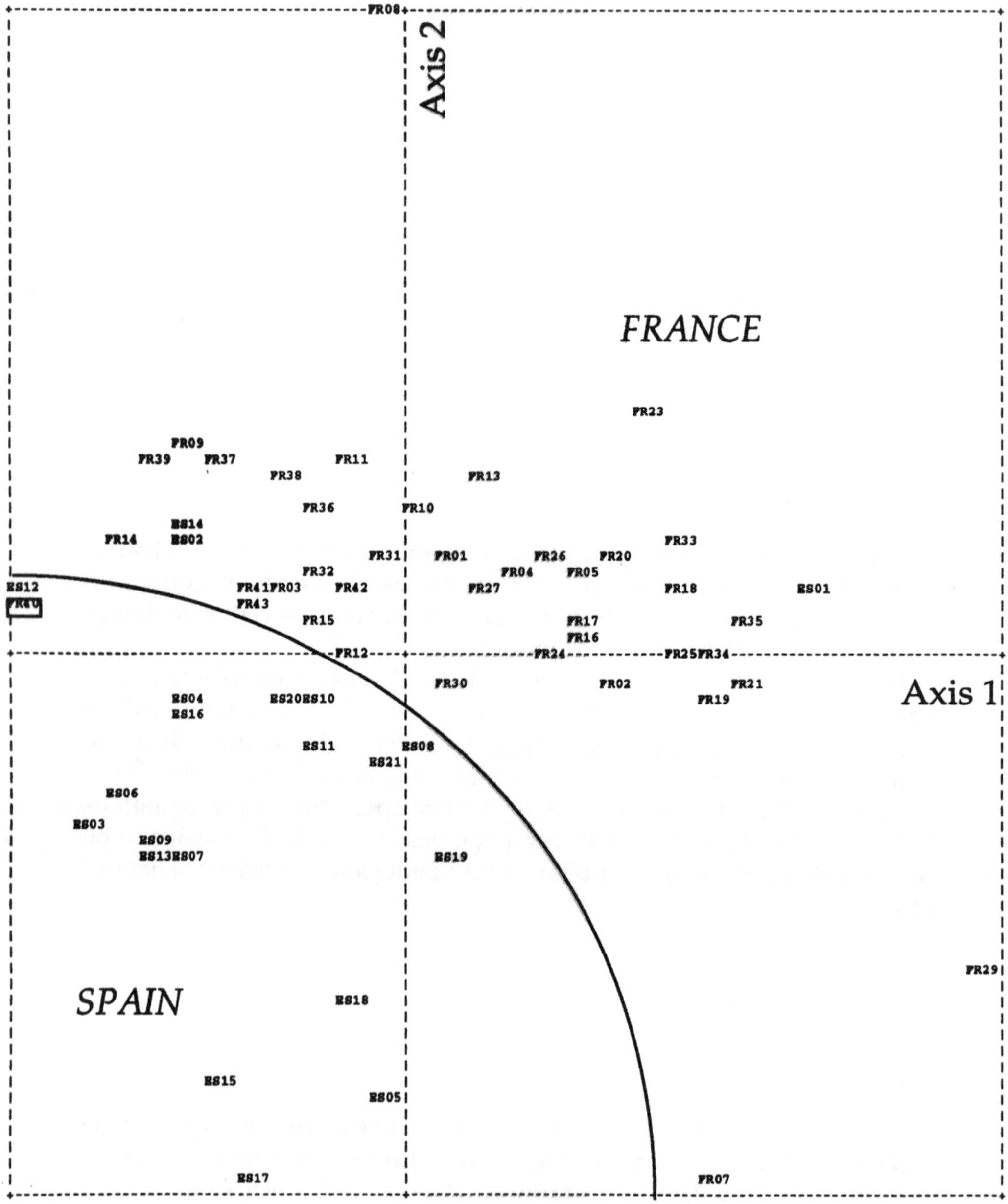

Figure 3.20 French and Spanich rosé wine separation using PCA on classical determinations and colour parameters. FR = French rosés; ES = Spanish rosés.

chromameter and absolute chromaticity expressed by L, a, b, C and H° (Garcia-Jares *et al.*, 1992). With these simple determinations and a representative set of rosé wines from both countries, a good distinction using PCA was made when colorimetric parameters predominate (L, a: axis 1 and b, ionisation index: axis 2). Wines were clearly separated into two groups (Figures 3.20). The French rosés have a tendency towards the colour red and Spanich ones towards orange.

3.6.1.2 Blending. The blending of wine is authorised in certain well-defined circumstances. It is a problem that needs to be evaluated as it is one of the primary frauds concerning wines.

According to EU regulations, Champagne rosé is the only authorised blend of wines of different colours. The blending of white and red table wine is permitted in Spain but the products may not be sold elsewhere (JOCE, 1994). In France, a distinction exists even between the 'rosé' wines and the 'clairet' category. Clairet wines, between red and rosé wines, are quite easily ascribed to their category by the analysis of their colour using tri-stimulus parameters L, a and b (Garcia-Jares *et al.*, 1992; 1993). However, a much more complex task is to distinguish a true 'rosé', made from only red grapes and a light maceration, from one produced by the blending of red and white grapes particularly when calculating the extent of blending. Garcia-Jares and Médina (1993b) assessed this difficult problem by associating a highly sophisticated mathematical construction model such as partial least squares and simple determinations such as the ultraviolet–visible spectrum. In the first stage, a model was built using the spectra of mixed red and white wines in different proportions (from 1–20% red wine) that simulate the visual characteristics of true rosés. The robustness of the model was then tested against wines of known mixed composition and the validity of the calculation of the proportion of red and white wine was estimated. A fairly good agreement was obtained; this method could be applied successfully to the detection of blending in the production of rosé wines (see Table 3.3).

3.6.2 Port wine

Pessanha (1987) tackled the problem of the authenticity of Port wines in a way that mixes several components. Part of the scope of this study was to differentiate genuine Port wine, imitations (mainly from Spain but not sold under this label) and wines made from similar techniques, for example Banyuls, a Vin Doux Naturel (VDN) from southern France. In all, 100 chemical determinations were made including:

- classical determinations
- 14 metals
- esters and fatty acids

Table 3.3 Prediction of white percentage with the PLS (two factors). The prediction error is expressed as the difference between predicted and true values

		White wine		
		Predicted %	True %	Prediction error
Test samples[a–c]				
BLA2		85.50	88.00	1.50
BLA3		84.75	84.00	0.75
BLC1		89.62	92.00	−2.38
BLE1		92.65	92.00	0.65
BLF3		83.90	84.00	−0.10
BLG2		78.80	88.00	−9.20
BLI1		95.96	92.00	3.69
BLI4		80.89	80.00	0.89
BLJ1		96.93	92.00	4.93
BLJ3		90.59	84.00	6.59
Unknown samples[d]	OD 280/320			
U1	1.69	76.65		
U2	1.93	85.17		
U3	1.42	83.58		
Predicted residual error sum of squares				4.22

[a]BL = blends of white and red wines.
[b]A to J = 10 red wines used for blending.
[c]1 to 4 = 8%, 12%, 16% and 20% respectively, red wines in the blend.
[d]U = undoubtful sample.

- superior alcohols and methanol
- phenolic substances (acids and neutral).

Owing to the limited number of imitations available, all the groups of chemicals (grouped by techniques) discriminated well.

For the imitations, an unknown substance (named ECLA in Figure 3.21d) was present almost universally in the high-pressure liquid chromatography determinations of the phenolic substances (Salagoïty and Bertrand, 1984).

With classical determinations, and because the imitations (unlike Port wines) are produced mainly with must fortified by alcohol with very little fermentation, the glucose/fructose ratio is above 0.9 for the imitations and below 0.8 for Port wines.

The most useful metals used in discrimination were rubidium, lithium, manganese, iron and aluminium. Rubidium permits an accurate separation of the imitations (Port wines having a mean value of 1 mg/l). A graphical representation of manganese against rubidium ensures good discrimination (Figure 3.21a). Among the metals used, manganese showed a clear distinction for the Banyul wines, where the soils have a very high manganese content; some toxicity for the wines was observed. The imitations also have a higher aluminium content (extensive use of bentonite?).

Fusel oil separated the genuine from the imitation wines equally well, and a graphical representation of amyl alcohol gives a clear view of the distinction (see Figure 3.21b).

When halting fermentation in Port wines, wine alcohol (76–78% vol) with very little rectification is used; the resulting product is consequently rich in higher alcohols (Andrade Perez, 1961). Isoamyl acetate and ethyl octanoate are the best indicators in the voltatile series (see Figure 3.21c). Figure 3.21c).

For clear and unsophisticated imitations, the situation is unambiguous and easy-to-apply rules were deduced from the above study (see Figure 3.21d).

3.6.3 Spirits

More recently, Cantagrel *et al.* (1993) proposed several methods, described in the following paragraphs, to distinguish the principal spirits.

The distribution of common components or their ratios in spirits, for example, higher alcohols, capylic acid, isopentanol ratio (M3B1/M2B1), syringaldéhyd/vanilin, is the first useful step. Using only six components, it is possible to separate almost perfectly (with a 0.6% error) Cognacs, Armagnacs and Brandies from different countries using Bayesian classification and PARVUS software (FORINA *et al.*, 1988). These components are meaningful in terms of wine-making, distillation process and ageing.

Discriminant analysis using 23 variables correctly separates 120 spirits into their main classes: Brandies (French, Spanish, Italian), Cognacs, Armagnacs, Calvados and Whiskies.

Isotopic analysis of the site-specific deuterium has proven useful in the distinction of 'chaptalisation' in wines but requires the use of a data bank because the natural distribution of the deuterium in the ethanol molecule is closely related to geographical origin (Martin *et al.*, 1983). Three isotopic variables, D/H1, D/H2 and R, the relative concentration of the deuterium on the first carbon of ethanol ($CH_2D\ CH_2\ OH$), the relative concentration on the second ($CH_3\ CHD\ OH$) and the intramolar ratio, permit good discrimination of Cognacs from other spirits; even the Whiskies are distinguished from the pure malt (Cantagrel *et al.*, 1993).

However if Brandies are made with wines of a different geographical origin the distinction is less obvious (e.g. for German Brandies produced with wines from the Cognac area).

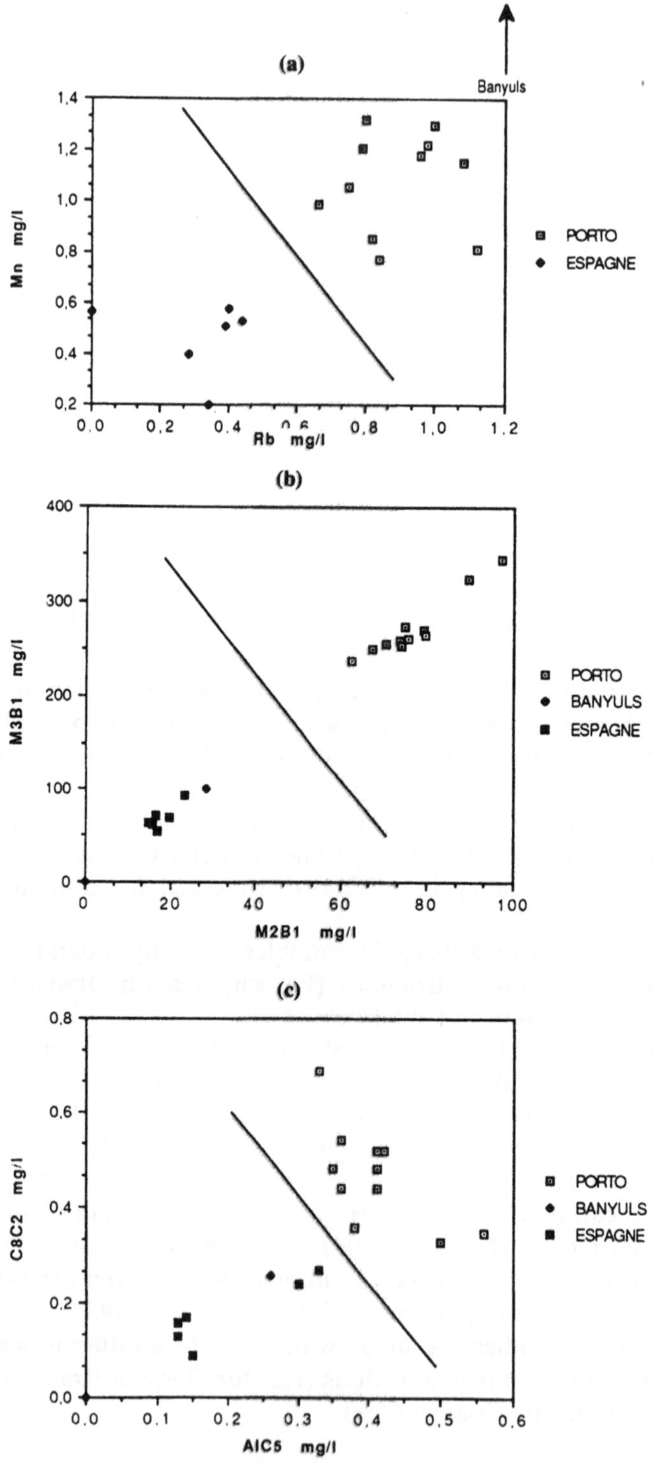
(a)
Banyuls
Mn mg/l
Rb mg/l
PORTO
ESPAGNE
(b)
M3B1 mg/l
M2B1 mg/l
PORTO
BANYULS
ESPAGNE
(c)
C8C2 mg/l
AIC5 mg/l
PORTO
BANYULS
ESPAGNE

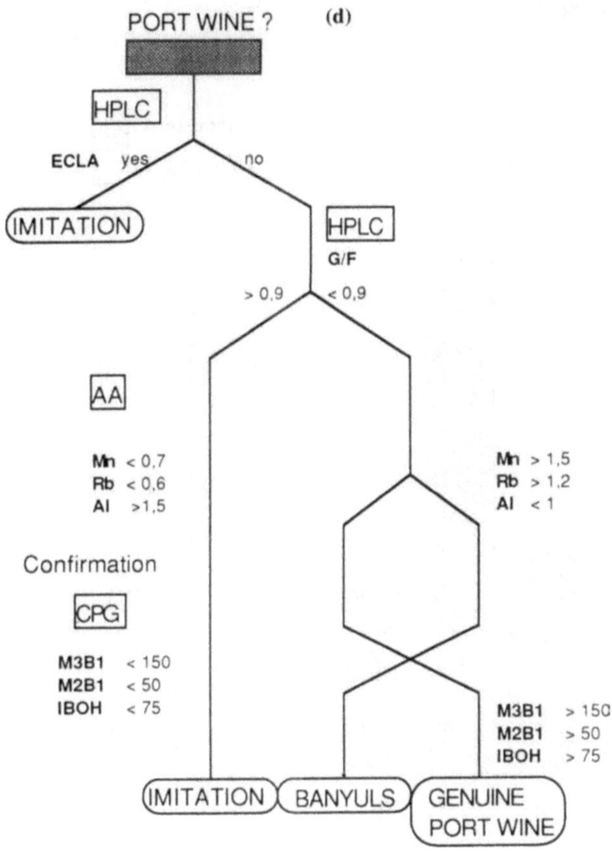

Figure 3.21 Characterisation of genuine port wines using: (a) minerals; (b) fusel oils; (c) volatile constituents. (d) Decision tree for the characterisation of genuine Port wines.

3.6.4 Champagne and sparkling wines

Recent EU regulations forbid the use of the word 'Champagne' on a label when produced outside the Champagne area, even when referring to the same process.

Tusseau *et al.* (1994) emphasised the fact that, after individually testing 100 analytical parameters, not a single one gives a satisfactory separation between the two groups. Using amino acids and multivariate analysis, 96% (50 samples) of the Champagnes but only 75% of the sparkling wines (60 samples) are classified correctly. A satisfactory classification is, however, achieved by adding 15 other parameters (e.g. oenological parameters, fusel oils, etc.) as shown by the linear discriminant function in Figure 3.22.

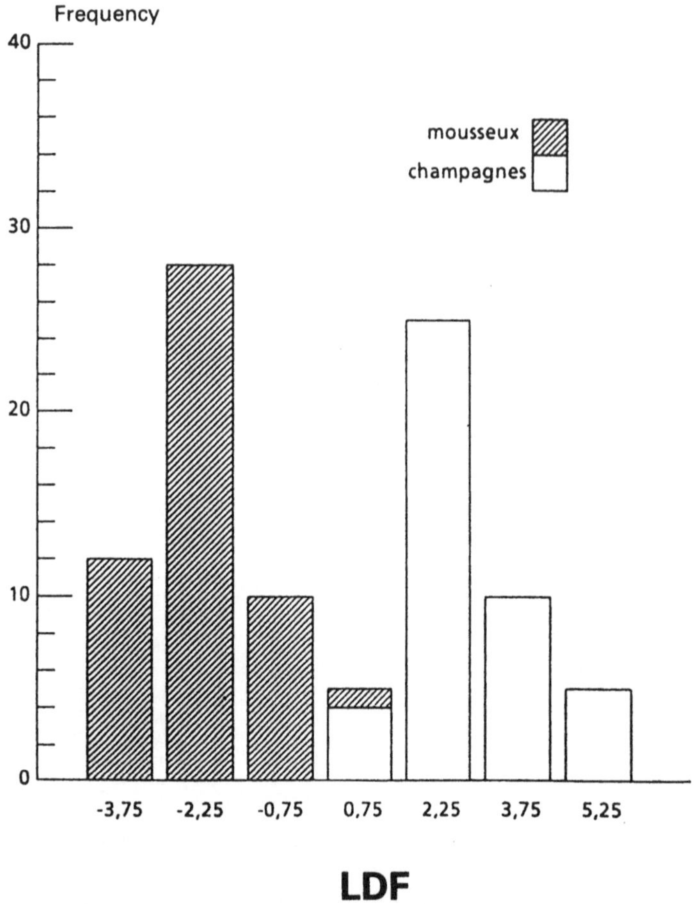

Figure 3.22 Linear discriminant function (LDF) for AOC Champagnes *versus* sparkling wines. Samples: 50 Champagnes; 60 sparkling wines. Features >100: minerals; amino-acids; volatiles; oenological parameters; sugars; polyphenolic compounds. Source: Tusseau *et al.* 1994.

Differentiation requires numerous new techniques but would be necessary in order to maintain the image of Champagne in the face of a wide range of worldwide imitations.

3.6.5 Botrytised sweet wines

Chauvet (1981) studied carefully the botrytised sweet wines of the Sauternes area. Using chemical determinations on wines and grapes, he developed an over-maturation index, which reflects the extent of growth of the so-called 'noble rot' – an essential step in the wine-making process for those wines. Using his data, a PCA was performed on wines from botrytised grapes; it showed that the different stages of the development of the 'noble rot' could be distinguished along axis 1 in Figure 3.23a. The plots for most of the commercial wines are found between the second stage and the third when projected as supplementary parameters on this graph. To build the first axis, magnesium and tartaric acids weighed the most. A two-dimensional plot of tartaric acid *versus* magnesium gave a very useful representation of the three groups (Figure 3.23b). The plane can be separated grossly into four quarters. It is well known that noble rot decomposes tartaric acid and, later, concentrates the grape. The decomposition is followed from group 1 (non-botrytised) to group 2 (fully botrytised). Concentration then occurs and the levels of magnesium rise to stage 3 ('roasted' stage). The fourth quarter (upper right) would correspond to grapes concentrated by the sun (no tartaric decomposition). The projection of commercial wines on this plot gives an assessment of the extent of the noble rot in the final product (Figure 3.23c).

3.7 Conclusion

The verification of wine authenticity may lay in the use of combined techniques such as those using isotopes and minerals. However, the difficult problem of 'chaptalisation' has been resolved by the use of a data bank; which may also be a possible solution to the problem of geographical origin. With the use of a database, however, problems such as representativity, authenticity, rules of interpretation, and the transposition of results from one laboratory to another arise. Authenticity and rules of interpretation are the easiest to solve through official action. The inter-laboratory variances require the use of robust methods and more importantly, the use of a reference material.

Isotope determination requires cumbersome and expensive techniques (e.g. NMR, isotopic mass spectrometry). Isotope techniques, although enticing, require the same sophisticated mathematical tools as the easier and more affordable means, which often work equally well.

Determination of the authenticity in wine calls for a pan-European effort, since any database completed in one laboratory will always be limited to an informal level (pilot level or feasibility level). Nevertheless it has been proven with consistancy over many years that this problem can be solved using the mineral content of wines (stable components with time), and that the easiest cases can be solved with inexpensive determinations (e.g. lithium, rubidium, manganese etc.). It has been solved for, what in our opinion is a much more difficult problem – 'chaptalisation'.

The mineral profile of wine is so characteristic that it is unique to a specific wine and can be used as proof of authenticity where a control

(a)

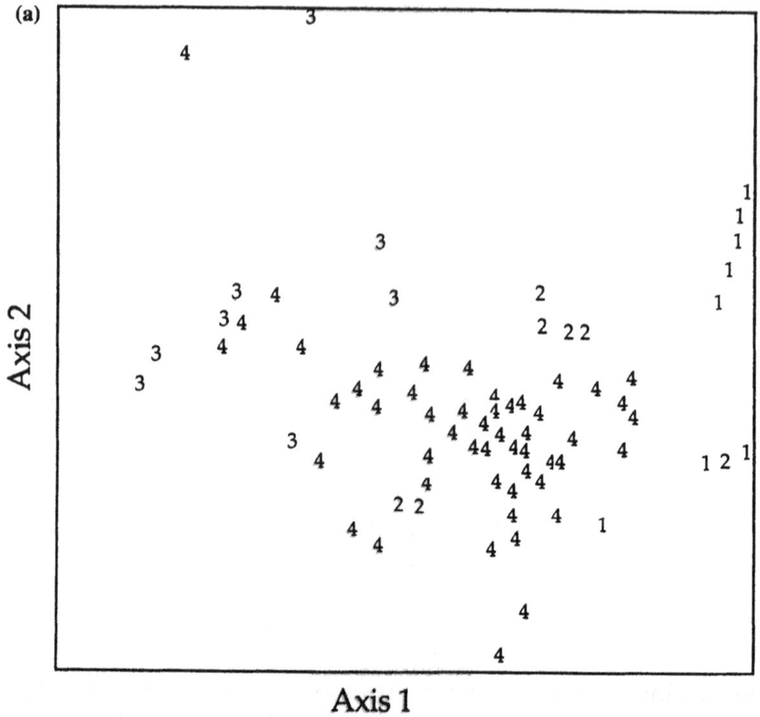
Axis 2
Axis 1

(b)

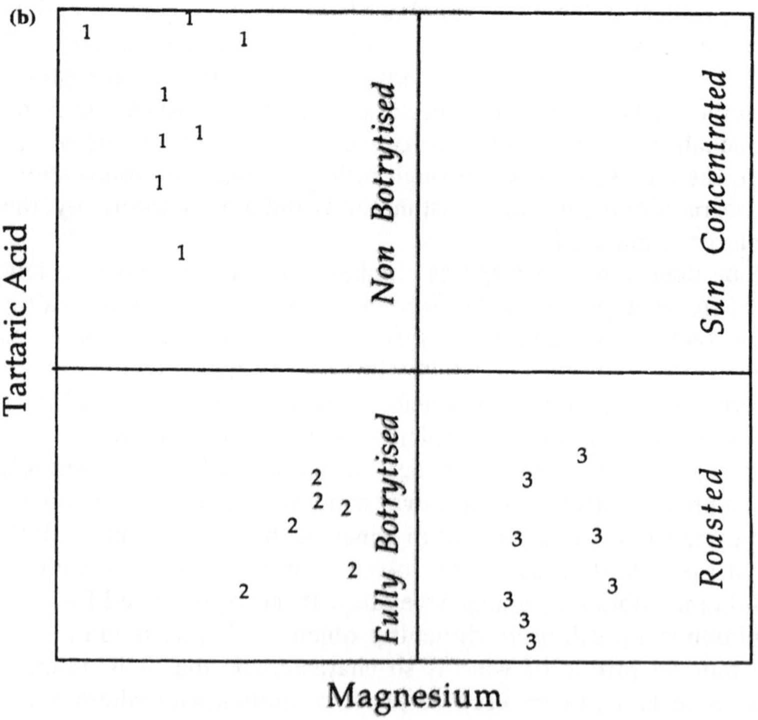
Tartaric Acid
Magnesium
Non Botrytised
Sun Concentrated
Fully Botrytised
Roasted

(c)

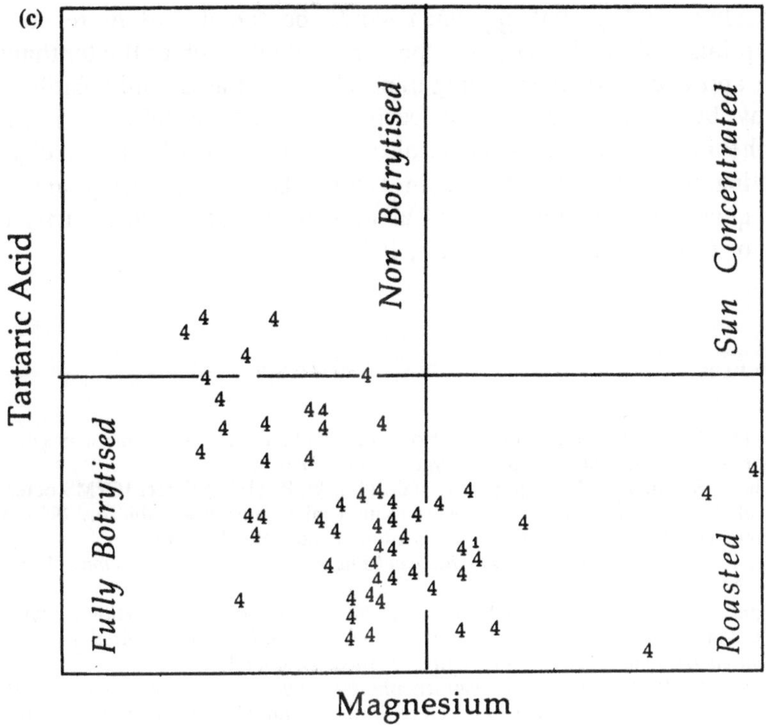

Figure 3.23 (a) PCA of botrytised sweet wines at different stages (28 measurements). (Commerical wines as supplementary individuals not included in the PCA calculation.) (b) Tartaric acid *versus* magnesium plot for reference wines elaborated at different botrytised stages. (c) Tartaric acid *versus* magnesium plot for reference wines and commercial wines, 1 = non-botrytised; 2 = fully botrytised; 3 = 'roasted'; 4 = commercial wines.

exists. It fares so much better than classical determinations. At any rate, it is possible to make accurate classifications of unknown wines and if, for the moment, no rational explanation is completely satisfactory (e.g. transfer of the minerals from the soil to the grape), the empirical solution is sufficient for practical purposes.

To determine the origin of a completely unknown wine, a global strategy could be conducted in three stages:

1. Lead isotope ratio determination to locate the origin of wine by continent.
2. ^{18}O or site-specific deuterium for the determination of latitude, north or south.
3. Metal determination such as lithium, rubidium or manganese for classification according to the sub-region.

An alternative global approach would be the use of more complex fingerprints and much more site-specific elements such as the lanthanides. Other approaches such as mixing minerals, amino acids and volatiles have already been used as a legal basis for detecting falsified products. Polyphenolic content of wine is also a good mark of authenticity (for detecting variety and region). In future, DNA sequencing and PCR techniques will probably be used as ultimate proof of authenticity (now used for forensic purposes) (Time, 1995).

References

ADDAD (1985) Association pour le développement et la diffusion de l'analyse des données. Laboratoire de Statistique, 4 place jussieu, 75005 Paris.

Augagneur, S., Médina, B., Szpundr, J. and Lobinski, R. (1996) Direct ICPMS determination of the rare earth elements in wine using a microconcentric nubilizer. 1996 Winter Conference on Plasma Spectrometry, Fort Lauderdale, FL, January.

Brooks, R.R. (1972) *Geobotany and Biogeochemistry in Mineral Exploration*, Harper and Row, New York.

Cantagrel, R., Vidal, J.P. and Snakkers, G. (1993) Le cognac: caractérisation analytique de son origine. *2nd European Symposium on Food Authenticity Isotope Analysis and Other Advanced Analytical Techniques*, Nantes, France, October 1993.

Chauvet, S. (1981) Application de l'Isotachophorèse au dosage des principaux acides du vin: contribution à l'étude de la relation entre l'intensité du développement de la pourriture noble et la composition chimique des vins blancs doux. Thèse doctorat d'Université de Bordeaux, France.

Chow, T.J. (1970) *Nature* **225**, 295.

Chow, T.J., Snyder, C.B. and Earl, J.L. (1975) Isotope ratios of lead as pollutant source indicators, United Nations FAO and International Atomic Energy Association Symposium Vienna, Austria (IAEA-SM-191/4) Proceedings pp. 95–108.

Day, M.P. (1993) Étude de la caractérisation de l'origine géographique des moûts et des vins par analyse conjointe de la composition en isotopes stables et en éléments minéraux. Thèse de doctorat Chimie Biologie Spécialité Chimie Analytique, Université de Nantes.

Diaconis, P. and Efron, B. (1983) Méthodes de calcul statistiques intensifs sur ordinateur. Pour la Science, July, 46–58.

Etchats, R. (1986) Essais de dosage de quelques éléments dans les vins par spectrométrie d'émission plasma DCP. Mémoire DEA Œnologie Ampélologie Université de Bordeaux II.

Etiévant, P., Schlich, P., Bouvier, J.C. *et al.* (1988) Varietal and geographic classification of French red wines in terms of elements, amino acids and aromatic alcohols. *J. Sci. Food Agric.* **48**, 25–41.

Fiorina, M., Lanteri, S. and Armanino, C. (1987) *Chemometrics in Food Chemistry*. Topics in Current Chemistry. Vol 141. Springer-Verlag, Berlin.

Fiorina, M., Leardi, R., Armanino, C. *et al.* (1988) Parvus: an extendable package of programs for data exploration classification and correlation. Elsevier Scientific Software, Amsterdam.

Forstel, H. (1995) Sous-Commission conventionnelle d'unification des méthodes d'analyse et d'appréciation des vins. 35 ème Session, 8–10 juin PARIS. OIV, 18 rue d'Aguesseau 75008 Paris.

Frank, I.E., Kowalski, B.R. (1984) Prediction of wine quality and geographic origin from chemical measurements by partial least-squares regression modeling. *Anal. Chim. Acta.* **162**, 241–51.

Garcia-Jares, C.M., Garcia-Martin, S. and Cela-Torrijos, R. (1995) Analysis of some highly volatile compounds of wine by means of purge and cols trapping injector capillary gas chromatography. Application to the differentiation of Rias Baixas Spanish white wines. *J. Agric. Food Chem.* **43**, 758–64.

Garcia-Jares, C.M. and Médina, B. (1993a) Caractérisation et différenciation des vins rosés d'appellation d'origine et de table français et espagnols. *J. Int. Sciences Vigne et Vin* **27(1)**, 35–51.

Garcia-Jares, C.M. and Médina, B. (1993b) Research on white and red wine blending in the production of rosés wines by means of the partial least squares method. *J. Sci. Food Agric.* **63**, 349–54.

Garcia-Jares, C.M., Médina, B. and Sudraud, P. (1992) Caractérisation des vins rosés et clairets par une mésure objective de leur couleur. *OIV, F.V.* **913**, 1915.

Garcia-Jares, C.M., Médina, B. and Sudraud, P. (1993) Mise au point d'une méthode pour la characterisation et la differentiation des vins clairets. *Rev. Française de'Œncologie* **140**, 19–24.

Glories, Y. (1978) Recherches sur la matière colorante des vins rouges. Thèse Doctorat d'Etat, Université de Bordeaux II.

Glories, Y. (1984) La couleur des vins rouges. II. Mesure, origine et interprétation. *Conne. Vigne Vin* **XVIII(4)**, 253–71.

Gonzales-Larraina, M., Gonzales, A. and Médina, B. (1987) Les ions métalliques dans la différenciation des vins rouges des trois appellation d'origine Rioja. *Conn. Vigne. Vins.* **2**, 127–40.

Goodacre, R., Kell, D.B., Bianchi, G. (1993) Rapid assessment of the adulteration of virgin olive oils by other seed oils using pyrolysis mass spectrometry and artificial neural networks. *J. Sci. Food Agric.* **63**, 297–307.

Grousset, F.E., Quétel, C.R., Thomas, B. *et al.* (1994) Transient Pb isotopic signatures in western European atmosphere. *Environ. Sci. Technol.* **28**, 1605–08.

Herrero-Latorre, C. and Médina, B. (1992) 'Utilisation de quelques éléments mineraux dans la differentiation de vins de Galice de ceux d'autres régions d'Espagne. *Conn. Vigne Vin.* **24**, 147–56.

Infometrix (1981) ARTHUR Infometrix Corporation, Seattle, Washington.

Jarvis, K.E. (1988) *Chem. Geol.* **68**, 31.

Jarvis, K.E. and Jarvis, I. (1988) *Geostand Newslett* **12**, 1.

JOCE (1987) Règlement CEE no 822/87 portant organisation commune du marché viti-vinicole.

JOCE (1994) Règlement 537/94 portant mesure transitoire en matière de campagne des vins de table en Espagne pour l'année 1994.

Kim, K.W. and Thornton, I. (1992) Influence of Ordovician uraniferous black shales on trace element composition of soils and food crops, Korea. *Appl. Geochem. (***Suppl 2**) 249–55.

Krauskopf, K.B. (1956) Sedimentary deposits of rare metals. *Econ. Geol* **50th anniversary volume** 411–63.

Kwan, W.O. and Kowalski, B.R. (1978) Classification of wines by applying pattern recognition to chemical composition data. *J. Food Sci.* **43**, 1320–23.

Kwan, W.O. and Kowalski, B.R. (1980a) Pattern recognition analysis of gas chromatographic data. Geographic classification of wines of *Vitis vinifera* cv. Pinot Noir from France and the United States. *J. Agric. Food Chem.* **28**, 356–59.

Kwan, W.O. and Kowalski, B.R. (1980b) Correlation of objective chemical measurement and subjective sensory evaluations wines of *Vitis vinifera* variety Pinot Noir from France and the United States. *Analytica. Chim. Acta.* **122**, 215–222.

Kwan, W.O., Kowalski, B.R. and Skogerboe, R.K. (1979) Pattern recognition analysis of elemental data; wines of *Vitis vinifera* cv. Pinot Noir from France and the United States. *J. Agric. Food Chem.* **27**, 1321–26.

Lacasta, F. (1982) Dosage de quelques métaux dans les vins par spectrométrie d'absorption et d'émission de flamme. Rapport de stage BTAO. DGCCRF 33405 Talence, France.

Latorre, M.J., Garcia-Jares, C., Médina, B. *et al.* (1994) Pattern-recognition analysis applied to classification of wines from Galicia (NW Spain) with certified brand of origin. *J. Agric. Food Chem.* **42**, 1451–56.

Latorre, M.J., Herrero, C. and Médina, B. (1992) Utilisation de quelques éléments minéraux dans la différenciation des vins de Galice. *Conn. Vigne Vin.* **24**, 147–56.

Legrand, V., Médina, B., Grenon, J.P. *et al.* (1991) Les contaminants minéraux des vins. *Analusis* **19(10)**, M39–M43.

MacKay, K. (1994) International Atomic Energy Agency, Marine research station Monaco, France. Personal communication.

Maarse, H., Slump, P., Tas, A.C. *et al.* (1987) Classification of wines according to type and region based on their composition. *Z. Lebensm. Unters. Forsch.* **184**, 198–203.

Manterola, J.J. (1985) Analyse de composés phénoliques neutres de vins rouges par chromatographie liquide haute pression, d'éléments minéraux par spectrométrie d'absorption atomique et d'émission, et traitement informatique des données, en vue de différencier les vins suivants les terroirs du Bordelais. Rapport de stage, Institut Supérieur d'Agronomique de Lille. DGCCRF 33405 Talence, France.

Martin, G.J. and Martin, M.L. (1983) *J. Chim. Phys.* **80**, 294.

Martin, G.J., Martin, M.L., Mabon, F. *et al.* (1983) A new method for the identification of the origin of ethanols in grain and fruit spirits: high-field quantitative deuterium nuclear magnetic resonance at the natural abundance level. *J. Agric. Food Chem.* **31**, 311–15.

Martin, G.J., Guillou, C., Martin, M.L. *et al.* (1988) Natural factors of isotope fractionation and the characterization of wines. *J. Agric. Food Chem.* **36**, 316–22.

Martinière, F., Gauthier, J.M., Sudraud, P. *et al.* (1979) Evolution de la radioactivité par le carbone 14 des vins de la Gironde. Application à la recherche des millésimes. *Ann. Fals Exper. Chim.* **72(775)** 263–74.

Masuda, A., Nakamura, N. and Tanaka, T. (1973) Fine structures of mutually normalized rare-earth patterns of chondrites. *Geochim. Cosmochim. Acta* **37**, 239–48.

Médina, B. (1978) Application de la spectrometrie d'absorption atomique sans flamme au dosage de quelques méteaux dans les vins. Thèse doctorat. Université Bordeaux.

Médina, B. and Van Zeller, A.L. (1984) Différenciation des vins de trois régions de France. *Conn. Vigne Vin* **18**, 225–35.

Mongereau, N. and Evin, J. (1993) Les applications du radiocarbone et de l'expertise. Études et documents. *CEJ* **5**, 105–10.

OIV (1995) Le Plomb, cahiers scientifiques et techniques. OIV 75008 Paris.

Pessanha Cunha Teixeira, M. (1987) Caractérisation chimique et sensorielle des Porto d'origine controlée. Analyse statistique des résultats et automatisation des dégustations. Mémoire Diplôme d'Etudes Approfondies Œnologie-Ampélologie, Université de Bordeaux II.

Rapp, A., Hastrich, H. and Engel (1979) *Vitis* **15**, 29.

Salagoïty-Auguste and Bertrand, A. (1984) Wine phenolics analysis of low molecular weight components by high-performance liquid chromatography. *J. Sci. Food Agric.* **35**, 1241–47.

Shirahata, H., Elias, R.W. and Patterson, C.C. (1980) Chronological variation in concentration and isotopic composition of anthropogenic atmospheric lead in sediments of a remote subalpine pond. *Geochimica Cosmochimica Acta* **44**, 149–62.

Siegmund, H. and Bächmann, K. (1977) Die Lagezuordnung von weinen durch bestimmung des spurenelementmusters. *Z. Lebensm. Unters.-Forsch.* **164**, 1–7.

Siegmund, H. and Bächmann, K. (1978) Z. Anwendung der numerischen taxonomie für die klassifizierung von weinen. *Lebensm. Unters.-Forsch.* **166**, 298–303.

Sudraud, P. (1958) Interprétation des courbes d'absorption des vins rouges. *Ann. Technol. Agric.* **2**, 203–08.

Sudraud, P. and Médina, B. (1993) Détermination de l'origine géographique d'un vin par analyse des éléments minéraux. *2nd European Symposium on Food Authenticity Isotope Analysis and Other Advanced Analytical Techniques*, Nantes, France, October 1993.

Thurin, E. (1984) Recherche de correlations entre l'analyse physico-chimique et l'analyse sensorielle. Diplôme National d'Œnologue, Institut d'Œnologie, Université de Bordeaux II.

Time Magazine (1995) Law: blood sweat and tears. All of the above could help British police to trace criminals when a new DNA bank is completed. **15 May**, 38.

Tusseau, D., Valade, M. and Moncomble, D. (1994) Control of the origin and nature of wines and spirits *108th AOAC International Meeting and Exposition, Portland, Oregon* 12–15 September. CIVC, BP135, 51204 Epernay, France.

Tusseau, D., Valade, M., Virion, M.C. *et al.* (1992) Controle de l'origine et de la nature des vins de champagne. *XXème Congrés Mondial de la Vigne et du Vin* OIV 72 ème Assemblée Générale, Madrid, 18–26 Mai 1996. Le Vigneron Champenois No 4 Avril 1993.

Vasconcelos, A.M.P., Higuinaldo, J. and das Neves, C. (1989) Characterization of elementary wines of vitis-vinifera varieties by pattern recognition of free amino acids profiles. *J. Agric. Food Chem.* **37**, 931–37.

Further reading

Andrade Perez, M.R. (1961–62) Doseamento dos alcools superiores no vinho do Porto e noutras biedbidas alcoolicas. Anais do Instituto do vinho do Porto.

Baldi, M., Riganti, V., Specchiarello, M. (1984) Caratterizzazione e classificazione di vini da tavola con metodologie statistiche. *Industrie della bevande.*, febbraio, 1–6.

Derde, M.P., Massart, D.L., Ooghe, W. *et al.* (1983) *J. Automatic Chem.* **5(3)**, 136–45.

Elbaz-Poulichet *et al.* (1984).

Larice, J.L., Archier, P., Rocheville-Divorne, C. *et al.* (1989) Composition anthocyanique des cépages. II Essai de classification sur trois ans par analyse en composantes principales et étude des variations annuelles de cépages de même provenance. *Revue. Fr. Œno* 121, Cahier Scientifique.

Lobinski, R., Witte, C., Adams, F.C. *et al.* (1994) Organolead in wine. *Nature* **370**, 24.

MacCurdy, E., Potter, D. and Médina, B. (1992) Rare earth elements in wine. *Lab. News* **9** 10–11.

Marais, J. Van Rooyen, P.C. and Du Plessis, C.S. (1981) Classification of white cultivar wines by origin using volatile aroma components. *S. Afr. J. Enol. Viticol.* **2(2)**, 45–9.

Médina, B. and Sudraud, P. (1979) Teneur des vins en lithium. *Ann. Fals. Exp. Chim.* **72(772)**, 65–71.

Médina, B. and Sudraud, P. (1980) Teneur des vins en chrome et nickel. Causes d'enrichissement. *Conn. Vigne. Vin.* **14(2)**, 79–96.

Rizzon, L. (1985) Incidence de la macération sur la composition chimique des vins. Thesis University of Bordeaux II.

Sudraud, F. (1985) Analyse de données se rapportant aux déterminations chimiques et gustatives de quelques vins blancs français. Mémoire Diplôme d'Etudes Approfondies Œnologie-Ampélologie. Université de Bordeaux II.

Vercauteren, J., Forveille, L. and Rutlegde, D. (1995) Les techniques d'identification progressent. La RMN reconnaît cépages et clones. *La Vigne* **mai**, 41–42.

4 Authenticity of meat and meat products

I.D. LUMLEY

4.1 Introduction

The terms 'meat' and 'adulteration' need to be defined for the purpose of this review. A typical dictionary definition of meat is 'the flesh of animals or animal tissue used as food', while the definition in the UK Meat Products and Spreadable Fish Products Regulations (1984) is 'the flesh, including fat, and the skin, rind, gristle and sinew in amounts naturally associated with the flesh used, of any animal or bird which is normally used for human consumption'. The Regulations go on to specify parts of the carcass that are to be regarded as meat and those that may not be used in uncooked meat products.

Typical dictionary definitions of adulteration are 'to debase, to make impure by addition of a foreign or inferior substance' and 'to prepare for sale by replacing more valuable with less valuable or inert ingredients'.

From these definitions it can be seen that adulteration of meat and meat products may involve, without informing the consumer, the addition of substances of a lower value and/or quality above the amounts required for normal or accepted processing or product formulation. For example, meat from a high-value species may be wholly or partially substituted with meat from a lower-value species; meat products may be extended with fat or rind so that these components are present in excess of the amounts naturally associated with the flesh used, or products may be extended by addition of water, non-meat proteins or other substances.

4.2 Components of meat products

4.2.1 Meat content

Food legislation differs across national boundaries but the requirement for manufacturers to state a minimum meat content on products is a common feature. Manufacturers know the amounts of ingredients in their products, therefore meat content can be calculated but the analyst is required to determine the meat content of a sample with little or no prior knowledge of ingredients, composition and processing history.

The Stubbs and More (1919) method, with modifications, is still used to estimate the meat content of meat products and is prescribed in a range of texts (e.g. Kirk and Sawyer, 1991; Commission Regulation (EEC), 1989). The method involves the determination of the fat-free meat content by determination of the total nitrogen content of the meat product under examination and the use of an appropriate 'nitrogen factor' to calculate meat content.

Nitrogen factors for a range of meats have been recommended by the Analytical Methods Committee (AMC) of the Royal Society of Chemistry over the years (see the summary in Kirk and Sawyer, 1991, p. 488). Reassessments of the factors for pork and beef have also been published (Analytical Methods Committee, 1991; 1993); while factors for lamb are currently under review. The species or origin of meat in a product must be known if the appropriate nitrogen factors are to be used.

Nitrogen factors are expressed as the percentage of nitrogen in raw meat on a fat-free basis. Agreed factors for cooked, processed or cured products are not available (this is not surprising, considering the range of cooking/processing practices), therefore analytically determined meat contents are expressed as raw meat.

The raw fat-free meat content (%) is calculated by:

$$\frac{\%\ \text{nitrogen content} \times 100}{\text{nitrogen factor}} = \%\ \text{fat-free meat} \qquad (4.1)$$

The total meat content (%) is given by:

$$\%\ \text{raw fat-free meat} + \%\ \text{fat} = \%\ \text{total meat} \qquad (4.2)$$

As the meat content so calculated is dependent upon the determination of nitrogen (by Kjeldahl) it is essential that corrections be made for contributions from non-meat sources of nitrogen, e.g. soya protein and cereal fillers.

The presence and nature of cereal fillers can be detected by light microscopy (presence of starch and other characteristic features) and the amount of carbohydrate (C), can be estimated by the difference, i.e.:

$$C = 100 - (\%\text{water} + \%\text{fat} + \%\text{protein} + \%\text{ash}) \qquad (4.3)$$

Values for the nitrogen content of cereal fillers have been published, for example 2.0% is recommended for wheat rusk. Potato starch and cornflour each have a negligible nitrogen content. If microscopy has indicated that the cereal filler is wheat rusk, then the nitrogen contribution (N_{filler}) from this source is given by:

$$N_{\text{filler}} = 0.02C \qquad (4.4)$$

where C is the carbohydrate content of the product obtained by difference.

The carbohydrate content can also be determined directly by chemical analysis. This can be achieved by selective enzymatic hydrolysis of starch followed by the determination of glucose by titrimetry, high-performance liquid chromatography (HPLC) or enzymatic analysis.

The presence of soya protein and milk protein can also be detected using microscopy. Methods for their estimation are described in this chapter, as are methods for the determination of wheat gluten.

Other corrections to the Stubbs and More calculation may have to be considered. The raw fat-free meat content can be obtained as described but the analyst may be required to calculate the 'lean meat' content of a product, which, in the UK, is defined as 'meat-free, when raw, of visible fat' (Meat Products and Spreadable Fish Products Regulations, 1984). Clearly allowance has to be made for intramuscular fat and the nitrogen contribution from the 'visible fat'. Formulae that can be used for calculation of lean meat content (Pearson, 1968; 1970; Cottam, 1975) allow for appropriate corrections to be made. A useful summary of meat content determination has been published by Pearson (1975).

4.2.2 Excess connective tissue

From the definition of meat previously quoted, it is clear that the analyst may have to assess if rind or connective tissue is present in a product in an amount above that expected to be naturally associated with the flesh used. The presence of 'excessive' rind or connective tissue will contribute nitrogen and therefore give a falsely high meat content (as determined by Stubbs and More) if a suitable correction is not made.

Several workers have suggested ways to make these corrections. Coomaraswamy (1972) published a procedure that involves the determination of hydroxyproline and calculations based on those of Pearson (1970), which enable the analyst to calculate meat contents corrected for various amounts of connective tissue. Lord and Swan (1984) similarly addressed the problem of connective tissue and presented a method for the calculation of connective tissue content. Data obtained from analysis of meat products, and a recommendation for the maximum connective tissue content associated with lean meat is also presented.

4.2.3 Excess gelatin

A similar problem to the presence of excess connective tissue is the presence of excess gelatin, particularly in canned ham products. The analyst may need to decide how much of the gelatin in such products should contribute nitrogen to the calculation of the meat content. One approach involves the determination of hydroxyproline in the total product, and comparison of the results against values published for the appropriate meat species and cut (e.g. Analytical Methods Committee, 1991), estimation of excess hydroxyproline, conversion of this figure to the gelatin equivalent, and calculation of the nitrogen contribution from the excess gelatin. Appropriate allowance can then be made in the calculation of meat content. An alternative procedure involves the determination of gelatin using a method similar to that published by Richardson and Tarassuk (1934). The gelatin is precipitated quantitatively from the product and the nitrogen content determined. The amount of 'excess' gelatin can be estimated and allowance made for its nitrogen contribution.

4.2.4 Added water

Analysts may be asked to determine the added water in a meat product; this is currently performed by determining the fat-free meat content and using the following relationship:

$$\%\text{ added water} = 100 - (\%\text{fat free meat} + \%\text{fat} + \%\text{NAA}) \quad (4.5)$$

where NAA is non-aqueous additions, such as soya protein, added gelatin and curing salts.

Analytical methods for the determination of nitrogen, ash, moisture, total fat and hydroxyproline are prescribed by the British Standards Institution (BS 4401). The uncertainties associated with these analytical determinations are known from a range of collaborative studies and should be taken into account when interpreting results. In addition, and perhaps more importantly, the analyst must be aware of the uncertainties associated with the use of published compositional data and his/her assumptions that are made when applying the information to a product under examination. If these facts are borne in mind, the results produced and conclusions drawn will allow the authenticity of meat products to be judged effectively.

4.3 Examination procedures

4.3.1 Microscopy

Microscopy has an important role to play in the examination and analysis of meats and meat products. One major advantage is that within a relatively short period of time the analyst can identify most major constituents in a meat product, obtain a rough estimation of the relative proportions of constituents and assess their form. This often allows a judgement to be made on the quality and authenticity of a product. It also suggests to the analyst the chemical analyses that subsequently may be necessary to perform.

The name of a meat product or the form in which it is presented to the consumer (e.g. steaklet or ribsteak) may suggest that it has been prepared from a particular cut, section or joint of meat. Examination of such products by microscopy can reveal if the muscle cell structure is essentially intact, that is in its natural form, or whether the structure is disrupted, which can indicate that the product has been prepared from restructured or reformed meat.

Microscopy can be used to identify the nature of meat used in products, for example whether it is present as cuts of muscle, or as pieces, a paste or mechanically recovered. Tissues and organs can sometimes be identified, and microscopy has been used to identify the presence of lung, kidney, intestine, pancreas, uterus, udder etc. in meat products.

Microscopy can also be used as a rapid screening technique to identify the presence of non-meat proteins and cereal additives, wheat rusk and the form of soya present, for example isolate, concentrate, or flour, can be identified, and gums and mucilages can be detected.

For the examination of meat products, cryosections of approximately 10–12 μm in thickness are required and are best prepared using a cryostat (freezing microtome). Chemical stains can then be applied to highlight particular features. Staining can also be supplemented by polarisation microscopy, as many ingredients when viewed between crossed polarising filters are birefringent, which can enhance their appearance.

One of the most useful stains for meat products is toluidine blue in an aqueous mountant (Flint and Firth, 1981). Each ingredient is coloured differently and when this is combined with unique structural features, ingredients such as muscle cells, collagen, elastin and added proteins (soya, milk, cereal) can all be located and identified. Other reagents are available for the detection of particular ingredients or adulterants. Starch from cereal additives can be stained blue with iodine in potassium iodide solution and liquid fat can be located using oil red O in aqueous propan-2-ol whilst associated crystalline fat is birefringent (Bayliss High, 1984). Raw and heat processed collagen (such as rind) can be distinguished from each other using picro-Sirius Red F3BA and polarisation microscopy (Flint and Pickering, 1984).

In products that have been finely comminuted, location and identification of characterising structural form by microscopy may be very difficult. Antibodies labelled with enzyme-detection systems or labelled with silver/gold visualisation (immunohistological staining) can be applied to cryosections of such meat products, and particular components recognised. However, this work is in its infancy and is limited by the antibodies available.

In summary, microscopy has an important role to play in the examination and characterisation of meats and meat products and the detection of adulteration or misdescription, and is an excellent complementary technique to chemical and immunological analysis.

4.3.2 *Identification of meat species*

The reliable identification of the species of origin of bulk meats, particularly after butchering, dicing or mincing, has been a problem faced by analysts for many years. Species identification is even more problematical in processed and heated products such as smoked and cooked sausages, canned and cooked meats. Identification of the major meat species present (by weight) has been possible but identification of low percentages of other meat species, which may be present in binary or ternary admixture, has been virtually impossible until recent years.

Traders or Customs and Health authorities may wish to confirm that consignments of meat and meat products being purchased or imported/exported are of the species declared, while enforcement authorities may wish to confirm that meats being purchased are of the nature, quality and substance demanded by the consumer and that label claims relating to composition are correct and not misleading. There have been various examples in the last decade of the substitution or dilution of an expensive meat species with meat from a cheaper species; horse and kangaroo substitutions make the headlines but substitution with 'acceptable' meats such as poultry and pork occurs, depending upon price differentials at a particular time.

Gross substitution can occur where the analyst would be looking for high percentages (e.g. >10%) of non-declared meat species, while adventitious contamination may mean that the analyst is looking for low percentages (e.g. 1–10% m/m) of non-declared species; this could be in raw or cooked products. If the analyst reports the presence of one or more undeclared species in a product, the question invariably asked by the customer is 'How much is present?'. Therefore when selecting analytical methods for meat-species identification, the analyst must consider whether:

- the method is sufficiently specific and sensitive;
- the method can be applied to cooked or heated meats and meat products;
- the method can be used quantitatively.

4.3.3 *Analytical methods available*

Methods for species identification must detect or determine a natural component of 'meat' that is a unique indicator of the species of origin. Lipids, proteins and, more recently DNA methods have provided the target for speciation studies. In this chapter, lipid analysis is mentioned in overview only, the main methods under review being electrophoresis, immunoassay and DNA analysis.

4.3.3.1 Lipid analysis. Examination of the fatty acid isomer profile by capillary gas chromatography, determination of fatty acids at the 2-position in triglycerides after enzymic hydrolysis and gas chromatography, and examination of triglyceride profile by gas chromatography and/or HPLC with mass detection can provide information that will allow tentative identification of some meat species (e.g. horse, pork, kangaroo) but the interpretation of data from raw meats in admixture is difficult. Interpretation of data from highly processed and cooked foods where more than one meat species is present, is almost impossible.

The composition of animal lipid varies depending upon many factors, including season and diet and, if lipid analysis is to be used for species identification, the analyst must have access to a comprehensive and reliable database of lipid compositional information. For example, beef and horse lipid can be differentiated by looking at the relative ratios of the fatty acid isomers 18:0 and 18.3Ω3 and 18:1Ω9 and 18:2Ω6 but differentiating fat from other species is much more difficult. Total fatty acid isomer profiles have to be considered and all ratios examined; multivariate statistical analysis is often required. Saeed *et al.* (1986) reported that the fatty acid 20:2Ω6 could be used as an indicator of the presence of pork fat in beef and lamb fat but this fatty acid occurs in beef, which limits the value of this indicator; it is rare that a single fatty acid index can be used for meat speciation. Lipid analysis is sometimes useful to provide complementary evidence in cases of gross substitution but detection limits are generally too high for detection of adulteration at the 0–15% level.

4.3.4 Electrophoresis

Many papers have been published since the 1970s detailing the use of electrophoresis for the identification of meat species. Although isoelectric focusing is now probably the most commonly used technique for this purpose because of its resolving power, Skrokki and Horni (1994) recently used a polyacrylamide-gel procedure for the detection of adulteration of minced meat. The main use was detection of pork in minced beef but resolution of proteins from beef, reindeer and mutton, however, appears less than ideal.

Kaiser *et al.* (1980) used isoeletric focusing of aqueous meat extracts (e.g. beef, pork, venison, roe, impala, chicken, turkey) followed by staining of the separated proteins with Coomassie Brilliant Blue to identify species in raw binary mixtures. The paper suggests that densitometric evaluation of the electrophoretograms allows quantitative analysis down to a minimum of 2–5% (m/m) of one of the components; however, this interpretation needs to be viewed with caution (see comments on quantitation in ELISA section). Sinclair and Slattery (1982) subsequently published a similar procedure for the differentiation of raw beef, pig, buffalo, horse and kangaroo; however, sheep and goat could not be distinguished using this method.

Some workers have described methods that involve the selective staining of specific enzymes, which have been separated from other proteins by isoelectric focusing. Coomassie Blue is a useful general stain for proteins but complex electrophoretograms can be obtained. Selective enzyme staining can make interpretation of electrophoretograms simpler because fewer bands require interpretation. The procedure involves pouring an agar solution, which contains the appropriate enzyme substrates, co-factors, coupling enzymes and tetrazolium dye, over the isoelectric focusing gel after completion of electrophoresis. The agar solidifies and, after incubation, isoenzyme bands from the focusing gel can be identified as a coloured band.

Heinert and Klinger (1980) and Prasad and Misra (1981) used esterase patterns to identify meat species while Slattery and Sinclair (1983) used muscle lactate dehydrogenase and esterase isoenzymes. King and Kurth (1982) stained isoelectric gels for phosphoglucomutase to detect pig, horse and buffalo in beef (detection at 1% m/m level), and later Kim and Shelef (1986) used agarose gel electrophoresis with identification of isoenzyme MM and myoglobin bands to identify binary mixtures of beef, pork, chicken and turkey (detection limits above 5%, clear differentiation at 25% m/m). Kauffer *et al.* (1990) also used isoelectric focusing with peroxidase staining of myoglobins for the detection of chicken meat mixed with beef and pork but detection limits were high, for example 25% (m/m) without concentrating sample extracts, and 10% (m/m) after a concentration step.

The papers referred to so far deal only with raw meats. King (1984) reported that if proteins were extracted from cooked meats with 6M guanidine hydrochloride and the extract dialysed against 1% triton X-100 the activity of adenylate kinase (AK) and creatinine kinase (CK) could be recovered sufficiently to allow isoelectric focusing, enzyme staining, and subsequent species identification. AK and CK activity could not be recovered if meats were heated above 120°C and 105°C respectively, which limits its application to some commercially heated products. King also states that CK dimers found during analysis can produce additional bands on electrophoretograms, which make interpretation difficult.

Hofmann and Bluchel (1986) identified the species of origin of raw meat by examination of myoglobin patterns after isoelectric focusing, and the sensitivity of the method was later improved by taking advantage of the pseudoperoxidase activity of myoglobin and applying peroxidase staining (Bauer and Hofmann, 1987). The method could not, however, be used on severely heated products or for detection of chicken and turkey, which contain low levels of myoglobin.

McCormick *et al.* (1992) investigated the identification of meat from game and domestic species using isoelectric focusing and enzyme staining for creatinine kinase and adenylate kinase. Pork and bison could be differentiated from all species examined but beef, elk, caribou, red deer and goat could not be differentiated from each other, neither could pronghorn, mule and white-tailed deer, sheep and moose; nevertheless the former group could be differentiated from the latter. The inability to individually identify many of the target species and also the lability of the enzymes on heating limit the application of the technique in this instance.

Janssen *et al.* (1990) developed an interesting alternative for myoglobin staining that was applicable to meats that had been heated to 120°C for 5 min. They extracted proteins from meats with 8M urea and 0.1% dithioerythritol and subjected the proteins to isoelectric focusing in ultra-thin gels that contained urea in addition to ampholytes. Once the proteins had been focused they were electroblotted onto nitrocellulose, then incubated with anti-human myoglobin serum, a linking antibody, peroxidase-antiperoxidase immune complex and enzyme substrate. Detection of less than 10% (m/m) of pork, horse, kangaroo and sheep meat was possible in beef that had been heated to an internal temperature of 120°C for 5 min.

This approach is interesting in that selective recognition of the species myoglobin bands is achieved using a combination of the non-species specificity of anti-human myoglobin serum in conjunction with the resolving power of isoelectric focusing.

4.3.4.1 Summary of electrophoresis. Isoelectric focusing is probably the most widely used form of electrophoresis for meat species identification, particularly if the analyst is required to examine samples for a wide range of adulterating species. Saline extraction of raw sample proteins, isoelectric focusing and subsequent chemical staining with Coomassie Blue can provide a considerable protein profile that can allow the differentiation of many species. This relatively simple approach can be applied to binary mixtures of meats but electrophoretograms become complex and interpretation may be difficult, if not impossible, for some mixtures. Less-complex electrophoretograms can be obtained if selective enzyme staining or immunoblotting techniques are used but again not all species in the admixture can be identified.

The majority of electrophoretic procedures have been applied to raw products but, as indicated, some have been applied to heated meats but their robustness must be evaluated critically if, or when, they are applied to commercially cooked products.

Some workers have attempted to use electrophoretic methods with densitometric scanning for semi-quantitative applications. Caution is required, and comments on quantitation in the immunoassay section of this chapter also apply to electrophoretic methods.

In summary, electrophoretic methods are very useful for screening purposes; are capable of detecting some species in binary or ternary meat species mixtures; and some may be applied to 'cooked' products and permanent records of protein profiles can be kept for evidential purposes. They also may be applied to 'exotic' species when antispecies antisera are not available and immunoassay is therefore not an option. The disadvantages are that significant analyst skills are required if reliable and reproducible results are to be obtained and not all species can be differentiated, particularly if in admixture.

4.3.5 Immunoassay procedures

Immunoassay procedures have been applied to problems of meat speciation for many years. In the 1890s they were used to identify horse proteins; in a 1942 text on food analysis (Nicholls, 1942) serological techniques were proposed as the best means of detection of horse meat in raw beef; and in the early 1950s the FDA in the USA used these techniques to detect beef adulteration with horse meat. Despite this long history of the application of immunoassay to meat speciation, robust methods have not yet been developed to meet all the needs of analysts.

4.3.5.1 Immunodiffusion methods. Immunodiffusion methods are ideal for routine species identification of raw meats. They are simple to use, inexpensive, provide a relatively rapid result and some commercial kits contain all the reagents required. Many papers have been published that give details of the production of species specific antisera by immunisation of various animals with a range of different and often unidentified antigens extracted from meats. Hayden (1978) used agar gel immunodiffusion (AGID) to detect serum proteins from pig, horse and rabbit flesh in extracts of ground beef. Antisera were raised in goats and rabbits by injection of whole species serum. Qualitative detection of each species in beef at the 1–3% level was claimed to be possible. In a subsequent paper Hayden (1979) describes the production of antisera to ovine, porcine and equine myoglobin. This antigen was chosen because it is relatively thermostable and could perhaps yield antisera that could be used to identify the species of origin of heated meats. Use of the antisera in AGID allowed the detection of the flesh of the corresponding species in raw and mildly heated products but assay reliability decreased with increased heating of the sample. Later, Hayden (1981) published details of an AGID procedure that used rabbit antisera raised against thermostable antigens of horse, pig, lamb and poultry adrenals and which could detect the flesh of the target species in sausages cooked to a core temperature of 71°C. Cross-reactivity was observed between beef and lamb, and chicken and turkey could not be differentiated. Detection levels were between 5% and 10% (m/m).

Kang'ethe *et al.* (1986) were another group who worked on the development of AGID methods to identify meat species in heated products. Antisera to a range of species of bovidae significant in Kenya were prepared, and antisera to cattle thermostable muscle antigens (TMAs) could be used to detect the presence of beef in meat products that had been commercially sterilised. However, applications did not extend to meats commonly consumed in Europe. Schweiger *et al.* (1983) developed an AGID procedure for the detection of chicken and turkey muscle in beef and pork products by raising antisera against the turkey muscle protein troponin. Results indicated that the presence of chicken or turkey could be detected in products cooked to a temperature of 70°C. This paper also reports the application of immunoelectrophoresis and counter-immunoelectrophoresis. It is interesting that this is one of the few reports of an immunological procedure designed to detect muscle as opposed to plasma or blood proteins, a significant fact when interpreting results of such examinations.

Swart and Wilks (1982) adopted the approach taken by many laboratories and used commercially available anti-species antisera in an AGID assay format (they prepared anti-kangaroo antiserum by immunising rabbits with kangaroo blood as this was not commercially available). Detection limits for sheep, horse and kangaroo meats in beef ranged from 5–20%, depending upon the species being detected, and the effects of cross-reacting antibodies were nullified by addition of the relevant species' proteins. The typical detection limit obtained in the authors' laboratory using AGID and commercial antisera is 3% (m/m).

Mageau *et al.* (1984) and Cutrufelli *et al.* (1986, 1987, 1988, 1989) published a series of papers describing AGID procedures for the detection of bovine, avian, porcine and ovine meats. They used reagent-impregnated filter paper discs to overcome the need for regeant handling (i.e. antisera and control species antigens), so that the kits could be used 'in the field'. The Overnight Rapid Bovine Identification Test (ORBIT) used pre-prepared agar-gel immunodiffusion plates and paper discs impregnated with rabbit, sheep or goat anti-bovine antiserum. Bovine serum albumin was used as the immunogen. The Poultry Rapid Overnight Field Identification Test (PROFIT) used goat anti-poultry antisera raised against chicken and turkey serum albumin, the Porcine Rapid Identification Method (PRIME) used goat anti-porcine antisera and the Serological Ovine Field Test (SOFT) used calf anti-ovine antisera. ORBIT and PROFIT were exposed to a collaborative study in 10 laboratories; Cutruffelli *et al.* (1987) analysed 30 raw meat samples and showed the techniques to be reliable and capable of detecting adulteration at the 10% (m/m) level.

An early reference to a dipstick assay was reported by Patterson and Spencer (1983). They developed a rapid dipstick test for meat speciation in which the solid phase was coated with species-specific antisera. A positive reaction was indicated by a colour change and 1% (m/m) of meat in admixture could be detected.

4.3.5.2 Commercially available AGID kits. Various AGID kits for meat speciation have been produced commercially but many do not appear to survive long in the marketplace. Rhone-Poulenc Diagnostics Ltd now sell the ORBIT, PROFIT, PRIME and SOFT test kits, which can be conveniently used to identify meat species present in raw meat products. Kits to detect deer and horse are now being produced. The Binding Site (Birmingham, UK) also market pre-prepared AGID plates and anti-species affinity purified antisera for meat species testing and anti-species antisera are also available from Sigma. If analysts purchase anti-species antisera for use in in-house AGID assays it is imperative that the species specificity of each batch purchased is evaluated against all possible animal species that may be present. It is often necessary to absorb out cross-reacting antibodies.

An alternative to AGID meat speciation is the FAST Immunostick Test Kit produced by Cortecs Diagnostics (UK), which can be used for beef, horse, pork, sheep/goat, poultry, rabbit and kangaroo detection. Immuno-reagents are immobilised on a dipstick and identification of meat species via a colour change can be achieved in less than 1 h. Detection limits for species in admixture are about 1% (m/m) for lean meat samples. FAST immunosticks for buffalo, goat, chicken, turkey and deer are under development.

4.3.5.3 Summary of AGID methods. A range of applications of AGID for meat speciation have been published during the last 16 years but no great advances have been made. Raw meats are identified via blood or plasma proteins rather than via muscle protein, and few methods can be applied to cooked samples; antisera described for this application are not commercially available. The analyst may prefer to use commercially available kits as all reagents are supplied and good quality control is exercised, they show good species specificity, are relatively cheap, easy to use and are suitable for screening large numbers of samples. Perhaps the one disadvantage is that the analyst is reliant on one or two suppliers who may or may not last long in the marketplace. In the author's laboratory, commercial kits are used but also in-house AGID methods have been validated and these employ commercially available anti-species antisera. This allows simple and cost-effective routine screening of raw meat samples using a suite of immunological reagents with no dependence on one particular commercial supplier. Commercially available AGID methods can only be applied to raw meats and they only detect blood or plasma proteins. There is a need for methods that can be applied to cooked samples to detect muscle proteins. Detection of muscle proteins implies that you are detecting 'meat' not perhaps just adventitious contamination of samples with blood from an undeclared species.

4.3.6 Enzyme-linked immunosorbent assay (ELISA)

4.3.6.1 Raw meats. Although ELISA methods have been used in biomedical research since the early 1970s, one of the first applications to food analysis was the determination of soya protein in meat products by Hitchcock *et al.* in 1981. Advantages of ELISA over AGID methods are that much less antisera is required, results can be obtained in several hours, large sample numbers can be processed and 'quantitative' results may be obtained with some assays. Two of the first reported applications of ELISA to meat speciation were published by Whittaker *et al.* (1982) and Kang'ethe *et al.* (1982). Both groups of workers used anti-species antisera raised in rabbits against serum albumins from the target species (e.g. beef, sheep, horse, kangaroo, pig and camel). Antigens were extracted from samples with buffer or saline and an indirect ELISA method used. Detection limits of 3% (m/m) were reported by Kang'ethe *et al.* (1982). Whittaker *et al.* (1983) later published a more comprehensive report of their ELISA procedure.

Patterson *et al.* (1984) published details of a double sandwich ELISA procedure capable of detecting below 1% (m/m) of meat species in admixture. The ELISA was capable of differentiating sheep and goat, and beef and buffalo. In a subsequent paper, Patterson and Spencer (1985) differentiated phylogenically related species using monospecific antisera produced by immunisation in target-related species; cattle were immunised with whole buffalo serum, sheep with whole goat serum and horses with whole donkey serum. Levels of 1% (m/m) or less of each target meat species could be detected using the previously described ELISA format.

In the same year, Jones and Patterson (1985) published an ELISA procedure that used antisera directed towards porcine serum albumin (PSA) raised in rabbits and sheep to detect pig meat in meat mixtures. Levels as low as 0.5% (m/m) pork in beef sausages could be detected. The same authors then reported a procedure (1986) in which commercially available anti-species antisera (e.g. anti-cow, anti-horse, anti-pig, anti-sheep/goat) could be used for meat speciation which incorporated integral assay inhibition of heterologous cross-reactivity by addition of heterologous serum albumins prior to ELISA. Similar methods were published by Ayob *et al.* (1989)

Griffiths and Billington (1984) used commercially available anti-species antisera for a competitive ELISA, and evaluated the procedure for the estimation of the apparent beef content of beef joints and model mixtures via determination of beef serum proteins. ELISA results agreed well with the known beef content of model mixtures (but calibration was from the same meats), whereas a wide variation in results was recorded for the fat-free beef content of a range of beef joints, which was attributed to differences in the amount of residual blood in the various joints. This is an observation that must be borne in mind when interpreting semi-quantitative data from this type of assay, where serum proteins are the target of the ELISA.

Following the Griffiths and Billington (1984) observation, Martin *et al.* (1988a) commented that most anti-species antisera are raised against whole blood serum or serum albumins and, because of the variable amounts of residual blood in meats, these antisera cannot be used to quantify amounts of undeclared meat species. They raised antisera to muscle proteins by extracting sarcoplasmic antigens from lean pig, horse, chicken and beef meat with isotonic saline and raised antibodies against the pig antigens in rabbits. The antisera were affinity purified against the sarcoplasmic extracts from the other meat species. The authors present evidence of the specificity of the anti-porcine sarcoplasmic protein (anti-PSP) antisera and state that a detection rate of 1–50% (m/m) pig substitution in raw lean beef was possible. A second paper (Martin *et al.*, 1988b) discusses the preparation of anti-chicken sarcoplasmic protein (anti-CHSP) antisera. The authors state that the sarcoplasmic extracts contain haemoglobin and chicken serum albumins. Can it be assumed therefore that the pork sarcoplasmic extracts also contain the corresponding proteins? The authors state that this should not affect assay response but no experimental evidence is provided to substantiate this claim. The essays are claimed to detect muscle but the effect of blood and plasma is not reported, neither is data to show the natural variation of sarcoplasmic proteins extracted from various muscles within and between animals. This is necessary if methods are to be used to estimate amounts of meat species present in admixture.

In a subsequent paper Martin *et al.* (1991) report on the preparation of hybridoma cell lines for preparation of monoclonal antibodies to CHSP. Calibration lines from 1% to 100% chicken are presented and quantification suggested. Martin *et al.* (1992) later described a procedure for the immunoaffinity isolation of pig-specific soluble muscle proteins that were used to develop a monoclonal antibody specific for porcine muscle proteins (Morales *et al.*, 1994).

The UK Laboratory of the Government Chemist, in collaboration with Beeches Biotech, have developed an indirect ELISA technique that is designed to detect and quantify lamb muscle fibre proteins in the presence of pork and beef meat. The polyclonal anti-lamb muscle fibre antisera do not detect blood proteins from lamb, pork or beef and therefore the assay can be said to detect muscle. Work has now been extended to include preparation of antisera against porcine muscle. It is hoped that these assays will enable the detection and determination of one meat in admixture with another, as opposed to the detection of serum proteins.

4.3.6.2 Cooked meats. All of the antisera used in the ELISA applications mentioned so far were raised against proteins that are thermally labile, and therefore can only be applied to raw meat products.

Berger *et al.* (1988) described a complex and lengthy procedure for the extraction of antigens from pork and chicken muscle that yielded potent species-specific antisera, which could be used to identify target antigens in cooked and canned products. The authors state that, although the immunising antigens are heat-resistant and species-specific, neither heat resistance nor specificity are absolute. The poultry ELISA showed virtually no cross-reaction with other meat species (e.g. horse, beef, pork, sheep, deer, kangaroo) but showed a strong reaction with chicken and turkey even after samples had been heated to 120°C for 15 min. The pork ELISA performed similarly, except that some cross-reactivity was shown with raw-tissue extracts. The authors state that the method cannot be used quantitatively.

Patterson and Jones (1989) also reported the development of an ELISA procedure for the detection of heated pig meat. The paper describes two ELISA methods designed to detect heated lean meats, particularly pork, in a variety of meat products. The protocol for production of the antigen is not described in detail because it is stated that the procedure is being developed commercially. An indirect ELISA produced a linear response of optical density *versus* percentage of pig meat in lean-meat mixtures with beef, lamb or chicken. The assay permitted the detection of 5% pork lean meat mixed with other lean meats. A significant variation in the response to individual pig muscles meant that the assay could not be used to evaluate absolutely the lean meat content of products. The paper does not report applications to commercially canned and cooked products; model mixtures only were used.

Sawaya *et al.* (1990a; 1990b) described the application of HPLC of triglycerides, gas chromatography (GC) of fatty acids, HPLC of ophidine dipeptides and ELISA for the detection of pork in processed meats. For ELISA, antigens were prepared from pork by extraction with saline followed by autoclaving at 120°C for 30 min. The filtered extract was used to immunise sheep. Mixtures of pork with beef and sheep meats and commercially canned meat products were homogenised with saline and autoclaved for 30 min to extract antigens. The sheep anti-porcine antisera was mixed with extracts from autoclaved beef, sheep and horse meats to block cross-reactions with heterologous species prior to a competitive

ELISA. The authors report pork could be detected at the 2% level in commercially canned products and in model meat mixtures that had been heated at up to 120°C for 30 min.

In one of the most recent papers on identification of heated meats, Sherikar *et al.* (1993) prepared thermostable antigens from adrenal glands and muscles by mincing, sonicating and heating (at 98°C for 15 min) in normal saline solution followed by centrifugation and autoclaving of the supernatant for 30 min. Buffalo, sheep, goat and pig were the species of interest. Antisera were raised in rabbits and specificity improved by absorption with antigens from other non-target species. The authors provide useful information on the electrophoretic study of the extracted thermostable antigens from the various species and identify them as troponin T (*cf.* Schweiger *et al.*, 1983). The antisera were used in AGID, counter-immunoelectrophoresis (CIE), unlabelled antibody peroxidase antiperoxidase (PAP) and ELISA methods. Meat mixtures were 'cooked' at 100°C for 15 min. In mixtures containing cattle, buffalo, sheep and goat, the AGID method had a limit of detection of 10% (m/m), whereas pork was detected at 5% (m/m) using AGID, and CIE; ELISA and PAP could detect the presence of 1% (m/m) of all species.

4.3.6.3 Commercial ELISA kits. Cortecs Diagnostics (UK) market ELISA kits designed to identify meat species in raw and cooked meats. The raw meat ELISA kit is capable of detecting cow, horse, pig, sheep, poultry and kangaroo meats in admixture, while the cooked meat ELISA kit is capable of detecting beef, pork, poultry and sheep. These kits are suitable for screening meat products and about 1–3% of each meat species in admixture can be detected. All reagents necessary for the assay are included in each kit. The kits cannot be used to estimate the amount of a particular meat species in a mixture.

4.3.6.4 Summary of ELISA techniques. A wide range of ELISA techniques have been applied to meat speciation since the early 1980s. Many are designed to detect blood plasma proteins and are suitable for the relatively rapid screening of raw meat samples; however, they cannot be applied to cooked products, and estimation of the amount of a meat in a mixture of meats is difficult. One advantage is that some anti-species antisera are commercially available and allow the development and use of in-house ELISA techniques; the disadvantage is that the range of anti-species antisera available is limited.

ELISA techniques that are designed to identify the species of origin of meats in cooked products usually involve a complex extraction of thermostable antigens and antisera must then be purified to achieve species specificity. Thermostability is not absolute and assay response and specificity diminishes with prolonged cooking or heating, and these assays cannot be used quantitatively. In addition, antisera are unavailable commercially.

Several ELISA techniques that are designed to detect muscle proteins have been reported and, with more investigation, may be capable of identifying and determining the amount of a meat species in a mixture of meats. Applications of these assays to cooked products have not been reported, and antisera are not yet commercially available.

The majority of papers report the production of antisera that are not commercially available and so where do these facts leave the food analyst who does not have access to animal house facilities? The analyst is most likely to rely on the use of commercially available ELISA kits for raw and cooked meat speciation. In-house ELISA techniques that use commercially available anti-species antisera to plasma proteins can be used successfully on raw meat products so long as antisera specificity is monitored with every assay and care is exercised if the analyst attempts to obtain semi-quantitative information from such assays.

Colleagues at the UK Laboratory of the Government Chemist 27 years ago (Hubbard and Pocklington, 1968) stated in a paper on meat speciation by lipid analysis that 'the wider application of serological techniques is limited by the restricted range of antisera available'. The situation has not significantly improved since that time. Many interesting reports on the production of antisera for ELISA have been published but most are research papers and commercially available antisera do not often materialise.

4.3.7 DNA methods for meat speciation

From the previous review of immunological methods, it is clear that identification of meat species present in highly processed and heated meat products is difficult and assay specificity a problem, particularly if wishing to differentiate between closely related species, such as sheep and goat, beef and buffalo. DNA is a relatively stable molecule and specific recognition of unique sequences of DNA from samples can provide the basis of a highly specific assay for species identification. The principle of DNA-based methods is relatively straightforward. For simple 'dot- or slot-blot' assays, DNA is extracted from the meat product, partially purified and, after denaturation immobilised on a solid surface, usually a nylon membrane. A species-specific segment of labelled DNA (a DNA probe) is then allowed to hybridise with any complementary sequences of DNA, which are immobilised on the membrane. If a species-specific labelled probe hybridises with its complementary sequence it will be bound to the membrane, and visualisation of the probe label allows identification of the species DNA present. The key to assay specificity is the specificity of the labelled DNA probe for the species of interest and the stringency of the hybridisation conditions used. Initially, DNA probes were radiolabelled but now colorimetric, fluorescent or chemiluminescent labels can be used. These are far more user-friendly for applications in food and enforcement laboratories.

The potential of DNA methods using such simple 'dot-' or 'slot-blot' assay formats on nylon membranes have been investigated by Bauer *et al.* (1987), Wintero *et al.* (1990), Chikuni *et al.* (1990) and Ebbehoj and Thomsen (1991a; 1991b). These workers used total genomic species DNA as the probe, but these probes contained highly repetitive base sequences, which were homologous between closely related species. Cross-hybridisation between related species thus occurred and methods were not sufficiently species-specific for general application. Pork and chicken were clearly identified but specific ruminant species identification was not possible. Ebbehoj and Thomsen (1991b) attempted to overcome this problem by adding unlabelled DNA from the cross-hybridising species, which allowed discrimination between some ruminants and some differentiation of sheep and goat.

Myer *et al.* (1993) reported a procedure that used the DNA amplification method, the polymerase chain reaction (PCR), for raw and cooked meat speciation using growth hormone gene-specific primers. A modification of this technique using sheep satellite I DNA primers for PCR amplification followed by Apa II restriction digestion of the product enabled differentiation of sheep and goat meats (Chikuni *et al.*, 1994).

Bartlett and Davidson (1992) and Forrest and Carnegie (1994) have reported on the use of forensically informative nucleotide sequencing (FINS) for identifying the animal origin of biological specimens and gourmet meats, respectively. The procedure involves extraction of DNA from raw and highly processed foods followed by PCR amplification of a specific segment of DNA, the mitochondrial cytochrome b gene, determination of the nucleotide sequence of the amplified DNA sequence, and phylogenetic analysis of the nucleotide sequence using a database, with subsequent matching of the most closely related species.

These PCR and sequence analysis-based techniques are relatively expensive, technically demanding and prone to contamination because of the extreme sensitivity of the reaction. This sensitivity is required for pathogen detection, for example, but is not required for detection of meat species in foodstuffs, which are present at percentage levels. A further drawback of this approach is that quantitative PCR has not been developed sufficiently at present to allow reliable quantification of species present in admixture.

There is clearly a need for a relatively simple and robust DNA-based procedure that can be used by food analysts to identify meat species present in raw and highly processed products. A simple slot-blot assay has been developed at the UK Laboratory of the Government Chemist, which has been applied successfully to the specific identification of pork, beef, rabbit, sheep and goat in admixture (Parkes *et al.*, 1994). Species-specific oligonucleotide probes, which recognise relatively short sequences of DNA; have been developed, therefore a considerable amount of DNA degradation can be accommodated without influencing assay specificity. Meat samples are washed with buffer to remove blood and plasma, the remaining muscle cells are lysed and total genomic DNA is extracted from authentic tissue or cooked and processed products with a buffer-SDS-EDTA mixture and, after incubation with ribonuclease A and proteinase K, DNA is extracted with phenol-chloroform and precipitated with ethanol. Sample DNAs are loaded onto nylon membranes and digoxigenin-labelled species-specific probes added for the hybridisation stage. After removal of non-specifically bound probes by a series of stringency washes, bound species-specific probes are visualised by chemiluminescent reaction and detection on radiographic film. The DNA detected is from lysed muscle cells, not from blood, which is washed out of the system. In consequence this method detects the presence of 'meat' and the defence that blood from another species adventitously contaminated the meat then becomes inapplicable.

Species-specific oligonucleotide probes were derived from satellite DNA sequences and oligonucleotides synthesised commercially. Goat and sheep can be differentiated using the assay procedure outlined and meat species in highly processed products, such as canned frankfurters, patés, fermented sausages and canned pet food can be identified. Work to date also suggests that quantitative information can be obtained from these assays. A range of assay refinements are underway, including the adoption of a microtitre plate format.

4.3.7.1 Summary of DNA methods. DNA-based methods for meat species identification provide a way to identify the species of origin in raw and cooked meat products. Total genomic probes can be used to identify distantly related species but tailored oligonucleotide probes need to be designed for differentiation of closely related species. This is now possible using a variety of techniques. Non-radioactively labelled probes can be used in relatively simple assay formats, which may mean that commercially available meat speciation kits based on DNA methods could soon be available for routine use by food analysts.

4.4 Detection and determination of non-meat proteins

Non-meat proteins are a useful ingredient for meat products because of their emulsifying and binding properties. Adulteration becomes an issue when non-meat proteins are added to extend the apparent meat content of the product in an attempt to overcome minimum meat content criteria and/or when the consumer is misled regarding the structure and form of the product. There is a need for methods that can detect and determine the amounts of non-meat proteins in meat products.

4.4.1 Soya protein

The detection and determination of soya protein in meat products has posed a problem to food analysts for many years. Microscopy is a good starting point. Examination of cryosections of meat products by light microscopy, after chemical staining, can yield a wealth of information. The presence of soya can be detected and the type identified by examination of the characteristic features associated with each form. Stereological point-counting techniques (Flint and Meech, 1978) have been used to estimate soya protein in meat products, although this method is time-consuming and not recommended.

Electrophoresis has been used to detect and estimate the amount of soya protein in meat products but this approach has essentially been overtaken by application of immunological methods.

Glyn (1939) used an immuno-precipitation test to detect soya flour in sausages, and the procedure was later refined by Ferguson *et al.* (1942) to make it semi-quantitative. Simple immunodiffusion methods using commercially available anti-soya antisera may be used for screening raw meat products for the presence of soya protein (detection limits of 1% (m/m) have been achieved in the author's laboratory); however most methods developed and validated since 1980 have been ELISA formats. ELISA has the advantage of offering the potential for quantification, which is important if meat content is to be determined by chemical analysis. Also most methods have been developed to detect and determine soya protein in cooked products.

Hitchcock *et al.* (1981) described an ELISA procedure based on antibodies raised to denatured-renatured soya proteins. Soya proteins were solubilised using urea and mercaptoethanol, the mixture was dialysed, which allowed proteins to renature, and the solubilised proteins used to immunise rabbits. Samples and standards were prepared in a similar manner but diluted as opposed to dialysed. The assay response to different commercial soya products was not consistent but specificity of the antibodies was good. It was found that the 7S soya protein fraction (β-conglycinin) was most antigenic after renaturation, the response of the whey and 11S fraction (glycinin) being 30% and 18%, respectively. Therefore soya protein composition, its pre-treatment, and efficiency of solubilisation and renaturation will have an effect on assay response.

The study by Hitchcock *et al.* (1981) used experimentally prepared immunoreagents that were in limited supply. To allow the procedure to be used by food analysts who require access to commercially available antisera, a modified method that used commercially available antisera raised against native and heat denatured (but not renatured) soya proteins was investigated. Model meat products, which contained soya protein isolate, and which were analysed after heating at 100°C for 30 min and 121°C for 30 min were used for the study. The extraction procedure for samples involved maceration with a sequence of organic solvents to remove fat prior to preparation of an acetone powder, from which proteins were extracted using urea and mercaptoethanol, as described by Hitchcock. The results of the study were published by Griffiths *et al.* (1984). The commercial anti-soya antisera reacted in a similar way to the experimental antisera in that it responded strongly to the 7S protein and relatively weakly to the 11S and whey proteins. It was therefore established that the antisera recognised renatured soya proteins, which meant that, although sample proteins needed to be renatured to enable sufficient solubilisation, there was no need to use the experimental antiserum to renatured soya proteins; indeed a commercial antiserum could be used. The results of this work were sufficiently promising for a collaborative study to be organised with 22 laboratories; the results were published by Crimes *et al.* (1984). The study samples were sausage-type products, which contained soya flour or textured soya concentrate at levels of incorporation between 0% and 7% soya protein. The repeatability values obtained were 30% and 60% of the determined value of soya protein for soya flour and soya concentrate, respectively. Reproducibility values were 70% of the determined soya protein content for both types of soya. These results were significantly better than those obtained from a study of a stereological procedure.

Rittenburg *et al.* (1987) published details of an improved indirect competitive ELISA, which was based on the earlier work of Hitchcock *et al.* and which provided a commercial assay kit containing standardised and stabilised reagents. The sample extraction procedure was modified to allow the use of an aqueous buffer as opposed to a series of solvents; this reduced considerably the time required for this stage of the analysis. Renaturation of soya protein after extraction with urea and dithiothreitol was achieved using cystine as opposed to dialysis. The majority of soya products examined gave a response, which was expressed as the percentage of soya protein measured by the immunoassay compared with the total protein content, of between 80% and 100%. In addition, cross-reactivity with other food proteins was low and trueness and precision appeared satisfactory.

The ELISA procedure in commercial kit format described by Rittenburg *et al.*, was tested collaboratively by 13 laboratories and the results published by Hall *et al.* (1987). The conclusions from the study were that using a single arbitrary soya standard as reference (supplied with the kit), the method permits the estimation of the levels of soya protein in raw or pasteurised meat products of unknown composition containing an unknown soya ingredient with a recovery of 80–100%. The procedure has repeatability and reproducibility values, expressed as a percentage of the mean, in the order of 45% and 60%, respectively. This information, and other data in the paper, are essential if the analyst is to interpret reliably the information from this ELISA kit, which is available from Cortecs Diagnostics, UK. The analyst should be aware that the recovery of soya protein from meat products that have been retorted at a temperature above the sterilisation temperature used in the study may be reduced.

The AOAC (1990) procedure for the determination of soya protein in raw and heat-processed meat products is based on the procedure of Hitchcock *et al.* (1981) but employs the use of commercial immunoreagents as reported by Griffiths *et al.* (1984) and Crimes *et al.* (1984). It is unfortunate that the quoted supplier of the antisera no longer produces the reagent. It is a worrying fact that food analysts often have to rely on a few, often only one or two, suppliers of suitable immunoreagents or immunokits. When these suppliers withdraw from the market, analysts can be left in a difficult position.

Ravestein and Driedonks (1986) adopted a different approach to the denaturation and solubilisation of soya protein. Glycinin subunits (11S protein) were isolated from soya, boiled in buffer containing sodium dodecylsulphate (SDS) and used for immunisation of rabbits to produce antisoya antisera. Meat products were extracted by boiling with buffer containing SDS and mercaptoethanol for 10 min; supernatents after centrifugation and dilution were used for ELISA. The authors state that quantitative analysis is possible down to 0.5% (m/m) and that the method can be applied to sterilised products (120°C, 30 min) without underestimation of soya content. The authors provide an interesting account of their method development and their views on the advantages of SDS as a denaturant and as a solvent over urea and mercaptoethanol. It is not known to the authors whether this method is still used, or whether the antisoya antisera is commercially available.

More recently, Yasumoto *et al.* (1990) described an ELISA procedure based on the antigenicity of tryptic digest peptide fragments that are characteristic of soya protein. It is stated that the assay is independent of soya cultivar and degree of denaturation of proteins in the sample. The suggested reason for assay independence from degree of denaturation (i.e. independence from changes in tertiary and secondary protein structure) is similar to the reasoning of Ravestein and Driedonks (1986).

4.4.1.1 Summary of soya protein. A range of ELISA procedures have been proposed for the detection and determination of soya protein in meat products but the majority of analysts will probably use a commercially available ELISA kit. This is the most convenient and reliable option for quantitative evaluation. Analysts who have access to an alternative source of suitable anti-soya antisera and a validated ELISA are perhaps in an enviable position in that they are not totally reliant on a single commercial kit supplier.

4.4.2 *Non-meat proteins other than soya*

4.4.2.1 Electrophoresis. Polyacrylamide gel (PAGE) has been used by many workers to detect non-meat proteins in meat products. However, problems can be experienced when examining heated products, and the effectiveness of a procedure is dependent largely upon the type of meat product being examined and the non-meat protein present. These methods often have limited scope and are not particularly robust. It is often necessary to develop a procedure for a particular application, for example a method for the detection of milk proteins may not be suitable for the detection of soya in the same, or a different meat matrix.

Dysseler *et al.* (1986) described a PAGE procedure for the detection of wheat gluten in cooked foods, Kim *et al.* (1987) used a SDS–PAGE method to detect soya and milk protein in chicken products while Woychick *et al.* (1987) used a similar approach for the detection of soya protein in frankfurters.

Janssen *et al.* (1986) used SDS electrophoresis coupled with nitrocellulose blotting and an immunoperoxidase staining system to detect soya protein in heated meat products and, in a second paper (Janssen *et al.*, 1987), applied a similar procedure to the detection of wheat gluten, whey protein, casein, ovalbumin and soya protein in heated meats. Eriksson and Yman (1992) described the quantitative analysis of casein and egg proteins by rocket immunoelectrophoresis on a PhastSystem (Pharmacia Ltd) with applications to meat products. The advantage of these approaches is that the separating power of electrophoresis is coupled with the selective staining ability of an immunoreagent; background proteins will not be recognised and interpretation of electrophoretograms is simplified.

Electrophoresis may be useful for confirmatory evidence, or application for the routine examination of similar products, but it is not ideally suited for the *ad hoc* assessment of a variety of meat products. Electrophoresis coupled with immunostaining is useful for the screening of samples and for evidential purposes but considerable practical skill is required if reproducible results are to be obtained. However, the procedure is applicable for cooked products where immunoassays may not be available.

4.4.2.2 Immunoassay. Since soya-derived products are the most common source of non-meat protein added to meat products, most effort appears to have gone into the development of methods to detect and determine this potential ingredient. Relatively few methods have been published for the detection and determination of other non-meat proteins, although detection of wheat gluten has attracted interest.

Immunoassays to detect and determine wheat gluten have been developed since the 1960s to detect its undeclared addition as a meat extender and because some people have an intolerance to wheat gluten and similar proteins from rye and barley. Skerrit and Hill (1990) published details of a sandwich-ELISA method based on monoclonal antibodies raised against omega-gliadins and related prolamines from wheat, rye and barley. The method could provide quantitative results on raw and cooked products and was exposed to a collaborative study involving 15 laboratories. The 13 samples in the study contained gluten over the range 0.02–10% (m/m); the meat products contained gluten at 2% (m/m) and less. The conclusion from the study was that the assay was semi-quantitative for the determination of gluten in the processed meat products; it was suggested that better precision was not obtained because of the possible inhomogeneity of the test samples. The repeatability and reproducibility data (RSD_r 14% and 26% and RSD_R 46% and 56%) were not dissimilar from those obtained from the collaborative study of the soya test kit (Hall *et al.*, 1987). This method is, however, a relatively simple means of estimating the gluten content of meat products. Gluten test kits are now available from Cortecs Diagnostics (UK) and Digen Limited (UK).

Turin and Bonami (1994) published details of the detection of egg-white proteins in dairy products by determination of ovalbumin. Their approach may be applicable to meat products. SDS–PAGE analysis was used to separate proteins from ice-cream and cheese and ovalbumin was detected by blotting onto nitrocellulose and immunochemical staining with commercially available rabbit anti-ovalbumin serum. An ELISA procedure was used to quantify the amount of ovalbumin present.

Immunoassay kits for the detection and semi-quantitative determination of bovine casein in foodstuffs are commercially available and can be applied to meat products.

A selection of antisera, which will recognise non-meat proteins such as soya, wheat gluten, bovine whey, bovine casein and chicken ovalbumin, are commercially available. They can be used in simple assay formats such as double or radial diffusion, in ELISA formats or for immunoblotting after electrophoresis.

Immunoassays are commercially available for the detection and determination of gluten and casein, and some antisera, which will recognise non-meat proteins if analysts need to develop their own methods of analysis, are commercially available. If analysts develop 'in-house' methods, they need to be fully validated before use, cross-reactivity of antisera checked and interpretation of results considered carefully in the light of comments and collaborative study data available for soya and gluten assays.

4.5 Mechanically recovered meat (MRM)

Traditional manual butchery procedures produce hand-deboned meat (HDM) but this practice results in residual meat remaining on the bones of a carcass. Machines that can remove this residual meat from animal bones and poultry carcasses are available, the product being called mechanically recovered meat (MRM). MRM has the appearance of highly comminuted meat; the types of mechanical recovery used destroy, to a greater or lesser degree, the natural muscle structure of HDM.

The undeclared use of MRM in meat products is a contentious issue that has stimulated much debate over the last 15 years or so. The UK Food Standards Committee (FSC/REP/72,1980) were of the view that MRM was not meat as would be understood by the consumer and recommended that if MRM was present in an amount more than 5% of the declared lean meat content of any meat product there should be a requirement for special reference to its use on the label.

There has been a continuing debate as to whether MRM is sufficiently different from HDM to warrant identification on food labels. There are several points to consider. Since MRM is disrupted muscle and is less expensive than whole-muscle meat it may be considered misleading to the consumer if it were incorporated into products described as being prepared from whole muscle or pieces of meat without an appropriate label declaration. The UK Food Labelling Regulations (1984) require that when any food ingredient has undergone any particular process or treatment, a reference to that process or treatment should be included in the name of the food and/or in the list of ingredients where its omission would otherwise mislead the consumer. The view is held by many that undeclared inclusion of MRM in meat products constitutes adulteration. The view in many member states of the EU is that MRM should be shown separately on ingredients lists of foodstuffs, a matter that the Commission is considering.

Although there remain differing views on the interpretation of these matters, it has been clear for a considerable time that additional information on the chemical composition of MRM has been required, and that methods for the detection of MRM in meat products are needed.

Meech and Kirk (1986) reported on the chemical composition of MRMs and assessed the data to see if there were compositional indices that could be used to indicate the presence of MRM in a product. Moisture, fat, protein, ash, bone content, fatty acids, sterols, pH, hydroxyproline and trace element contents were determined. Despite some compositional differences between MRM and HDM, the variations in composition between MRMs was large and the authors concluded that a chemical index for the presence of MRM would be difficult to determine.

A series of four papers were published which describe work funded by the UK Ministry of Agriculture Fisheries and Food, designed to further investigate MRM chemical composition and methods for MRM detection.

Crosland *et al.* (1995) determined the fat, moisture, nitrogen, ash and collagen, calcium, iron and purine content in beef, lamb, pork, chicken and turkey MRMs and also for HDM trimmed from close to the bone. Effects of machine type and operating conditions were included in the study. The overall conclusion was that there were several differences between MRM and HDM but, generally, MRM was not consistently or significantly different from the HDM and that the components determined in the study could not be used for the reliable detection of MRM in meat products.

Savage *et al.* (1995) describe the examination of MRMs from different machines and animal species by gel electrophoresis. Mechanical recovery of meat subjects the bones to high pressures and it was thought that this could result in the release of bone fluids into MRM, which could then be used as markers to differentiate MRM from HDM. SDS and water extracts of samples were examined using SDS–PAGE. Raw beef, raw lamb and raw pork HDM and MRM could be differentiated by comparison of myoglobin, haemoglobin and other protein bands on the electrophoretograms. Poultry MRM and HDM could be distinguished but with more difficulty than the red meats. Machine type did not appear to have a large influence on protein profile. It was possible to identify MRM in minced meat samples and an order of percentage incorporation. The authors state that, with further development, it may be possible to use the technique to detect and possibly quantify MRM in meat products and apply the method to cooked products.

Pickering *et al.* (1995a; 1995b) report on the application of light microscopy to the differentiation of MRM and HDM. Again samples from different species and MRMs produced by different machines and processing conditions were used. The samples examined were the same as those used by Savage *et al.* (1995). Cryosections of samples were made, stained with toluidine blue and examined by light microscopy for the presence of hyaline cartilage. Hyaline cartilage is a dense connective tissue that is associated with bone rather than muscle and it was thought that this cartilage may be released into MRM by the high pressure involved during MRM production. Hyaline cartilage was observed in almost all MRM samples but not in the HDM samples, confirming that its presence in a sample could be used as an indicator of the presence of MRM. Mixtures of HDM and MRM in various proportions were examined to test the sensitivity of detection. Beef MRM could be detected at the 10% (m/m) level and chicken at the 25% (m/m) level but it was difficult to detect pork MRM in HDM. Hyaline cartilage was observed in some commercially canned meat products, indicating that the method can be applied to cooked products.

Although the results demonstrate that the method is rapid and useful as a screening procedure for detection of MRM, the authors state that the method needs to be evaluated further on a greater number and range of samples and on commercially prepared meat products.

In the fourth paper of the series Pickering *et al.* (1995b) describe an immunological approach to the detection of bone marrow antigens, the premise being that bone marrow will be exuded into MRM during production and that it could be used as a marker for MRM in meat products. Since chicken MRM is the most commonly used MRM in meat products in the UK, studies were performed on chicken bone marrow. Electrophoresis was used to examine proteins from MRM, HDM and chicken bone marrow cells. Proteins present in MRM and bone marrow, but not in HDM were identified, and used as the immunogen for polyclonal antiserum production. A non-competitive solid-phase ELISA technique was developed from the antiserum produced. The antiserum showed high immunoreactivity with chicken and turkey MRM, slight reactivity with turkey HDM but low reactivity with the MRM and HDM of other species. Blood, skin and liver showed a response to the assay that was higher than that for HDM, and these components could lead to an overestimation of MRM. When the assay was applied to 'blind' samples of MRM in HDM, the amount of MRM was overestimated; this was attributed to the effect of residual blood in the HDM used. The authors state that the assay requires further refinement and the selectivity of the antiserum requires improving; however, the results show that an immunological approach to the detection of MRM is possible; quantification may also be achievable.

It is evident from the results presented, that light microscopy may provide a rapid screening method for the detection of MRM and that electrophoresis and immunoassay may provide detection and perhaps quantification of MRM; however, the methods require further development, particularly if they are to be applied to cooked products.

4.6 Summary

In this chapter, detail of the development of methods to detect the adulteration of meat and meat products has been described in overview to give the reader an appreciation of some of the analytical developments to date, some of the problems still faced by analysts, and the scope of methods available. If more information is required the original references should be consulted.

Each of the methods discussed should not be considered in isolation. They are all complementary and, when used as a suite of techniques, will provide the analyst with the information required to make informed decisions regarding the authenticity of meats and meat products.

Trends in food adulteration change depending upon many factors including the relative costs of ingredients, changes in food-processing techniques, and developments in analytical science. The analysts skills are, and probably will be, continually challenged by the adulteration and misdescription of meats and meat products.

References

Analytical Methods Committee (1991) Nitrogen factors for pork: a re-assessment. *Analyst* **116**, 761–66.

Analytical Methods Committee (1993) Nitrogen factors for beef: a re-assessment *Analyst* **118**, 1217–26.

AOAC Official Methods of Analysis (1990) Soy protein in raw and heat processed meat products. Enzyme-linked immunosorbent assay. (15th edn.) **988.10**, 942–45.

Ayob, M.K., Ragab, A.A., Allen, J.C. *et al.* (1989) An improved rapid ELISA technique for detection of pork in meat products. *J. Sci. Food Agric.* **49**, 103–16.

Bartlett, S.E. and Davidson, W.S. (1992) FINS (Forensically Informative Nucleotide Sequencing): a procedure for identifying the animal origin of biological specimens. *Bio Techniques* **17(1)**, 408–11.

Bauer, F. and Hofman, K. (1987) Elektrophoretische Tierartbestimmung. Steigerung der Empfindlichkeit durch Peroxidasefarbung der Myoglobine. *Fleischwirtschaft* **67**, 861.

Bauer, C., Teifel-Greding, J., Liebhardt, E. (1987) Species identification of heat denaturized meat samples by DNA analysis. *Archiv für Lebensmittelhygiene* **38**, 149–76.

Bayliss-High, O. (1984) Lipid histochemistry. Royal Microscopical Society Microscopy Handbooks, Oxford University Press.

Berger, R.G., Mageau, R.P., Schwab, B. *et al.* (1988) Detection of poultry and pork in cooked and canned meat foods by enzyme-linked immunosorbent assays. *J. Assoc. Off. Anal. Chem.* **71(2)**, 406–9.

British Standards Institution, BS 4401. Analytical Methods for Meat and Meat Products. Determination of ash: Part 1, 1980: ISO 936:1978. Determination of nitrogen: Part 2, 1980: ISO 937:1978. Determination of moistures: Part 3, 1970: ISO/R 1442. Determination of total fat: Part 4, 1970: ISO 1443:1973. Determination of L(–)-hydroxyproline: Part 11, 1979 ISO 3496:1978.

Chikuni, K., Ozutsumi, K., Koishikawa, T. *et al.* (1990) Species identification of cooked meats by DNA hybridisation assay. *Meat Sci.* **27**, 119–28.

Chikuni, K., Tabata, T., Kosugiyama, M. (1994) Polymerase chain reaction assay for detection of sheep and goat meats. *Meat Sci.* **37**, 337–45.

Commission Regulation (EEC) No 226/89. *Official Journal of the European Communities* 31/1/89, NOL 29/11–12.

Coomaraswanay, M. (1972) The calculation of meat content in the presence of excess rind and/or connective tissue. *J. Assoc. Public Analysts* **19(2)**, 33–43.

Cottam, J. (1975) A note on the calculation of lean meat content of sausages. *J. Assoc. Public Analysts* **13**, 133–34.

Crimes, A.A., Hitchcock, C.H.S. and Wood, R. (1984) Determination of soya protein in meat products by an enzyme-linked immunosorbent assay procedure: Collaborative study. *J. Assoc. Public Analysts* **22**, 59–78.

Crosland, A.R., Patterson, R.L.S., Higman, R.C. *et al.* (1995) Investigation of methods to detect mechanically recovered meat in meat products – I. Chemical composition. *Meat Sci.* **40**, 289–302.

Cutrufelli, M.E., Mageau, R.P., Schwab, B. *et al.* (1986) Development of poultry rapid overnight field identification test (PROFIT). *J. Assoc. Off. Anal. Chem.* **69(3)**, 483–87.

Cutrufelli, M.E., Mageau, R.P., Schwab, B. *et al.* (1987) Detection of beef and poultry by serological field screening tests (ORBIT and PROFIT): Collaborative study. *J. Assoc. Off. Anal. Chem.* **70(2)**, 230–33.

Crutrufelli, M.E., Mageau, R.P., Schwab, B. *et al.* (1988) Development of porcine rapid

identification method (PRIME) by modified agar gel immunodiffusion. *J. Assoc. Off. Anal. Chem.* **71(2)**, 444–45.
Cutrufelli, M.E., Mageau, R.P., Schwab, B. *et al.* (1989) Development of serological ovine field test (SOFT) by modified agar gel immunodiffusion. *J. Assoc. Off. Anal. Chem.* **72(1)**, 60–61.
Dysseler, P., Jacqmain, D. and Karydas, J. (1986) Electrophoretic determination of gliadins in cooked foods. *Belgian J. Food Chem Biotechnol.* **41(6)**, 143–49.
Ebbehoj, K.F. and Thomsen, P.D. (1991a) Species differentiation of heated meat products by DNA hybridisation. *Meat Sci.* **30**, 221–34.
Ebbehoj, K.F. and Thomsen, P.D. (1991b) Differentiation of closely related species by DNA hybridisation. *Meat Sci.* **30**, 359–66.
Eriksson, A. and Yman, I.M. (1992) *Food Safety and Quality Assurance: Applications of Immunoassay Systems* (Eds M.R.A. Morgan, C.J. Smith, P.A. Williams). Elsevier Applied Science, Amsterdam, pp. 65–68.
Fergusson, C.S., Racicot, P.A. and Rane, L (1942) Study of use of precipitin test for determination of soy flour in sausages. *J. Assoc. Off. Agric. Chem.* **25**, 533.
Flint, F.O. and Meech, M.V. (1978) Quantitative determination of texturised soya protein by a stereological technique. *Analyst* **103**, 252–58.
Flint, F.O. and Firth, B.M. (1981) A toluidine blue stain mountant for the microscopy of comminuted meat products. *Analyst* **106**, 1242–43.
Flint, F.O. and Pickering, K. (1984) Demonstration of collagen in meat products by an improved picro-sirius red polarisation method. *Analyst* **109**, 1505–06.
Food and Drug Administration (FDA), USA. (195X).
Forrest, A.R.R. and Carnegie, P.R., (1994) Identification of gourmet meat using FINS (Forensically Informative Nucleotide Sequencing). *Bio Techniques* **17(1)**, 24–26.
Glynn, J.H. (1939) Quantitative determination of soy bean protein in sausage and other protein mixtures. *Science* **89**, 444.
Griffiths, N.M. and Billington, M.J., (1984) Evaluation of an enzyme-linked immunosorbent assay for beef blood serum to determine indirectly the apparent beef content of beef joints and model mixtures. *J. Sci. Food Agric.* **35**, 909–14.
Griffiths, N.M., Billington, M.J., Crimes, A.A. *et al.* (1984) An assessment of commercially available reagents for an enzyme-linked immunosorbent assay of soya protein in meat products. *J. Sci. Food Agric.* **35**, 1255–60.
Hall, C.C., Hitchcock, C.H.S. and Wood, R. (1987) Determination of soya protein in meat products by a commercial enzyme immunoassay procedure: collaborative trial. *J. Assoc. Public Analysts* **25**, 1–27.
Hayden, A.R. (1978) Determination of residual species serum albumin in adulterated ground beef. *J. Food Sci.* **43**, 476–78, 492.
Hayden, A.R. (1979) Immunochemical detection of ovine, porcine and equine flesh in beef products with antisera to species myoglobin. *J. Food Sci.* **44(2)**, 494–500.
Hayden, A.R. (1981) Use of antisera to heat-stable antigens of adrenals for species identification in thoroughly cooked beef sausages. *J. Food Sci.* **46**, 1810–13.
Heinart, H.H. and Klinger, A. (1980) Tierartpeczifische Eiweissdifterentzierung: Protein und Enzymmuster bei Reh und Hirsch. *Fleischwirtsch* **60**, 1682–85.
Hitchcock, C.H.S., Bailey, F.J., Crimes, A.A. *et al.* (1981) Determination of soya proteins in food using an enzyme-linked immunosorbent assay procedure. *J. Sci. Food Agric.* **32**, 157–65.
Hofman, K. and Bluchel, E. (1986) Bestimmung der Tierart von rohem Muskelfleisch auband der Myoglobinmuster im pH Gradienten Gel. *Fleischwirtschaft* **66**, 916.
Hubbard, A.W. and Pocklington, W.D. (1968) Distribution of fatty acids in lipids as an aid to the identification of animal tissues. *J. Sci. Food Agric.* **19**, 571–77.
Janssen, F.W., Voortman, G. and Baaij, J.A. (1986) Detection of soya proteins in heated meat products by blotting and dot-blot. *Z. Lebensm Unters Forch.* **182**, 479–83.
Janssen, F.W., Voortman, G. and Baaij, J.A. (1987) Detection of wheat gluten, whey protein, casein, ovalbumin and soy protein in heated meat products by electrophoresis, blotting and immunoperoxidase staining. *J. Agricol. Food Chem.* **35**, 563–67.
Janssen, F.W., Hogele, G.H., Voorpostel, A.M.B. *et al.* (1990) Myoglobin analysis for determination of beef, pork, horse, sheep and kangaroo meat in blended cooked products. *J. Food Sci.* **55(6)**, 1528–30, 1563.

Jones, S.J. and Patterson, R.L.S. (1985) Double-antibody ELISA for detection of trace amounts of pig meat in raw meat mixtures. *Meat Sci.* **15**, 1–13.

Jones, S.R. and Patterson, R.L.S. (1986) A modified indirect ELISA procedure for raw meat speciation using crude anti-species antisera and stabilised immunoreagents. *J. Sci. Food Agric.* **37**, 767–75.

Kaiser, K.P., Matheis, G., Knita-Durrmann, C. *et al.* (1980) Qualitative and quantitative analysis of raw binary meat mixtures by means of isoelectric focussing. *Z. Letensm. Unters. Forch.* **171**, 415–19.

Kang'ethe, E.K., Jones, S.J. and Patterson, R.L.S. (1982) Identification of the species of origin of fresh meat using an enzyme-linked immunosorbent assay procedure. *Meat Sci.* **7**, 229–40.

Kang'ethe, E.K., Gathuma, J.M. and Lindqvist, K.J. (1986) Identification of the species of origin of fresh, cooked and canned meat and meat products using antisera to thermostable muscle antigens by Ouchterlony's double diffusion test. *J. Sci. Food Agric.* **37**, 157–64.

Kauffer, K., Beneke, B. and Bentler, W. (1990) Detection of comminuted chicken meat in native and heated meat mixtures by means of isoelectric focussing in polyamide gels. *Fleischwirtsch.* **70(10)**, 1188–94.

Kim, H. and Shelef, L.A. (1986) Characterisation and identification of raw beef, pork, chicken and turkey meats by electrophoretic patterns of their sarcoplasmic proteins. *J. Food Sci.* **51**, 731–41.

Kim, Y.H., Park, G.B., Lee, M.H. *et al.* (1987) Studies on the determination of non-meat proteins in chicken products. *Korean J. Animal Sci.* **29(9)**, 416–19.

King, N.L. (1984) Species identification of cooked meats by enzyme staining of isoelectric focussing gels. *Meat Sci.* **11**, 59–72.

King, N.L. and Kurth, L. (1982) Analysis of raw beef samples for adulterant meat species by enzyme-staining of isoelectric focussing gels. *J. Food Sci.* **47**, 1608–12.

Kirk, R.S. and Sawyer, R. (1991) *Pearson's Composition and Analysis of Foods*, 9th edn. Longman, Harlow, UK.

Lord, D.W. and Swan, K.J. (1984) Connective tissue in beefburgers. *J. Assoc. Public Analysts* **22**, 131–40.

Mageau, R.P., Cutrufelli, M.E., Schwab, B. *et al.* (1984) Development of an overnight rapid bovine identification test (ORBIT) for field use. *J. Assoc. Off. Anal. Chem.* **67(5)**, 949–54.

Martin, R., Azcona, J.I., Casas, L. *et al.* (1988a) Sandwich ELISA for detection of pig meat in raw beef using antisera to muscle soluble proteins. *J. Food Protection* **51(10)**, 790–94.

Martin, R., Azcona, J.I., Tormo, J. *et al.* (1988b) Detection of chicken meat in raw meat mixtures by a sandwich enzyme immunoassay. *Int. J. Food Sci. Technol.* **23**, 303–10.

Martin, R., Wardale, R.J., Jones, S.J. *et al.* (1991) Monoclonal antibody sandwich ELISA for the potential detection of chicken meat in mixtures of raw beef and pork. *Meat Sci.* **30**, 23–31.

Martin, R., Azcona, J.I., Hernandez, P.E. and Sanz, B., (1992) Partial purification of pig specific soluble muscle proteins. *Fleischwirtshaft* **72(6)**, 899–900.

McCormick, R.J., Collins, D.A., Field, R.A., Moore, T.D. (1992) Identification of meat from game and domestic species. *J. Food Sci.* **57(2)**, 516–17, 520.

Meat Products and Spreadable Fish Products Regulations (1984).

Meech, M.V. and Kirk, R.S. (1986) Chemical characterisation of mechanically recovered meats. *J. Assoc. Public Analysts* **24**, 13–26.

Meyer, R., Candrian, U. and Luthyl, J., (1993) Species differentiation and detection of soya in heated meat products by the polymerase chain reaction (PCR). *Mitt. Gebiete Lebensm. Hyg.* **84**, 112–21.

Morales, P., Garcia, T., Gonzalez, I. *et al.* (1994) Monoclonal antibody detection of porcine meat. *J. Food Protection* **52(2)**, 146–49.

Nicholls, J.R. (1942) *Aids to Analysis of Food and Drugs*. Baillière, Tindall and Cox, London, p. 275.

Parkes, H.C., Hunt, D.J. and Lumley, I.D. (1994) Quantitative and specific species identification in raw and cooked meat products employing oligonucleotide probes. Prepared for publication.

Patterson, R.L.S. and Jones, S.J. (1989) Species identification in heat processed meat products. Proceedings of 35th International (Copenhagen) Congress on Meat. *Sci. Technol.* **2**, 529–36.

Patterson, R.M. and Spencer, T.L. (1983) A rapid on-site test for speciation of meat. *Australian Veterinary J.* **60(12)**, 381–82.

Patterson, R.M. and Spencer, T.L. (1985) Differentiation of raw meat from phylogenically related species by enzyme-linked immunosorbent assay. *Meat Sci.* **15**, 119–23.

Patterson, R.M., Whittaker, R.G., and Spencer T.L. (1984) Improved species identification of raw meat by double sandwich enzyme-linked immunosorbent assay. *J. Sci. Food Agric.* **35**, 1018–23.

Pearson, D. (1968) The calculation of the lean meat content of pork sausages. *J. Assoc. Public Analysts* **6**, 129–34.

Pearson, D. (1970) A general formula for the calculation of the lean meat content. *J. Assoc. Public Analysts* **8**, 22–24.

Pearson, D. (1976) The examination of meat products with special reference to the assessment of meat content. *Analyst* **100**, 73–81.

Pickering, K., Evans, C.L., Hargin, K.D. *et al.* (1995a) Investigation of methods to detect mechanically recovered meat in meat products. III. Microscopy. *Meat Sci.* **40**, 319–26.

Pickering, K., Griffin, M., Smethurst, P.L. *et al.* (1995b) Investigation of methods to detect mechanically recovered meat in meat products. IV. Immunology *Meat Sci.* **40**, 327–36.

Prasad, V. and Misra, D. (1981) Differentiation of meats of different species of animals by muscle esterase pattern in different age and sex groups. *Indian J. Anim. Sci.* **51**, 211–14.

Ravestein, P. and Driedonks, R.A. (1986) Quantitative immunoassay for soya protein in raw and sterilised meat products. *J. Food Technol.* **21**, 19–32.

Richardson, G.A. and Tarassuk, N.P. (1934) Detection of gelatin in cultured buttermilk and cottage cheese. *J. Assoc. Off. Anal. Chem.* **17**, 314–19.

Rittenburg, J.H., Adams, A., Palmer, J. *et al.* (1987) Improved enzyme-linked immunosorbent assay for determination of soy protein in meat products. *J. Assoc. Off. Anal. Chem.* **70(3)**, 582–87.

Saeed, T., Abu-Dagga, F. and Rahman, H.A. (1986) Detection of pork and lard as adulterants in beef and mutton mixtures. *J. Assoc. Off. Anal. Chem.*, **69(b)**, 999–1002.

Savage, A.W.J., Richardson, R.I., Jolley, P.D. *et al.* (1995) Investigation of methods to detect mechanically recovered meat in meat products. II. Gel Electrophoresis. *Meat Sci.* **40**, 303–17.

Sawaya, W.N., Saeed, T., Mameesh, M. *et al.* (1990a) Detection of pork in processed meat: experimental comparison of methodology. *Food Chem.* **37**, 201–19.

Sawaya, W.N., Mameesh, M.S., El-Rayes, E. *et al.* (1990b) Detection of pork in processed meat by an enzyme-linked immunosorbent assay using antiswine antisera. *J. Food Sci.* **55(2)**, 293–97.

Schweiger, A., Baudner, S. and Gunther, H.O. (1983) Isolation by free-flow electrophoresis and immunological detection of troponin T from turkey muscle: an application in food chemistry. *Electrophoresis* **4**, 158–63.

Sherikar, A.T., Karkare, U.D., Khot, J.B. *et al.* (1993) Studies on thermostable antigens, production of species-specific antiadrenal sera and comparison of immunological techniques in meat speciation. *Meat Sci.* **33**, 121–36.

Sinclair, A.J. and Slattery, W.J. (1982) Identification of meat according to species by isoelectric focussing. *Austr. Vet. J.* **58**, 79–80.

Skerrit, J.H. and Hill, A.S. (1990a) Monoclonal antibody sandwich enzyme immunoassay for determination of gluten in foods. *J. Agric. Food Chem.* **38**, 1771–78.

Skerrit, J.H. and Hill, A.S. (1990b) Enzyme immunoassay for determination of gluten in foods: Collaborative study. *J. Assoc. Off. Anal. Chem.* **74(2)**, 257–64.

Skrokki, A. and Hormi, O. (1984a) Composition of minced meats: Part A. Methods. *Meat Sci.* **38(3)**, 497–501.

Skrokki, A. and Hormi, O. (1984b) Composition of minced meat: survey of commercial ground meat. *Meat Sci.* **38(3)**, 503–9.

Slattery, W.J. and Sinclair, A.J. (1983) Differentiation of meat according to species by electrophoretic separation of muscle lactate dehydrogenase, esterase isoenzymes and isoelectric focusing of soluble muscle proteins. *Austr. Vet. J.* **60**, 47–51.

Stubbs, G. and More, A. (1919) The estimation of the approximate quantity of meat in sausages and meat pastes. *Analyst* **44**, 125–27.

Swart, K.S. and Wilks, C.R. (1982) An immunodiffusion method for the identification of the species of origin of meat samples. *Austr. Vet. J.* **59**, 21–22.
Turin, L. and Bonomi, F. (1994) Immunochemical detection of ovalbumin in dairy-based foods. *J. Sci. Food Agric.* **64**, 39–45.
UK Food Labelling Regulations (1984) Statutory Instrument 1984 No. 1305. HMSO, London.
UK Food Standards Committee (1980) Report on Meat Products FSC/REP/72. HMSO, London.
Whittaker, R.G., Spencer, T.L. and Copland, J.W. (1982) Enzyme-linked immunosorbent assay for meat species testing. *Austr. Vet. J.* **59**, 125.
Whittaker, R.G., Spencer, T.L. and Copland, J.W. (1983) An enzyme-linked immunosorbent assay for species identification of raw meat. *J. Sci. Food Agric.* **34**, 1143–48.
Wintero, A.K., Thomson, P.D. and Davies, W. (1990) A comparison of DNA-hybridisation, immunodiffusion, countercurrent immunoelectrophoresis and isoelectric focusing for detecting the admixture of pork and beef. *Meat Sci.* **27**, 75–85.
Woychick, J.H., Happich, M.C., Trinh, H. and Seilers, R. (1987) Quantification of soya protein in frankfurters by gel electrophoresis. *J. Food Sci.*, 52–6, 1532–4.
Yasumoto, K., Sudo, M. and Suzuki, T. (1990) Quantitation of soya protein by enzyme-linked immunosorbent assay of its characteristic peptide. *J. Sci. Food Agric.* **50**, 377–89.

5 Authenticity of fish

I.M. Mackie

5.1 Introduction

Whenever the distinguishing features such as shape or skin pigmentation of fish are removed on processing, there can be doubt about the authenticity of the species if only a fillet or portion of flesh, as in a fish finger, is available for examination. If it is then necessary to establish the species unequivocally to check, for example, that it is as declared on the label of a product or that it is as specified in a commercial contract, a reliable means of identification other than by sensory criteria is required. For the purposes of this review, the term 'fish' will be assumed to include shellfish and crustacea.

In the UK there is general legislation protecting the consumer from misrepresentation (Trade Descriptions Act 1968; Consumer Protection Act 1987) but additionally, for fish, there are legal requirements for the use of specific names for commercial species, namely the Food Labelling Regulations 1984, and The Food Labelling (Scotland) Regulations 1984. These regulations require that prescribed names for fish species are used, for example, that sole is *Solea solea* and not Canary sole (*Solea canariensis*); that *Squalus* spp. are called dog fish and not rock salmon; that *Anarichas* spp. are called catfish and not rock turbot. Scampi, otherwise known as Norway lobster or Dublin Bay prawn, is required to be the flesh of *Nephrops norvegicus* but not that of its near relatives from the Pacific, the *Metanephrops* spp. Similarly, rainbow trout (*Oncorhynchus mykiss*) – the widely farmed species of trout, which was originally introduced from North America, cannot be substituted for the native European trout (*Salmo trutta*), more commonly known as sea trout or brown trout.

The substitution of cheaper for more expensive foods and other commodities is not, of course, a new problem. It has no doubt been practised by unscrupulous operators ever since trading began but today, as more species of fish are processed into products almost completely devoid of the morphological characteristics of the animal, there are possibly more opportunities for substitution or adulteration than in earlier times. When fish species are closely related, the flesh is often similar in appearance (and taste) and, although trained taste panellists and other connoisseurs of fish gastronomy may be able to identify the species some of the time, their pronouncements or judgements have, however, on many occasions been shown to be faulty and little better than the rest of the apparently less discriminating sector of the population. To be absolutely certain of the species, reliable objective methods are required.

Species of the *Gadidae* (the cod family) such as saithe (*Pollachius virens*), lythe (*Pollachius pollachius*), haddock (*Melanogrammus aeglefinus*), and whiting (*Merlangius merlangus*) for example, are all white-fleshed fish species of low fat content. Nevertheless, there are large differences in price between whiting and cod for example (UK Sea Fisheries Statistics 1991 and 1992), and therefore incentives for the unscrupulous trader to substitute cheaper for more expensive species. Also, because of the general shortage of cod and haddock, there is the possibility of imported white fish such as Alaska pollack (*Theragra chalcogramma*) or hoki *Macruronus novaezelandiae* from New Zealand being substituted.

Substitution rather than adulteration is more likely to be practised as the fish processing industry, in the UK at least, is still largely based on processing whole fish into fillets. Opportunities for adulteration of more expensive species such as cod with cheaper species such as whiting or ling can, however, arise in the production of the frozen fish blocks. These are usually made by filling rectangular trays or moulds with fillets of fish, which are subsequently used for the manufacture of fish fingers. Products made from this raw material although containing other 'white fish' of equivalent nutritional value when labelled as 'cod fish fingers' would be contravening the UK Trade Descriptions Act 1968 and the Food Safety Act (1990). Similarly for crustacean products such as breaded scampi, which are often prepared from comminuted tails of the Norway lobster, opportunities can arise for adulteration with cheaper crustacean species such as imported Tropical shrimps or substitution by the Pacific *Metanephrops* spp.

Although not a product of the UK fish-processing industry, surimi, the raw material used for the manufacture of shellfish analogues or substitutes such as 'crab sticks' (Piggot, 1986; Lanier and Lee, 1992; Mackie, 1992a; 1994a) has the potential for adulteration and substitution. This material has been used in Japan for many centuries to manufacture the traditional fish-gel products of that country namely 'kamakoko', and since the early 1960s it has been the basis of a huge largely sea-based processing industry, using Alaska pollack (*Theragra chalcogramma*) – the white-fleshed species of the North Pacific – as the raw material for both these and the fish sausage type of gel products. Alaska pollack is now used by Japan, USA and other countries, for the manufacture of shellfish substitutes of which 'crab sticks' is the most widely known product (Piggot, 1986; Anon, 1987a). Other products are mock lobsters claws, crab claws and scallop substitutes, all of which could be substituted for authentic shell fish. The surimi is essentially the minced flesh of Alaska pollack, which has been washed free of all water-soluble components, including the water-soluble proteins, and mixed with relatively high concentrations of mono or disaccharides (e.g. sucrose), to protect the proteins from denaturation on frozen storage as a 'surimi block' (Mackie, 1992). When mixed with salt and heated the block forms a gel, with the properties of shellfish. As pressure on the Alaska pollack fishery has increased, the annual catch has decreased and, as so often happens in all fisheries throughout the world, what was once a cheap raw material is now no longer so. As a result, there have been moves to find alternative species for surimi manufacture and as the product is required to have specifications for quality and composition to comply with the laws of the countries in which the shellfish substitutes are sold, means of identifying the component species are required. For 'crab sticks' there are often requirements that the fish component is declared and that the content of crab is also given (Anon, 1987a). Reliable methods for determining the species of fish used in such products are thus required if standards are to be set and maintained.

Examples of substitution of cheaper for more expensive species of fish are not difficult to find. In the fresh fish trade, the cheaper flat-fish species, such as megrim and witch, can be substituted for plaice or lemon sole and black halibut or Greenland halibut (*Reinhardtius hippoglossoides*) can be passed off as the true halibut *Hippoglossus hippoglossus*. For this latter species, substitution can be unintentional as it is often assumed that a fish named Greenland halibut is indeed the true halibut (*Hippoglossus hippoglossus*) from Greenland. In actual fact it is a much smaller species with a black upper skin, often called such in other languages (e.g. flétan noir or hipogloso negro). The texture of the flesh is much softer than that of halibut and it is generally agreed that this fish is of inferior eating quality to halibut. Another fish that is sometimes found masquerading as a more expensive species is rainbow trout (*Oncorhynchus mykiss*). The species when farmed in the sea, often develops a silvery appearance and this has encouraged the unscrupulous seller to substitute it for seatrout or salmon trout (*Salmo trutta*). It is not unknown for it to undergo a transformation into 'smoked salmon' (i.e. smoked *Salmo salar*). Again, because of the similarity in appearance of all smoked salmon, the Pacific salmon (*Onchorhynchus* spp.) can also be found as a substitute for *Salmo salar*, which is the mainstay of the Scottish and Norwegian fish-farming industries. It is worthwhile pointing out here that care must be taken in understanding the difference in meaning between 'Scottish smoked salmon' and 'smoked Scottish salmon' – the latter product is likely to have been farmed and smoked in Scotland, while the former could mean smoked in the Scottish manner but farmed elsewhere.

To move on to the other problem of differentiating wild from farmed salmon brings one into another controversial subject. Strong views are expressed on the superior eating quality of wild salmon over farmed salmon (and prices in restaurants reflect this perceived difference). However, the claimed superior eating quality is often attributed to its superior physical fitness acquired on its migration to and from Greenland waters and, while this may often be the case, examples are also found of wild salmon in much poorer nutritional state than their farmed cousins. In general, farmed salmon are more likely to be of a more consistent eating quality but with a higher fat content than the wild counterpart (Torry Research Station, 1985). To differentiate wild and farmed salmon by sensory criteria is not possible and as the wild stocks have interbred to varying extent with escaped farmed fish, the authenticity of the wild population itself may be under question.

Canned fish can also present problems of identity when only steaks, fillets or minced flesh is available. The potential for substituting one species for another, already discussed in relation to smoked salmon, also applies to canned salmon, as it is usually produced in the form of steaks which have few if any species-identifying features. Similarly canned steaks or minced flesh of tuna and bonito cannot be identified by sensory criteria and as some of the species are much cheaper than the other (e.g. bonito and skipjack tuna), reliable means of differentiating them from the other tuna species are required. Bonito, in particular, has a lower import tariff than the tunas and, not surprisingly, it is believed that some of the canned tuna entering the EU is more likely to be bonito than tuna. For commercial reasons there is also a need to differentiate the true tuna species within the *Thunnus* family: Big eye (*Thunnus obesus*), Bluefin (*Thunnus thynus*), Yellowfin (*Thunnus albacares*) and Albacore (*Thunnus alalunga*) are all sold by name.

Problems of identity can also arise with sardine and anchovy, as it is known that sardine fillets can be taken through the salted anchovy process to give a similar product to 'anchovy'. Differences in appearance are not usually sufficiently reliable for establishing the identity, particularly when only broken fillets of fish are produced, as on a pizza.

Other products that require objective methods of species identification are preserved fish eggs or roe. Of particular interest from a substitution point of view is the true caviar, most of which is produced from three species of sturgeon by the countries bordering the Caspian Sea – that is, Russia and Iran. The familiar black eggs are obtained from the species, *Sevruga*, *Oscietra* and *Beluga* of which those from *Beluga* are the most expensive (Anon, 1987b). With a product such as caviar, which is both expensive and exclusive, there must be pressure to substitute the cheaper for the more expensive species, particularly when differentiation is based on sensory criteria such as colour of the yolk and pigmentation of the 'skin' of the egg. Substitutes for caviar itself are also produced from other species of fish and are marketed as 'mock caviar' (e.g. lumpfish eggs). Here the difference in egg size is likely to rule out any considerations of substitution for true caviar.

5.2 Objective methods of fish species identification

For methods to be of value in identifying the species of a portion of muscle tissue, they have to be based on the analysis of components of animal cells that are as unique to the species as are the more obvious morphological features of the live animal. As these very features are themselves a reflection of genetically determined differences in their constituent proteins, it is evident that analytical techniques that can differentiate the constituent proteins are likely to be suitable for species identification. Fortunately physicochemical differences in size and net charge of proteins are revealed as differences in mobility when proteins of animal tissues are subjected to electrophoresis and related techniques (Mackie, 1980; Durand *et al.*, 1985; Mackie, 1990; Rehbein, 1990; Andrews, 1992; Yman, 1992; Sotelo *et al.*, 1993; Mackie, 1994). Thus, when proteins of a species of fish are separated by electrophoresis, they yield profiles that are unique to the species and which are the basis of this method of species identification.

Other systems that can recognise the specific nature of the flesh of a species of fish use the specificity of interactions between antibodies and antigens – in other words, immunoassay systems (Patterson, 1985; Hitchcock, 1988). These methods depend upon the prior production in an animal of an antibody to a particular foreign protein or polysaccharide. When this foreign molecule (antigen) is introduced into the body of the animal usually by injection, the animal responds by producing antibodies as part of its defence mechanism. Each antibody interacts specifically with the antigen that caused its production in the first place and effectively removes it from the animal's body. Antibodies, so raised against a particular protein or group of proteins of a species of fish can then be used to identify that species in a standard immunoassay procedure. One major disadvantage of this method of species identification is its poor sensitivity to cooked products, since the antibodies are normally raised against solutions of native proteins isolated from raw flesh. As the native structure is often destroyed on heating, the specificity of the antigen–antibody reaction is also lost. The extent of denaturation of proteins to heat varies considerably and, if certain species have more-stable proteins, there will be a greater likelihood of success using this method. In general, however, the immunoassay system is only suitable for the identification of raw species of fish. Another disadvantage relates to the large number of species of fish available commercially compared with the three or four main 'meat' animals of beef, pork, sheep and possibly goat that are likely to be involved in species substitution. This raises questions of the practicability of producing antibodies to so many likely species and, unlike the protein electrophoretic method, there is no way of comparing a profile of a sample with those of reference species to establish, for example, whether it is a mixture of two or more species or a single species. The assay gives no indication of the presence or absence of antigens of other species as it is specifically an analysis for a pre-selected species for which an antibody has been produced. Its greater value possibly lies in detecting and determining the extent of adulteration of a product with a known adulterant (e.g. soya protein or horse meat in a beef burger). Immunoassay test kits for such adulterants are available commercially.

The third method of fish species identification is based on the analysis of deoxyribonucleic acid (DNA), the basic genetic material of all life. Techniques such as the widely known DNA fingerprinting analysis can be applied equally well to the identification of fish species as to individual humans but prior knowledge of the sequences of DNA of many commercially important species of fish is required. Compared with databases available on humans and higher animals, little information is available on fish, except for certain families such as tuna and salmon, where extensive sequence databases have been built up because of phylogenetic interest in closely related species (Foote *et al.*, 1989; Thomas and Beckenback, 1989; Bartlett and Davidson, 1991; Bartlett and Davidson, 1992; Block *et al.*, 1993; Carnegie, 1994; Russell and Carnegie, 1994). DNA technologies require specialised training in the particular techniques used and, as equipment is still relatively expensive, the analytical systems are still far removed from routine use in an analytical laboratory. Nonetheless, they are likely to be of value for more difficult problems in fish species identification, such as the identification of heat-processed (canned) tuna, salmon or sardine species. DNA is less affected by heat than are proteins and the information provided is potentially greater when analysing the genetic material itself rather than an expression of it, namely the proteins.

5.3 Principles of electrophoresis and isoelectric focusing

Electrophoresis was first introduced as an analytical procedure by the Swedish scientist, Tiselius in 1930. The detailed theory is complicated but, for the purpose of this review, it is sufficient to say that it refers to the separation of charged molecules, such as proteins or DNA, in an aqueous medium under the influence of an electric field applied between positive and negative electrodes. The fundamental requirement for separation by electrophoresis is that the molecules under study must have either a net positive or net negative charge under the pH of the buffer selected for the analysis. Charged species with a high net charge will tend to move more quickly than those with a lower net charge and, when molecules have the same net charge, the smaller ones will tend to move more quickly, in other words, movement is dependent on both charge and size of the molecules being separated.

If a protein is taken as an example of a molecule with both positive and negative charges, it is evident that the net charge will be dependent on the pH of the solution, as it will be determined by the degree of ionisation of its constituent amino and carboxyl groups. At low pH values, the free carboxyl group will not be ionised and the amino groups will be protonated, giving a net positive charge, the actual value of which will be dependent on the relative proportion of the two chargeable groups in that particular protein. Conversely, at high pH values, the free carboxylic acid groups will be ionised but the amino groups will not be protonated giving as a result a net overall negative charge. At intermediate values of pH, some of the free side-chain groups will be protonated, while others will be neutral depending upon their individual pK values. There will be one particular pH at which a protein molecule will have a net zero charge, that is when there is a balance of positive and negative charges. This value, designated pI, is called the isoelectric point of the protein, and is characteristic of each individual protein.

5.3.1 Electrophoretic systems

When first developed, electrophoresis was carried out in free solution in a moving boundary system (Tiselius, 1930) and, while giving very accurate values for fundamental mobility measurements of charged particles it was not, however, useful as a routine analytical tool. It has been largely superseded by systems in which the liquid medium is supported in a gel, to reduce diffusion. Usually polyacrylamide or agar gels are used depending upon the nature of the molecules being separated. Earlier systems employed starch as the supporting medium but this has been largely replaced by polyacrylamide and agar because of their superior resolving power and more robust handling characteristics.

5.3.1.1 Polyacrylamide gels. Polyacrylamide gels are co-polymers formed from acrylamide monomer $CH_2 = CH\ CONH_2$ and the dimer N,N′–methylenebisacrylamide, the crosslinking agent. The reaction requires an initiator and a catalyst. Ammonium persulphate is a commonly used initiator and tetramethylenediamine (TEMED) is usually the catalyst. The reaction takes place via vinyl polymerisation and gives a randomly coiled gel structure. Since the polymerisation reaction is very sensitive to certain chemical impurities that may be present in the components of the mixture, it is important to use highly purified reagents.

Gels of widely varying pore size, and hence sieving properties, are obtained simply by altering the total concentration of the monomer and crosslinking agent. These compounds are usually present in a ratio 95:5 which in a 7.5% gel (total of monomer and dimer) gives a pore size of about 5 nm whereas in a 30% gel it is 2 nm. The pore size can be even more closely tailored to the dimensions of the proteins being separated by altering the ratio of monomer to dimer.

It is the flexibility in design of the characteristics of this support medium that makes polyacrylamide gel electrophoresis such a powerful tool in the analysis of proteins. These gels, unlike those used for chromatography, are not granular but are formed into continuous columns or slabs depending upon the mould used. The gels are, in effect, a network of polymer molecules surrounded and penetrated by buffer, with the spaces within this framework being the 'pores'. When an external voltage is applied between the electrodes, protein molecules, together with the anions and cations of the buffer solution, migrate towards the appropriate electrode. As they move they experience a frictional resistance to their movement, which is related to the relative sizes of the gel pores and the radii of the charged species (i.e. there is a sieving effect). In addition, the net charge and mass effects also influence the separation of the proteins.

Once adequate separation of proteins has taken place, the gel is removed from the system. It is then necessary to 'fix' the proteins by precipitating them with either a solution of perchloric acid or trichloroacetic acid (Mackie, 1980; Laird *et al.*, 1982; Mackie, 1990). These 'fixed' proteins must then be visualised by staining the gel with a suitable dye, such as Coomassie Blue in a solution of ethanol, acetic acid and water typically in a ratio of 25:8:67. Subsequent washing of the stained gel with the same solvent removes unbound dye, giving blue bands of proteins, which are protein separation profiles against a water-clear background. Fixing and staining procedures are required for all electrophoretic systems that use gel supports as proteins that meet the requirements of electrophoresis must be in solution, and for satisfactory fixing and visualisation they have to be removed by a precipitant and then made visible by a suitable staining procedure. Subsequent examination of the developed gels can be carried out visually or by using a densitometer or image analyser to measure positions and concentrations of the proteins in the profiles or separation patterns.

5.3.1.2 Agarose gels. Agarose is a linear polymer of D-galactose and 3,6 anhydro-D-galactose residues and is obtained from seaweed (*Agar agar*). It is readily dissolved in boiling water and, on cooling to about 38°C, it forms a gel through the formation of hydrogen bonds. As the pores are relatively large compared with those of acrylamide, agarose gels are more suitable as a support medium for the separation of larger molecules, such as DNA and RNA, which can range in size from 10–60 000 base pairs (Sambrook *et al.*,

1989). Generally flatbed systems are preferred as the support provided from below allows lower concentrations of agarose to be used. All polynucleotides such as DNA and RNA contain one phosphate group per base residue and, since phosphate groups form a negative charge under the alkaline conditions required to dissolve DNA, they move towards the anode according to their molecular mass. In contrast to proteins, which are ampholytes and can be in solution in both acidic and alkaline media, DNA is in effect a polyanion, soluble only under alkaline conditions. Detection of the separated DNA polymers is usually achieved by incorporating a fluorescent dye, ethidium bromide into the gel. At the end of the electrophoretic run, the separated DNA molecules are then identified by photography using ultraviolet light.

Once the photograph is developed, comparison with reference markers or other DNA residues can then be made as with protein separation profiles.

5.4 Separation systems

5.4.1 Zone electrophoresis

5.4.1.1 Gel rods. For the gel rod system, polyacrylamide gels are prepared in glass or plastic cylindrical tubes of internal diameter about 5 mm and length 70–100 mm. Typically six tubes are mounted through grommets in an upper electrolyte compartment, with their bottom ends dipping into the buffer solution of the lower electrolyte compartment to give continuity of ionic strength and pH throughout the system. The same buffer solution (typically TRIS–glycine pH 8.0) is used for both upper and lower electrolyte compartments. The sample is then applied by pipette to the top of the gel in a sucrose solution to minimise diffusion into the upper electrolyte compartment and reference samples are applied to the other gels. After the electrophoretic run has been completed, the gel is removed from the tubes by irrigation with water from a syringe, fixed with trichloroacetic acid and perchloric acid and stained with a suitable dye (e.g. Coomassie Blue). Comparison of the sample profile with those of reference samples is then made to identify the species (Mackie, 1969).

5.4.1.2 Slab gels. Polyacrylamide slab gels have now largely replaced the rods as supports for electrophoresis (Mackie, 1980). They have distinct advantages over the former system in being able to analyse several samples at the same time on the same gel and therefore under the same conditions of electrophoresis. As the gel is mounted on a cooling plate, the removal of heat produced during the electrophoresis is more uniform, giving superior resolving power and greater flexibility. Gels of a wide range of thicknesses can be prepared, mounted on plastic backing or glass. Gel dimensions can be altered readily to suit the requirements and samples can be applied to any position on the gel. For some analyses, the quality of resolution is dependent on whether the sample is applied at the anodic or cathodic end.

5.4.2 SDS polyacrylamide gel electrophoresis

When protein molecules are dissolved in solutions of the anionic detergent, sodium dodecyl sulphate (SDS), their individual charges are lost and a net negative charge is formed as a result of the anion dodecyl sulphate complexing with the protein present (Weber and Osborn, 1969). The uptake of detergent is the same per unit mass for all proteins and, in consequence, the mobility on electrophoresis is proportional to the charge. Separation on the basis of molecular mass of protein provides an alternative method of analysis of flesh proteins and thereby of identifying fish species. As will be discussed in section 5.6.2, this method of electrophoresis is used to identify the species of samples of cooked fish, since the proteins, although denatured are extractable in 2% SDS solution to give species-specific profiles on electrophoresis in the SDS solution.

Depending upon the size of proteins to be separated, a concentration of acrylamide is selected to optimise the sieving effect (Rehbein and Karl, 1985; Scobbie and Mackie, 1988). In addition, a stacking gel of lower acrylamide concentration is used to allow rapid movement of the sample to the top of the main separating gel, and thereby to form a narrow zone of proteins prior to entry into the main separating gel. Other refinements, such as concentration gradients of acrylamide, can be used to provide increased resolution of closely separating proteins.

5.4.3 Isoelectric focusing

Isoelectric focusing differs from the previously described systems in that a stable pH gradient is created between the anode and cathode by the focusing or movement to positions of electrical neutrality, of a mixture of synthetic ampholytes selected to provide a range of pH values at which they are electrically neutral (i.e. a pH gradient). Narrow-range pH gradients (e.g. pH 4–6 or pH 5–8) can be used to amplify a particular part of a separation profile, while wide range ones (e.g. pH 3.5–9.5) are often used for more general separations. When proteins are present in such a medium they, together with the ampholytes, migrate and focus at those positions on the pH gradient where they have no charge (Figure 5.1). The resolving power of isoelectric focusing is extremely high, as differences of as little as 0.01 between pIs of proteins can be separated. Once a steady state has been attained, the gel is removed, fixed, stained and washed as for other polyacrylamide gels. The synthetic ampholytes themselves are

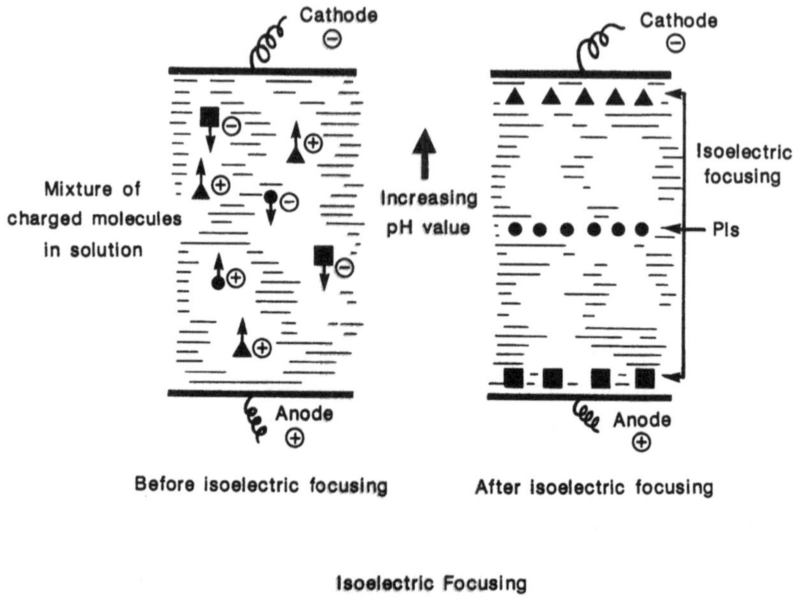

Figure 5.1 Diagramatic presentation of isoelectric focusing.

washed out during the initial fixing step and do not interfere with the uptake of dye by the proteins. By reference to standards of known pI, which are run at the same time, the pIs of the proteins of the species being examined can be established (Rehbein *et al.*, 1995).

5.4.4 Capillary electrophoresis

Capillary electrophoresis (CE) is a relatively recent development in electrophoresis (Ewing *et al.*, 1989; Novotny *et al.*, 1990; Kleparnik and Bocek, 1991). It is based on free-zone electrophoresis in buffer-filled capillaries of typical dimensions (e.g. 50 μm internal diameter and 100 cm length), which are used as the separation chambers. Potentials of as high as 30 kV are applied and promote rapid migration of proteins while minimising zone spreading and the effect of other charged entities. Owing to the small sample volumes of solutions used, however, it is necessary to have highly sensitive detection systems. Ultraviolet and fluorescence detectors have been developed for CE systems but as the signal-to-noise ratio is low for ultraviolet detection of proteins applied in these small volumes, a limitation is placed on its use for their analysis. Capillary electrophoresis offers the potential of a fully automated instrumental system as an alternative to present electrophoretic methods, which are highly demanding both of operator skills and time. At its present state of development, however, it cannot be considered as an alternative analytical procedure for fish species identification.

5.5 Nature of fish-flesh proteins

Before discussing the analysis of the proteins of fish flesh, some understanding of their nature and of the effects on them of treatments such as post-mortem storage in ice, frozen storage and heat processing is required. Fish flesh itself is contractile muscle and, like that of other animals, it comprises specialised cells or fibres containing thick and thin filaments, which interact with one another in response to nerve impulses and produce the contraction and relaxation required for movement of the animal. In all animals, these muscle cells or fibres are arranged parallel to one another and are held together by connective tissue of collagen or elastin. In mammals and birds, this membrane connects with the tendons, also of connective tissue, which anchor the muscle cells to the skeleton of the animal. In fish, however, a very different arrangement of muscle cells is found: instead of being connected to tendons, the cells are bound together one cell deep to form segments or mytomes of muscle. These bundles of muscle cells are shaped like the letter 'W' and are orientated across the mid-plane of the fish, with the central parts directed towards the head of the animal. The ends of the cells are attached to sheets of connective tissue called 'myocommata', which separate one block of mytomes from another. Since myocommata are composed largely of collagen, they break down on heating to release the mytomes, which are still in their original shape and more commonly recognised as the characteristic 'flakes' of cooked fish. Compared with mammalian and avian muscle cells, those of fish are short, generally less than 20 mm, even in large fish. In mammals and birds they can run the whole length of the muscle and may be as long as 300 mm (Love, 1970; Hultin, 1992).

Fish muscle consists of two types of cells, white and red, with the latter always present as the minor component. White muscle as in the white fish species cod (*Gadus morhua*) has an aerobic metabolism. It depends upon glycogen for energy and is well supplied with glycolytic enzymes, which permit intense contraction during periods of flight. Red muscle, on the other hand, has an aerobic metabolism and is rich in haemoglobin, myoglobin and mitochondria. In fatty species, particularly the surface-swimming pelagic fish of which tuna and mackerel are typical examples, red muscle is well developed to provide energy for sustained high swimming speeds. It is the white muscle, however, that is of interest as far as fish species identification is concerned and, for the purposes of this review, it will be assumed to be what is implied by the term 'fish flesh'. As there are differences in the separation profiles of the two types of muscle, it is important that white muscle is free of red muscle and blood if reproducible results are to be obtained.

As with all structural and contractile muscle, fish flesh contains three main groups of proteins: (i) the sarcoplasmic or water-soluble proteins; (ii) the myofibrillar proteins; and (iii) the connective tissues proteins (Table 5.1).

These proteins are readily separated by fractional extraction technique employing water and strong salt solutions (Mackie, 1993).

5.5.1 *Sarcoplasmic proteins*

The sarcoplasmic or water-soluble proteins make up 20–35% of the total protein content of muscle depending upon the species (Mackie, 1993). They are present within the muscle cells usually at concentrations as high as 20% and are largely enzymes reponsible for metabolism within the cell. Since they are of low molecular weight (40–60 kDa), readily extractable with water and, of course, electrically charged, they are ideally suited for separation by electrophoresis. It would appear also that genetic differences between species are more developed in this than in the other groups of proteins. The unique nature of the sarcoplasmic proteins is such that the separation patterns or profiles obtained on electrophoresis or isoelectric focusing can be used for the unequivocal identification of the species.

5.5.2 *Myofibrillar proteins*

The myofibrillar proteins are responsible for the structure of the muscle and, although there are fewer differences between species, they can be of importance in the identification of the species of heat-denatured flesh, as for example in the profiles obtained on SDS electrophoresis of SDS extracts of whole muscle. The myofibrillar proteins, of which myosin, actin, tropomyosin and the troponins are the major components, account for 65–75% of the total protein content of muscle and are responsible for functional properties such as gelling ability, water-holding capacity and textural eating characteristics.

Table 5.1 Main groups of proteins in the flesh of fish and mammals

	Composition (%)	
Protein group	Fish	Mammals
Sarcoplasm	20–35	30–35
Myofibrils	65–75	52–56
Connective tissue	3–10	10–15

5.5.3 Connective tissue proteins

The connective tissue proteins represent only 3–10% (Mackie, 1983; Mackie, 1994a) of the total proteins in fish compared with 10–15% in mammals (Goll *et al.*, 1977). As with the myofibrillar group, the connective tissue proteins are likely to contribute to differences in the profiles of total muscle extracts such as those used for SDS electrophoresis of heat-denatured flesh as they also are extracted in SDS solution.

5.6 Experimental procedures for electrophoretic methods

For as long as the sarcoplasmic proteins remain in the native state, they are readily extractable with water or dilute salt solutions, making them in a suitable form for separation by isoelectric focusing or electrophoresis. However, this condition does not apply when the fish is cooked, as the sarcoplasmic proteins together with the structural proteins are denatured and precipitated by heating. They are no longer in their native forms and they cease to be extractable in water. For these and even more severely heat-denatured products, such as heat-sterilised canned fish, alternative methods of analysing species-specific proteins, now denatured, have to be considered.

It is convenient to consider fish species identification under the following three categories of processing: (i) raw fish flesh; (ii) cooked but not autoclaved fish flesh; and (iii) heat-sterilised, autoclaved fish flesh.

5.6.1 Raw fish flesh

The species of raw fish is normally readily established by comparing the isoelectric focusing (IEF) profile of an aqueous extract with those of aqueous extracts of an authentic species. Although the separation profiles of individual fish, even of the same species show some degree of variation they, nonetheless, have sufficient number of major zones in common, which are different from those of closely related species, to enable unequivocal identification of the species to be made.

The procedure used varies to some extent between laboratories but, in general, a portion of fish muscle is extracted with water in the ratio 1:2. The mixture is homogenised for 30 s to 1 min and the homogenate centrifuged at low centrifugal force (e.g. 8100 g), for a few minutes. The supernatant solution is then stored in a refrigerator for no more than 2 days until it is used for IEF. For IEF, a large gel system (e.g. 245 × 110 × 1 mm) or a small gel slab system (e.g. Phast 43 × 40 × 0.35 mm) is used. A wide pH range (e.g. 3.5–9.5) or a narrow-range pH gradient (e.g. pH 5–8, pH 4–6) is used depending upon the isoelectric points of the main proteins to be separated. A recent collaborative experiment among eight European laboratories (Rehbein *et al.*, 1995) investigated the effect of small differences in procedures. In this study, frozen samples of reference materials and unknown samples were sent to seven laboratories by the co-ordinating laboratory, as a preliminary exercise aimed at identifying those steps in the methodology (either procedures for preparation of samples, conditions of IEF or subsequent staining procedures) that were critical for the reproducibility of the results. These preliminary studies showed that water was satisfactory as the extractant and that the position of application of the sample solution to the IEF gel had a great influence on the band pattern. Other variables between laboratories, such as type of ampholyte, protein content of sample, volt hour product and staining procedure, were also found to affect the protein profile obtained. These results support the widely accepted view that, to have full confidence in the identification of a species, it is necessary to have samples of authentic species run on the same gel. It is not sufficient to rely on profiles obtained under apparently the same conditions of IEF. Continuing studies among the laboratories are aimed at optimising the procedure so that data on pIs of the proteins of a single species can be measured from marker proteins of known pI and used as the basis of identification of species from a computer database of pI and the corresponding concentration of proteins specific to species of fish.

IEF profiles of raw fish of the same species have been shown to be remarkably constant, although minor differences attributable to polymorphism are sometimes found. Extensive studies on the effects of geographical location of fishing ground have shown that there is little variation in the profiles of a species. These results give confidence in the use of the method for identification of fish species.

An example of a typical isoelectric focusing experiment obtained on the large gel system is shown in Figure 5.2. In this case, a narrow-range of pH gradient (pH 4.0–7.0) has been used as it gives greater resolution of the anodic proteins that are used to differentiate the species. The gel shows the species-specific nature of the profiles and their reproducibility when replicates are run. Comparison of the profiles of the gadoid species shows a degree of commonalty of the anodic proteins, as might be expected within a family.

Separation profiles obtained with the Phast system are shown in Figure 5.3. Although the gels are considerably smaller than the standard large gels, the definition obtained is more than adequate for differentiating even those closely related species. Automatic staining and destaining reduces operator input considerably and enables the result to be obtained in 2 h. One disadvantage of the Phast system is that it tends to lose the anodic proteins, the parvalbumins, which are useful in differentiating gadoid species, in the anodic solution. Consequently, differentiation of these species has to be based more on proteins of pIs between pH 5 and 7.

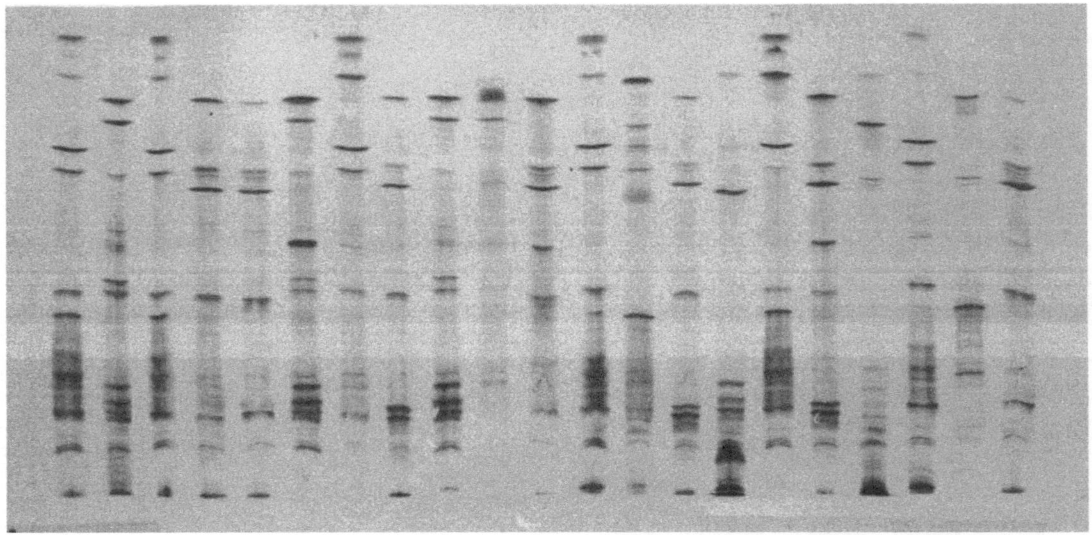

Figure 5.2 Isoelectric focused profiles of aqueous extracts of raw fish (L–R): ling (*Molva molva*); cod (*Gadus morhua*); ling; saithe (*Pollachius virens*); cod; ling; saithe; cod; Alaska pollack (*Theragra chalcogramma*); saithe; ling; turbot (*Psetta maxima*); saithe; sea bass (*Dicentrachus labrax*); ling, saithe, halibut (*Hippoglossus hippoglossus*); ling; Greenland halibut (*Reinhardtius hippoglossoides*); and saithe. pH range = 4.0–7.0.

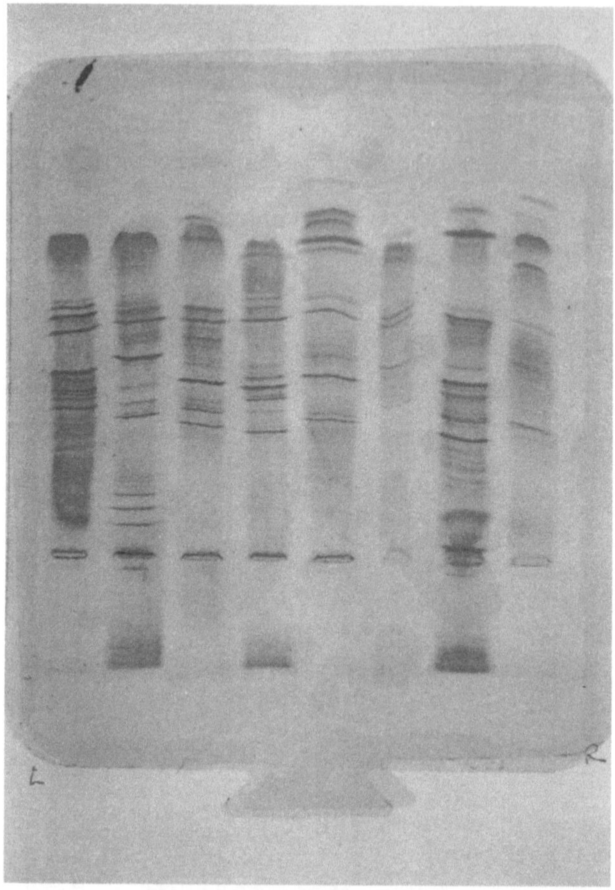

Figure 5.3 Isoelectric focused profiles of aqueous extracts of raw fish obtained on the Phast system (L–R): cod (*Gadus morhua*); haddock (*Mellanogrammus aeglefinus*); whiting (*Merlangius merlangus*); saithe (*Pollachius virens*); ling (*Molva molva*); Alaska pollack (*Theragra chalcogramma*); hake (*Merluccius merluccius*); and hoki (*Macruronus novaezelandiae*). pH range 3.5–5.9)

5.6.1.1 Influence of iced storage. During iced storage, changes in the composition of the sarcoplasmic proteins may occur as a consequence of proteolysis by endogenous and bacterial enzymes. In addition, proteins can be leached out by melting ice. Studies at the Torry Research Station and elsewhere have shown that the profile remains remarkably constant even to the stage of spoilage when the fish has become unfit for human consumption.

5.6.1.2 Frozen storage. The effect of frozen storage has been shown to have a minimal effect on the profiles. This is not surprising as it is well established that sarcoplasmic proteins are more resistant to denaturation during frozen storage, compared with the myofibrillar proteins (Mackie, 1993). For this reason it is considered that samples of flesh of authentic species can be stored at −30°C for at least 1 year without any significant quantitative loss in the sarcoplasmic proteins or any alteration in the IEF profile of a species. If the flesh is minced, however, and stored at relatively high temperature (e.g. −10°C) considerable loss of the sarcoplasmic proteins will occur. This is believed to be caused by mincing, which breaks cell walls and releases potential reactants, such as trimethylamine oxide (TMAO) and the enzyme trimethylamine oxide demethylase. Storage at the higher temperature results in more rapid rates of reaction and consequently a greater degree of protein denaturation.

5.6.2 Cooked but not autoclaved fish

When proteins are denatured on heating, an irreversible process of aggregation leading to the formation of precipitates usually takes place. These aggregates are inextractable in the normal protein solvents such as water or strong salt solutions, as the native properties of the proteins have been destroyed. It is believed that the bonds that are formed are largely non-covalent, hydrophobic and hydrogen bonds, possibly with some degree of electrostatic interaction between the charged groups of the denatured proteins (Mackie, 1993; Mackie, 1994a). Some –S–S– bonds may also be formed. It has been shown that heat-denatured proteins can be dissolved in a concentrated solution of SDS, a solvent that splits hydrogen bonds. A reducing agent such as dithiothreitol is often added to break any –S–S– bonds that may be present. The extracted protein residues can then be separated using SDS electrophoresis and identification of the species can be made by comparing the protein profile of the unknown with those of authentic samples that have been treated in the same way. Instead of using SDS solutions as solvents, high concentrations of urea (8 M) have been used (An *et al.*, 1988). Electrophoresis has also been carried out in urea (Mackie, 1983; Rehbein, 1990) but problems can arise because of the crystallisation of the urea. For these reasons SDS electrophoresis is the preferred system.

While there are fewer differences between species (largely because of the major contribution to the profile by the myofibrillar and connective tissue proteins), it is nonetheless possible to differentiate very closely related species such as those of the cod family. This point is illustrated in Figure 5.4, where an SDS profile of an unknown (reformed extruded scampi) is compared with those of authentic species of gadoids, which could have been used to extend or adulterate the product. In this case, the profiles of the gadoids can be differentiated from one another. The test sample corresponds to authentic scampi with no detectable presence of any other species. For these analyses, silver staining (Blum *et al.*, 1989; Craig *et al.*, 1995) has been used as it gives more clearly defined profiles and higher sensitivity than Coomassie Blue. Although scampi is a raw product, the SDS procedure is preferred to IEF of aqueous extracts as more species-related proteins are obtained (Craig *et al.*, 1995). It also has an advantage in detecting the adulteration of scampi with other crustacea such as 'warm-water prawns'. With IEF of aqueous extracts, only a few zones of similar profile are obtained for scampi and potential adulterants, whereas when using SDS, clear differences in the mobility of one or two species-specific zones are seen (Craig *et al.*, 1995).

Typical SDS separation profiles of some crustacea and white flesh species obtained with silver staining are shown in Figure 5.5. Species differences are often more evident in the low molecular weight proteins as many of the higher molecular weight protein residues are common to all

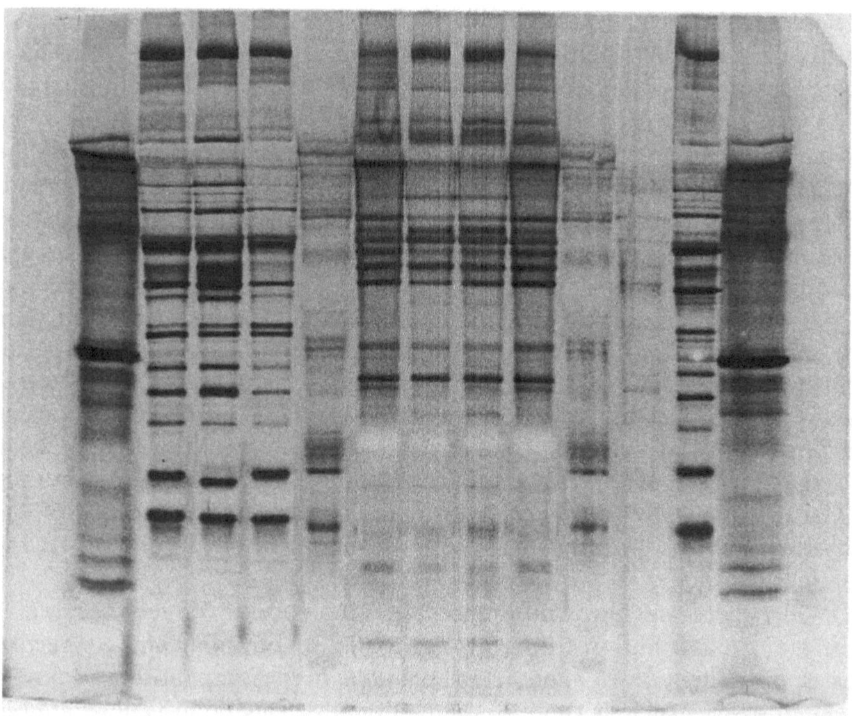

Figure 5.4 Separation profiles obtained on SDS electrophoresis of extracts of cooked (L–R): cod (*Gadus morhua*); haddock (*Melanogrammus aeglefinus*); whiting (*Merlangius merlangus*); soya protein, scampi (*Nephrops norvegicus*); unknown; scampi, unknown, soya protein, monkfish (*Lophius piscatorius*); saithe (*Pollachius virens*). Standard molecular weight markers are on L and R.

species. The profiles show that even with closely related species such as scampi (*Nephrops norvegicus*) and its counterparts from the Pacific (*Metanephrops* spp.) there are sufficient differences in profiles to differentiate the species. The profiles, in turn, are readily distinguishable from those of shrimps. The technique is of value in differentiating cooked flesh; in general and specifically when salted smoked fish are being analysed it is often preferable to compare the IEF profiles of aqueous extracts, which have previously been dialysed to remove salt, with those obtained using SDS electrophoresis. The identification of smoked Greenland halibut is established by running authentic samples of true halibut (*Hippoglossus hippoglossus*) and Greenland halibut (*Reinhardtius hippoglossoides*) on either side.

SDS electrophoresis has been used successfully to analyse surimi-based gel products that are manufactured into shellfish substitutes. Sufficient differences in the SDS profiles of what are essentially sarcoplasmic protein-free muscle proteins have been obtained to enable the identity of the

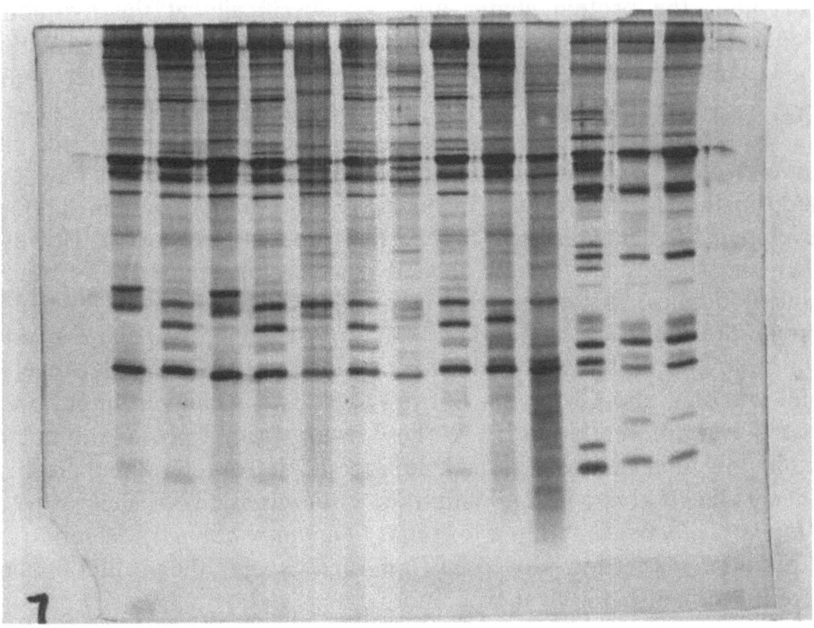

Figure 5.5 Separation profiles obtained on SDS electrophoresis of (L–R) shrimp (*Macrobrachium rosenbergii*); sample 1; *Macrobrachium rosenbergii*; scampi 2; scampi (*Nephrops norvegicus*); sample 1; Korean scampi (*Metanephrops* spp.); sample 2; shrimp (*Penaeus indicus*); shrimp (*Penaeus monodon*); halibut (*Hippoglossus hippoglossus*); smoked Greenland halibut (*Reinhardtius hippoglossoides*) and Greenland halibut. Samples 1 and 2 were hilsa (*Clupea elisha*).

species of fish used in the production of surimi to be established (Torry Research Station, 1985).

5.6.3 Heat-sterilised and autoclaved products

When fish flesh is subjected to the high temperatures and pressures of the canning process, the constituent proteins undergo further denaturation, possibly in the form of increased covalent-bond formation. The nature of these changes is not well understood. However, evidence for continued denaturation of proteins in the canning process is seen in the loss of protein zones on SDS electrophoresis and increased background staining (Mackie *et al.*, 1992). These features of SDS electrophoresis of SDS extracts have meant that the method has little, if any, application to canned fish products; the few protein zones that are observed are often common to all the species examined and the strong background staining adds to the problems of differentiation.

Nevertheless, a procedure using cyanogen bromide in formic acid was shown by the author (Mackie and Taylor, 1972) to give species-specific profiles on electrophoresis of protein hydrolytic fragments. In this procedure, the protein chains are split specifically at the methionine residues on reaction with cyanogen bromide in formic acid (Gross, 1966). The protein fragments obtained can then be dissolved in water or 6 M urea, and when separated by IEF, they give species-specific profiles that can be compared with those of authentic species. Since the first demonstration of the application of this reaction, modifications have been made in the procedure and, by using IEF, a suitable system has been developed for heat-sterilised canned fish products in general. However, when the species are closely related, such as members of the tuna or salmon families, the differences in profiles obtained are insufficient to enable differentiation to be made. Within the salmon species examined, two groupings have been made, namely Group 1 – Chum, Sockeye and Pink salmon – and Group 2 – Spring, Coho and Atlantic salmon (*Salmo salar*) – but differentiation below these levels was not possible. Similarly, within the tuna family, Skipjack tuna could be differentiated from the others (Mackie *et al.*, 1992). Within the *Thunnus* family of Yellowfin tuna (*Thunnus albacares*), Big eye tuna (*Thunnus obesus*), Bluefin tuna (*Thunnus thynnus*) and Albacore (*Thunnus alalunga*), the profiles obtained were indistinguishable from one another.

In Figure 5.6, the IEF profiles of the cyanogen bromide peptides of canned salmon and other species are given to illustrate the overall similarity of the separation profiles of all species. In the middle pH range, however differences between species are sufficient to allow identification to be established provided that the species are well separated zoologically, e.g. pilchard, sardine and tuna.

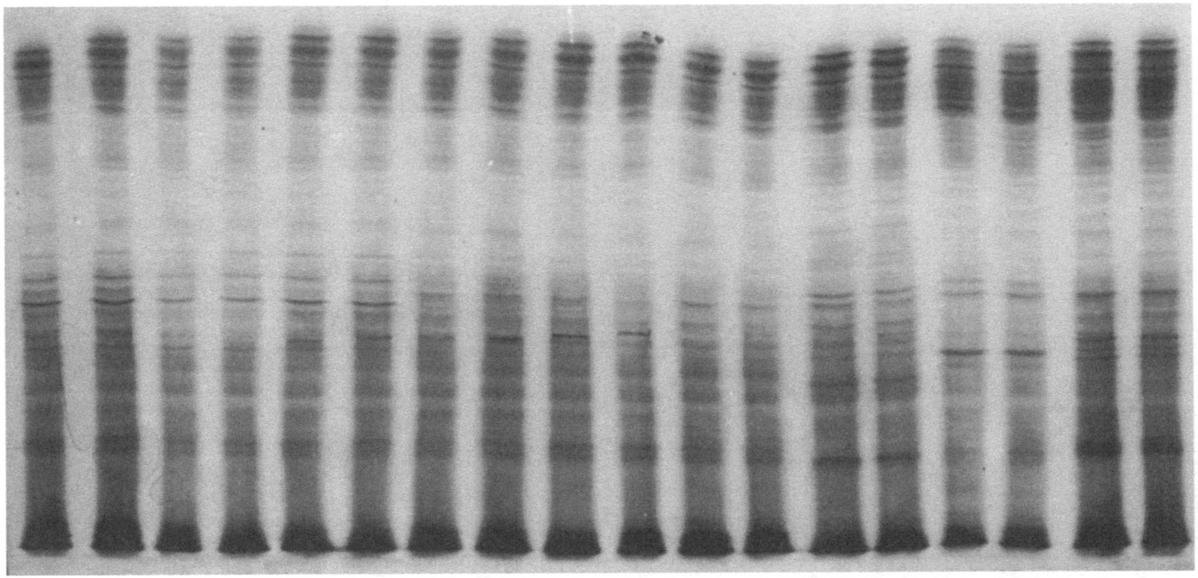

Figure 5.6 Isoelectric focused profiles of cyanogen bromide peptides of heat-sterilised canned fish (L–R): Chum salmon (*Oncorhynchus keta*); Sockeye salmon (*Oncorhynchus nerka*); Pink salmon (*Oncorhynchus gorbuscha*); Spring salmon (*Oncorhynchus tschwytscha*); Coho salmon (*Oncorhynchus kisutch*); Atlantic salmon (*Salmon salar*); pilchard (*Sardina pilchardus*); herring (*Clupea harengus*) and bonito (*Sarda sarda*). pH range = 3.5–9.5.

5.7 Alternative protein-based methods of fish species identification

5.7.1 Immunoassay procedures

Immunoassay systems have not been used to any great extent for fish species identification; however, two recent studies using the ELISA procedure have been reported. An *et al.* (1990) developed a simple ELISA technique to verify the identity of rock shrimp (*Sicyonia brevirostris*) and to detect and quantify it in a mixture of raw and heat-processed seafood. In this study, monoclonal antibodies were raised to a heat-denatured protein that was specific to rock shrimp. Another approach (Taylor and Jones, 1992) has been the development of an ELISA technique to identify canned sardine (*Sardina pilchardus*) in the presence of mackerel, herring and sild. In this study, a polyclonal antiserum was raised against a crude extract of water-soluble proteins from authentic canned sardine; however, problems with cross-reaction with other species were not resolved fully. The robustness of these methods has not been fully determined, however, and further testing and modification is required.

5.7.2 Capillary electrophoresis

While the potential of capillary electrophoresis has been recognised, its application to authenticity problems in fish and other foods is still very limited (Ewing *et al.*, 1989). A review of the methodology of capillary electrophoresis for protein separation has been published by Novotny *et al.* (1990), and Chen *et al.* (Chen, 1991; Chen *et al.*, 1992) have established potential conditions for the analysis of model proteins, serum and milk proteins. More recently, Le Blanc *et al.* (1994) showed that clear differences in profiles of species of the gadoid family could be obtained using a high ionic strength phosphate buffer. The profiles, however, are very subject to pH and, unlike conventional electrophoresis, changes attributable to frozen storage were obtained. Nonetheless, separation of the sarcoplasmic proteins was achieved in less than 9 min and with high reproducibility of migration times. The potential of this method for fish species identification is now established but further work is needed to determine the reliability of the system.

5.7.3 High-performance liquid chromatography (HPLC)

Like capillary electrophoresis, HPLC uses a detection system of either ultraviolet absorbance or fluorescence. As these detectors also respond to non-protein substances present in muscle extracts (e.g. nucleotides and amino acids), they lack the protein specificity of dye-binding as used for the detection of proteins after conventional gel electrophoresis. For this reason, caution is required in interpreting perceived differences in separation profiles. Two papers that have been published (Osman *et al.*, 1987; Armstrong *et al.*, 1992) emphasise the simplicity and speed of the analysis but they also point out that intra-species variation is found and that there is loss of data through interference from the internal standard, which is required to establish relative retention times of the peaks. Change in peak retention times, width and resolution arise with even slight variation in the chromatographic parameters, such as mobile-phase composition, gradient generation and temperature. Further work, including comparative studies with conventional electrophoresis, is required before any comment can be made on its reliability with unknown samples of fish.

It has not yet been established to what extent factors such as physiological condition of the fish and post-mortem changes during iced and frozen storage affect the separation profiles.

5.8 DNA techniques of fish species identification

There is interest in the application of DNA technologies to identify species of fish that cannot be identified using existing methods. As with proteins, DNA is also degraded on heating but procedures are now available to make it possible to isolate species-distinguishing fragments from such degraded material. Basic to all of the procedures is the polymerase chain reaction (PCR), which enables a species-distinguishing sequence of DNA to be targeted and amplified. Most of the studies have concentrated on the cytochrome *b* gene of mitochondrial DNA for which, even for fish, there are several sequences available for comparison. Those of the tuna species have been the subject of extensive phylogenetic studies (Block *et al.*, 1993). DNA probes rely on the specific interaction between the sequence of the probe and the complimentary sequence in the segment of the DNA of interest. On subsequent separation by electrophoresis on agar, the amplified DNA product can be identified and isolated for further analysis by an ever increasing range of technologies. The sequence of nucleotide residues can be determined and compared with those in a computer database or the product can be digested with enzymes (endonucleases), which act on specific nucleotide sequences. The digested DNA can then be subjected to electrophoresis on agar gel and examined under ultraviolet light to reveal a series of fluorescent bands, which, as for protein profiles, would be specific for the species assuming that the correct enzymes had been selected. The great advantage of the DNA approach is that, although the DNA is considerably degraded as a result of the canning process, there is sufficient material to be recognised by the primers. However, it is well known that there is considerable variation in the degree of heating in commercial practice and as it is usually well in excess of what is required for sterilisation, any evaluation of the DNA technology for the identification of species of canned fish must take these variables into account. The post-mortem history of the fish prior to canning should also be considered as degradation of DNA is known to take place to some extent on iced and frozen storage.

Typical separation profiles of DNA isolated from canned and raw tuna are shown in Figure 5.7. The DNA isolated from all the canned samples is severely degraded and shows only DNA residues of the order of 100 base pairs. By selecting primers specific for a known sequence of the cytochrome *b* mitochondrial gene, it has been possible to isolate sufficient quantities of the PCR products of the seven commercially important species when canned (Figure 5.8) and to obtain their sequences. By reference to a database, each of them has been identified from their

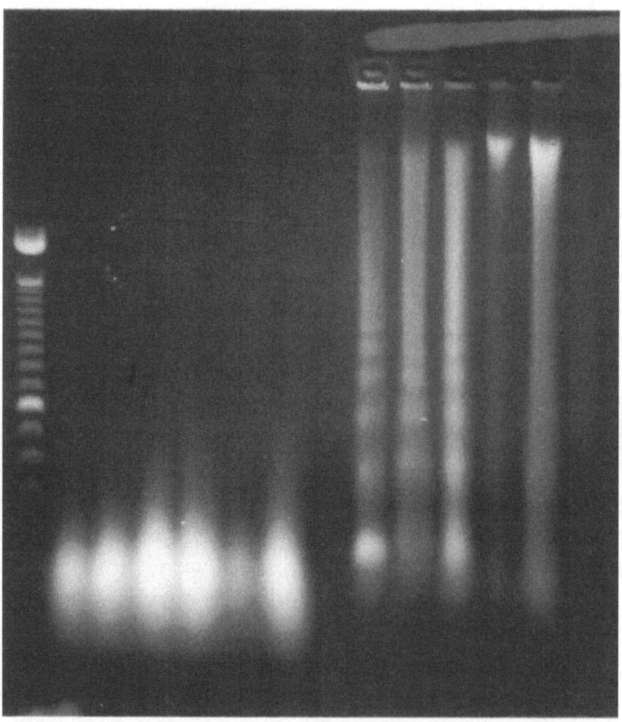

Figure 5.7 Agarose gel electrophoresis. DNA isolated from canned and raw tuna samples (L–R): 100 base pair DNA ladder; Bluefin tuna (*Thunnus thynnus*) – canned; albacore (*Thunnus alalunga*) – canned; skipjack tuna (*Katsuwonus pelamis*) – canned, yellowfin tuna (*Thunnus albacares*) – canned; bonito (*Sarda sarda*) – canned; Bigeye tuna (*Thunnus obesus*) – canned; Bluefin tuna – raw; albacore – raw; skipjack tuna – raw; yellowfin tuna – raw; bonito – raw.

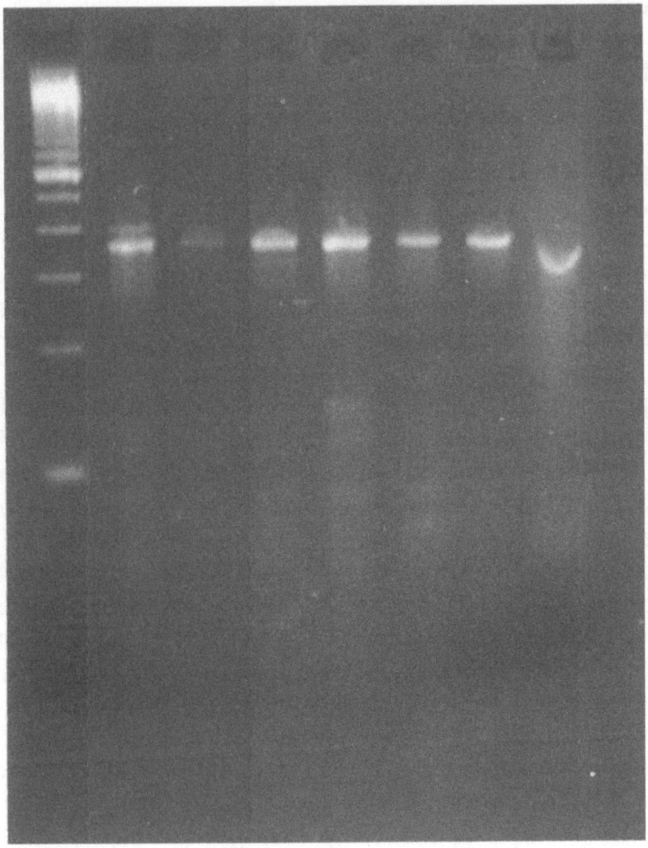

Figure 5.8 Agarose gel electrophoresis: 350 base pair PCR product of mitochondrial cytochrome *b* gene from canned tuna samples (L–R): 100 base pair DNA ladder; skipjack tuna (*Katsuwonus pelamis*) – canned; little tuna (*Euthynnus alleteratus*) – canned; bluefin tuna (*Thunnus thynnus*) – canned; albacore (*Thunnus alalunga*) – canned; Bigeye tuna (*Thunnus obesus*) – canned; yellowfin tuna (*Thunnus albacares*) – canned; and bonito (*Sarda sarda*) – canned.

specific sequences. However, because of the great variability of conditions of storage of tuna prior to canning and of the heat treatments of the canning process itself, validation of the method is required on commercially canned products.

The use of restriction enzymes has also been investigated, and separation profiles obtained on electrophoresis has established that this simpler approach can be used for the routine identification of species of canned tuna and bonito.

5.9 Fish eggs

The identification of species of fish egg or roe is unreliable if it is based on sensory criteria alone (e.g. visual appearance). As with fish flesh, problems arise, particularly when the species are closely related. For example, eggs of haddock, cod and saithe cannot be identified or differentiated by marine biologists when present in plankton trawls. Similarly, the roe of rainbow trout (*Oncorhynchus mykiss*) is visually indistinguishable from that of Atlantic salmon (*Salmo salar*). However, when individual eggs are analysed using SDS electrophoresis, it has been shown that species-specific profiles are obtained (Scobbie and Mackie, 1990). Of particular interest is the use of the technique to identify salmon eggs (Scobbie and Mackie, 1995) the possession of which is illegal in Scotland for historical reasons connected with the preservation of salmon fishing (Salmon Fisheries (Scotland) Act 1868). It was not unknown for poachers to use salmon eggs as fish attractants and to escape the arm of the law when caught for lack of unequivocal evidence.

The key to the use of SDS electrophoresis for this analysis has been the adoption of gradient gel (Figure 5.9). This has improved the resolution over that obtained in a uniform 12.5% gel. Even the eggs of closely related species such as haddock and whiting can be readily differentiated as can those of salmon, rainbow trout and brown trout. As the species-specificity of SDS profiles has now been established for the eggs of more commercially important species of fish, it is likely that this procedure will be of value in differentiating the eggs of the three species of sturgeon – Beluga (*Huso huso*), Oscietra (*Acipenser gueldenstaedli*) and Sevruga (*Acipenser stellatus*). The preliminary results shown indicate that it is possible to differentiate the eggs of those species by SDS electrophoresis.

5.10 General conclusions

Isoelectric focusing of aqueous extracts of fish muscle is the preferred procedure for the identification of species of raw fish. Wide experience of its use throughout the world has led to its general acceptance as the final arbiter in cases of dispute when the species needs to be established unequivocally. With recent advances in the use of densitometry and image analysis, data on pI values and peak height and areas can now be held in computer databases. As this information is built up it will have the potential of enabling species identification to move from simple visual comparison to a quantifiable computer-based analysis. These developments will make it more likely that the system will identify mixed species products and give an estimate of the quantitative composition of the respective species.

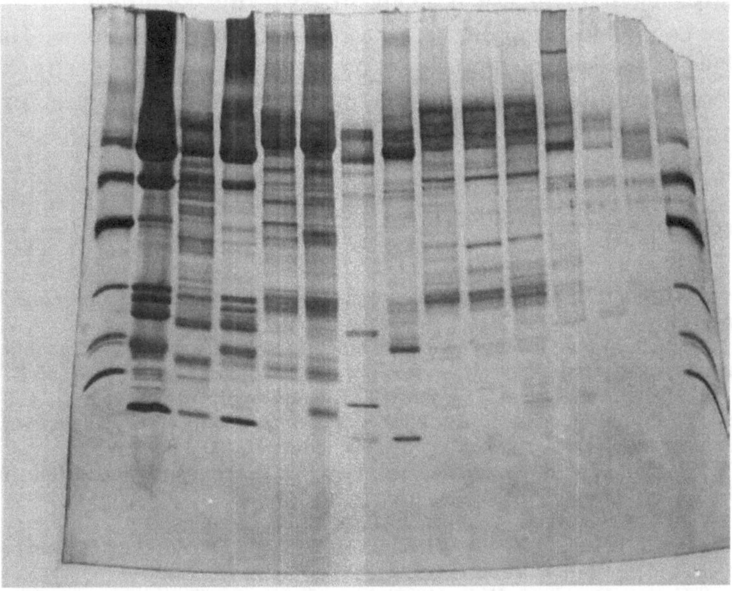

Figure 5.9 Separation profiles obtained on SDS electrophoresis (gradient gel) of SDS extracts of eggs of the following species of fish (L–R): protein marker; salmon (*Salmo salar*); rainbow trout (*Oncorhynchus mykiss*); brown trout (*Salmo trutta*); cod (*Gadus morhua*); haddock (*Melanogrammus aeglefinus*); whiting (*Merlangius merlangus*); herring (*Clupea harengus*); sturgeon (*Oscietra*); sturgeon (*Sevruga*); sturgeon (*Beluga*); flying fish (*Exocoetus* spp.); mackerel (*Scomber scombrus*); shrimp (*Pandalus borealis*) and protein markers. A 5–30% gradient gel was used.

Similarly, SDS electrophoresis is likely to remain the preferred procedure for species identification of cooked fish products and raw fish products such as scampi, where there is potential adulteration by other crustacea. Further develoments in optimising the procedure will enable corresponding data on molecular weight and area of protein zone to be determined by image analysis and stored on a computer database.

Although immunoassay systems have not been applied to any significant extent to the identification of fish species, there may be particular problems, such as the detection and estimation of crab meat content of surimi products where they could have application. For instance it may be anticipated that an antibody to a specific crab protein could be prepared and an ELISA technique developed for detecting and quantifying the specified component.

As far as DNA techniques are concerned, it is unlikely that they will replace the protein-based systems in the immediate future. This is because they require a high degree of practical skill and access to computer databases of DNA sequences and the equipment required is also expensive. Their main use is likely to be in the identification of heat-sterilised canned fish products, such as salmon, tuna and sardine. They are also likely to be used to complement data obtained by SDS electrophoresis in determining the species of fish eggs for either biological or food authenticity purposes.

References

Andrews, A.T. (1992) Electrophoresis measurements in food and agriculture. *Food Technol. Internat. Eur.* **XX**, 219–23.

An, H., Marshall, M.R., Otwell, W.S. and Wei, C.I. (1988) Electrophoretic identification of raw and cooked shrimp using various protein extraction systems. *J. Food Sci.* **53(2)**, 313–18.

An, H., Klein, P.A., Kao, K.J., Marshall, R.R., Otwell, W.S. and Wei, C.I. (1990) Development of monoclinal antibody for rock shrimp identification using enzyme-linked immunosorbent assay. *J. Agricol. Food Chem.* **38**, 2094–100.

Anon (1987a) *Surimi – It's American Now*, Fisheries Development Foundation Inc., The Alaska Writers' Group, Anchorage, Alaska, USA.

Anon (1987b) Caviar. *Seafood Internat.* **XX,** 35–39.

Armstrong, S.G., Leach, D.N. and Wyllie, S.G. (1992) The use of HPLC protein profiles in fish species identification. *Food Chem.* **44**, 147–55.

Bartlett, S.E. and Davidson, W.S. (1991) Identification of *Thunnus* tuna species by the polymerase chain reaction and direct sequence analysis of their mitochondrial cytochrome *b* genes. *Can. J. Fish Aquat. Sci.* **48**, 309–19.

Bartlett, S.E. and Davidson, W.S. (1992) FINS (Forensically informative nucleotide sequencing): a procedure for identifying the animal origin of biological specimens. *Biotechniques* **12(3)**, 408–11.

Block, B.A., Finnerty, J.R., Stewart, A.F.R. and Kidd, J. (1993) Evolution of endothermy in fish: mapping of physiological traits on a molecular phylogeny. *Sci.* **260**, 210–14.

Blum, H., Beiser, H. and Cross, H.J. (1989) Improved silver staining of plant proteins, RNA and DNA in polyacrylamide gels. *Electrophoresis* **8**, 93–99.

Carnegie, P.R. (1994) Quality control in the food industries with DNA technologies. *Austr. Biotechnol.* **4(3)**, 146–49.

Chen, F.T.A. (1991) Rapid protein analysis by capillary electrophoresis. *J. Chromatogr.* **559**, 445–53.

Chen, F.T.A., Kelly, L., Polmieri, R., Bieler, R. and Schwartz, H. (1992) Use of high ionic strength buffers for the separation of proteins and peptides by capillary electrophoresis. *J. Liq. Chromatogr.* **15**, 1143–61.

Consumer Protection Act, Great Britain (1987), HMSO, London.

Craig, A., Ritchie, A.H. and Mackie, I.M. (1995) Determining the authenticity of raw reformed breaded scampi (*Nephrops norvegicus*) by electrophoretic techniques. *Food Chem.* **52**, 451–54.

Durand, P., Laudrein, A., Quéro, J-C. (1985) *Catalogue Electrophorétique des Poissons Commerciaux*, IFREMER, Nantes, France.

Ewing, A.G., Wallingford, R.A. and Olefirowiez, T.M. (1989) Capillary electrophresis *Anal. Chem.* **61(4)**, 292A–303A.

Food Safety Act, Great Britain (1990) HMSO, London.

Foote, C.J., Wood, C.C. and Withler, R.E. (1989) Biochemical genetic comparison of sockeye salmon and kokanee, the anadromous and non-anadromous forms of *Oncorhynchus nerka. Can J. Fish Aquat. Sci.* **46**, 149–58.

Goll, D.E., Robson, R.M. and Stromer, M.H. (1977) Muscle proteins. In *Food Proteins*, (Eds J.R. Whitaker and S.R. Tannenbaum). Avi, Westford, Connecticut, USA. p. 121–74.

Gross, E. (1966) The cyanogen bromide reaction. In *Methods in Enzymology XI* (Ed. C.W.H. Hirs). Academic Press, New York, pp. 238–55.

Hitchcock, C.H.S. (1988) Opportunities and incentives for developing food immunoassays.

In *Immunoassays in Veterinary and Food Analysis I.* (Eds B.A. Morris, M.N. Clifford and R. Jackman). Elsevier Applied Science, London, pp. 3–16.
Hultin, H.O (1992) Biochemical deterioration of fish muscle. In *Quality Assurance in the Fish Industry; Developments in Food Science, 30* (Eds H.H. Huss, M. Jakobsen and J. Liston). Elsevier Science, Amsterdam, pp. 125–38.
Kleparnik, K. and Bocek, P. (1991) Review. Theoretical background for clinical and biochemical applications of electromigration techniques. *J. Chromatogr.* **569**, 3–42.
Laird, W.M., Mackie, I.M. and Ritchie, A.H. (1982) Differentiation of species of fish by isoelectric focusing on agarose and polyacrylamide gels – a comparison. *J. Ass. Public Analysts* **20**, 125–35.
Lanier, T.C. and Lee, C.M. (1992) *Surimi Technology*. Marcel Dekker, New York.
LeBlanc, E.L., Singh, S. and LeBlanc, R.J. (1994) Capillary zone electrophoresis of fish muscle sarcoplasmic proteins. *J. Food Sci.* **59(16)**, 1267–70.
Love, R.M. (1970) *The Chemical Biology of Fishes*. Academic Press, London.
Mackie, I.M. (1969) Identification of fish species by a modified polyacrylamide disc electrophoresis technique. *J. Assoc. Public Analysts* **7**, 83–87.
Mackie, I.M. (1980) A review of some recent applications of electrophoresis and isoelectric focusing in the identification of species of fish and fish products. In *Advances in Fish Science and Technology* (Ed. J.J. Connell). Fishing News Books, Farnham, pp. 444–51.
Mackie, I.M. (1983) New approaches in the use of fish proteins. In *Developments in food Proteins, 2nd edn.* (Ed. B.J.F. Hudson). Applied Science, Barking, Essex, pp. 215–62.
Mackie, I.M. (1990) Identifying species of fish. *Analytical Proc. Royal Soc. Chem.* **27**, 89–92.
Mackie, I.M. (1992) Surimi from fish. In *The Chemistry of Muscle-based Foods*, (Eds D.E. Johnston, M.K. Knight and D.A. Ledward). Royal Society of Chemistry, Cambridge, pp. 200–07.
Mackie, I.M. (1993) The effects of freezing on flesh proteins. *Food Rev. Internat.* **9(4)**, 75–610.
Mackie, I.M. (1994a) Fish protein. In *New and Developing Sources of Food Proteins* (Ed. B.J.F. Hudson). Chapman and Hall, London, pp. 95–143.
Mackie, I.M. (1994b) Fish speciation. *Food Techno. Internat. Eur.* **XX**, 177–80.
Mackie, I.M. and Taylor, T. (1972) Identification of species of heat-sterilised canned fish by polyacrylamide disc electrophoresis. *Analyst* **99**, 609–11.
Mackie, I.M., Chalmers, M., Reece, P. *et al.* (1992) The application of electrophoretic techniques to the identification of species of canned tuna, in *Pelagic Fish: The Resource and Its Exploitation* (Eds J.R. Burt, R. Hardy and K.W. Whittle). Fishing News Books, London, pp. 200–07.
Novotny, M.Y., Cobb, K.A. and Liu, J. (1990) Recent advances in capillary electrophoresis of proteins peptides and amino acids. *Electrophoresis* **11**, 735–49.
Osman, M.A., Ashoor, S.H. and Worsch, P.C. (1987) Liquid chromatographic identification of common fish species. *J. Assoc. Off. Anal. Chem.* **70(4)**, 618–25.
Patterson, R.L.S. (Ed.) (1985) *Biochemical Identification of Meat Species*. Elsevier, London.
Piggot, G.M. (1986) Surimi – the 'high-tech' raw material from minced fish flesh. *Food Res. Internat.* **2**, 213–46.
Rehbein, H. (1990) Electrophoretic techniques for species identification of fishery products. *Z. Lebensm, Unters-Forsch* **191**, 1–10.
Rehbein, H. and Karl, H. (1985) Identification of fish muscle proteins with buffers containing sodium dodecyl sulphate. *Z. Lebensm Unters-Forsch* **180**, 373–83.
Rehbein, H., Etienne, M., Jerome, M. *et al.* (1995) Influence of variation in methodology on the reliability of the isoelectric focusing method of fish species identification. *Food Chem.* **52**, 193–97.
Russell, A.R. and Carnegie, P.R. (1994) Identification of gourmet meat using FINS (Forensically informative nucleotide sequencing). *Biotechniques* **17(1)**, 24–26.
Salmon Fisheries (Scotland Act) 1868. Revised to 31 October 1978. HMSO, London.
Sambrook, J., Fritsch, E.F. and Manciatis, T. (1989) *Molecular Cloning. A Laboratory Manual*, 2nd edn. Cold Spring Harbour Laboratory Press, New York, USA.
Scobbie, A.E. and Mackie, I.M. (1988) The use of sodium dodecyl sulphate polyacrylamide gel electrophoresis in fish species identification – a procedure suitable for cooked and raw fish. *J. Sci. Food Agric.* **44**, 343–57.

Scobbie, A.E. and Mackie, I.M. (1990) The use of sodium dodecyl sulphate-polyacrylamide gel electrophoresis in species identification of fish eggs. *Comp. Biochem. Physiol.* **96B(4)**, 743–46.
Scobbie, A.E. and Mackie, I.M. (1995) The use of sodium dodecyl sulphate-polyacrylamide gel electrophoresis to confirm the presence of salmon (*Salmo salar*) eggs in illegal fish bait. *Electrophoresis* **16**, 306–07.
Sotelo, C.G., Pineiro, C., Gallardo, J.M. *et al.* (1993) Fish species identification in seafood products. *Trends Food Sci. Technol.* **4**, 395–401.
Taylor, W.J. and Jones, J.L. (1992) An immunoassay for verifying the identity of canned sardines. *Food Agric. Immuno.* **4**, 169–74.
'The Food Labelling Regulations 1984', S.I.No. 1305, HMSO, London.
'The Food Labelling (Scotland) Regulations 1984', S.I.No. 1519 (S.128), HMSO, London.
Thomas, W. and Beckenbach, A.T. (1989) Variation in salmonid mitochondrial DNA: Evolutionary constraint on mechanisms of substitution. *J. Mol. Evol.* **29**, 233–45.
Tiselius, A. (1930) *Nova Acta Reg. Soc. Sci, Uppsaliensis.* Ser IV, **7**, 1–107.
Torry Research Station (1985) *Internal report.*
Trade Descriptions Act, Great Britain (1968), HMSO, London.
UK Sea Fisheries Statistics (1991) and (1992), HMSO, London.
Weber, K. and Osborn, M. (1969) The reliability of molecular weight determination by dodecyl sulphate–polyacrylamide gel electrophoresis. *J. Biol. Chem.* **244**, 4406–12.
Yman, I.M. (1992) Meat and fish species identification. *Eur. Food Drink Rev.* **Autumn 1992**, 67–72.

6 Authenticity of cereals

B.G. OSBORNE

6.1 Introduction

Cereal grains are the fruits of cultivated grasses that are processed in order to make them into palatable foods or drinks. The processing quality, hence consumer acceptability of the product, depends on physical and chemical characteristics of the grain, which are at least partly genetic in origin. Different species of cereals have widely different uses; for example, wheat (and to some extent rye) can be used to make leavened bread but there is also a wide range of qualities within varieties of each species such that not every variety of wheat produces bread of acceptable quality to the consumer. Sometimes, there are perceived differences in quality between lots of the same variety when grown in different locations, for example Basmati rice. Inevitably, there exists a price differential when some varieties enable greater added value through processing. The need to establish authenticity of cereals therefore stems from the potential for cheaper but inferior substitutes to circulate in international and national trade and become incorporated intentionally or otherwise in food and drink products. The main authenticity issues for cereals that have been addressed so far are the identification of wheat and barley varieties, the detection of common wheat in durum semolina and pasta and the authenticity of Basmati rice.

6.2 Overview of techniques

The techniques available for identification of different species and varieties (cultivars) of cereals depend on the uniqueness of the DNA in each cereal. This may be determined directly, or indirectly by examination of the physical or chemical characteristics that the DNA controls (Figure 6.1).

New developments in molecular biology, notably the polymerase chain reaction, have enabled techniques to be developed to discriminate between species, varieties or even individuals on the basis of their DNA. Such methods are extremely specific, sensitive and can be used even on highly processed foods. However, they are not yet in widespread use in the food industry and are confined to specialised laboratories. Furthermore, it has

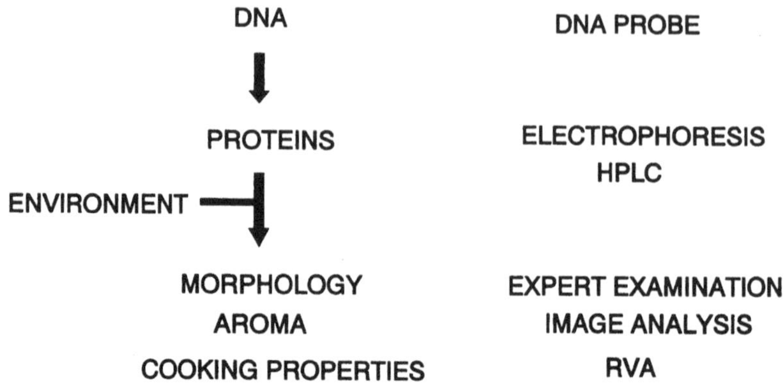

Figure 6.1 Methods overview.

not been established how direct DNA-based methods can resolve authenticity issues where origin is as important as species and variety (e.g. Basmati rice).

The function of DNA is to provide the code that defines the structure of the proteins made in the plant's cells. Thus, an alternative and widespread method of determining genetic information is to identify which proteins are contained in a sample. The most suitable for this purpose are the nitrogen storage proteins from the endosperm of the grain, known as gliadins in wheat and hordeins in barley. These have the merit of being both abundant and having many allelic forms that are readily separated into fingerprint patterns by polyacrylamide gel electrophoreis (PAGE), high-performance liquid chromatography (HPLC) or capillary electrophoresis (CE). Unfortunately, rice and oats have adopted a different nitrogen-storage mechanism in which this type of protein (prolamines) plays a minor role. The abundant glutelin proteins in rice and oats do not give fingerprint patterns that can be used for variety identification. Nevertheless, procedures for identification of rice varieties by PAGE or HPLC separation of endosperm proteins have been reported (Lookhart *et al.*, 1987; Hussain *et al.*, 1989; Huebner *et al.*, 1990). However, these procedures require very high resolution and sensitivity, which is difficult to achieve in routine operation.

Proteins control growth, which partially determines properties such as morphology, aroma, and cooking characteristics. These may be assessed by expert examination, image analysis (IA), near infrared spectroscopy (NIR), gas–liquid chromatography (GLC) or specialised equipment such as the Rapid Visco-Analyser (RVA). However, these characteristics are also strongly influenced by environmental factors. This may be viewed both as a disadvantage in that it makes the manifestation of genetic factors less clear, or an advantage where origin is an important authenticity issue.

6.3 Wheat and barley variety identification

Modern wheat and barley varieties are bred for specific agronomic and processing characteristics. Efficient farming and processing of cereals depends, to a large extent, on the ability to select the best varieties for the intended market. As a consequence, it becomes important to verify the authenticity of the chosen variety. Thus, all sectors of the cereals industry, from plant breeders through variety and seed-testing authorities and certification agencies, seed merchants and farmers, and ultimately to grain buyers, processors and consumers, may need at some stage to determine varietal identity and purity. The principal methods for cereal variety identification have been reviewed by Cooke (1988).

6.3.1 Morphology

The genetic information in the DNA determines how much and in what order the structural materials are deposited when synthesised by the enzymes. For grains with the same genetic code (i.e. the same variety) the development of structural materials will be common to all grains. Conversely, grains of different varieties exhibit several characteristic morphological features (Plates 1–9) that can be used by a trained operator to identify the variety. This provides a very quick and cheap means of checking a sample but the result should be supported by a more objective method.

A more objective means of measurement of features, such as grain shape and size, is available through the use of IA (Figure 6.2). IA refers to the automatic acquisition of quantitative visual data and its subsequent computer processing to compare images objectively (Bertrand *et al.*, 1991). This is achieved by means of a microcomputer interfaced to a monochrome video camera. The continuous television signal is first converted into an array of points, called pixels, on a rectangular grid. Each pixel can be identified and related to its neighbours to define the boundaries of recognisable images. Colours are resolved as many (usually 128) shades of grey. The digitised image can be corrected for variation in background illumination before being transformed into binary form and then subjected to statistical processing. The binary image is created by defining all pixels darker than an arbitrarily chosen grey level as black and defining as white all those lighter than this. In this way, the features of interest become black and the background white.

A typical IA system for cereal variety identification (Draper and Keefe, 1989) consists of a sample presentation device to present the grains one at a time and a charge-coupled device (CCD) video camera to record the images. The camera is linked to a frame-grabber, to digitise the image and store it in a RAM, and a computer that pre-processes the raw digitised

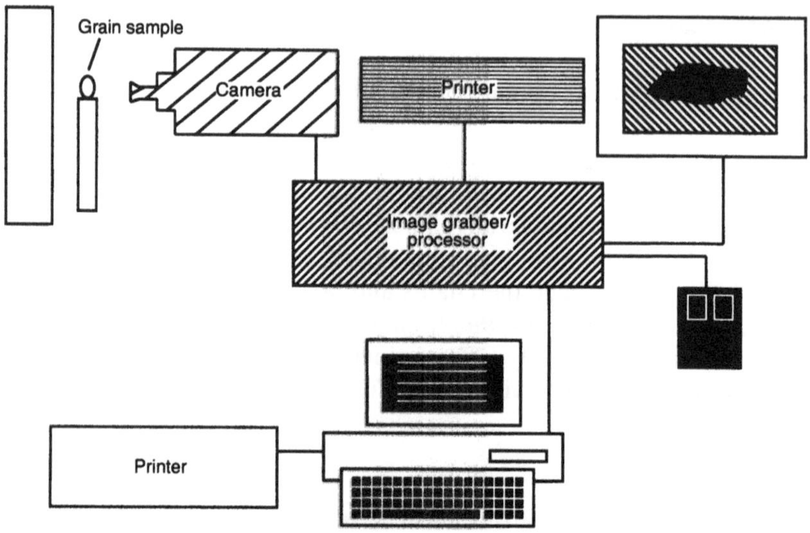

Figure 6.2 Schematic diagram of a typical image analysis system.

images to reduce noise and identify the relevant image for further processing (Figure 6.3). Since IA uses a television image, it is restricted to two-dimensional parameters. Fortunately, the wheat grain is bilaterally symetrical so that an approximate description can be achieved from projected images of two views, one from the side and one from the top. The principal features of the stored image are extracted mathematically for comparison with a stored database using pattern recognition. A database of 67 descriptors has been found sufficient to recover virtually all the shape information offered by a two-dimensional image of a wheat grain.

6.3.2 *Separation of endosperm storage proteins*

The most widely used methods for wheat and barley variety identification are based on the separation of the endosperm storage proteins, gliadins and hordeins respectively. These are relatively easy to extract and may be separated by one of several complimentary techniques described below. These components are controlled directly by the plant's DNA and so are highly characteristic and stable across different growing locations and seasons.

6.3.2.1 Polyacrylamide gel electrophoresis (PAGE). The most widely used technique for protein separation in cereal variety identification is electrophoresis. Applications of electrophoresis include distinctness testing and seed certification, quality control in flour milling and malting and

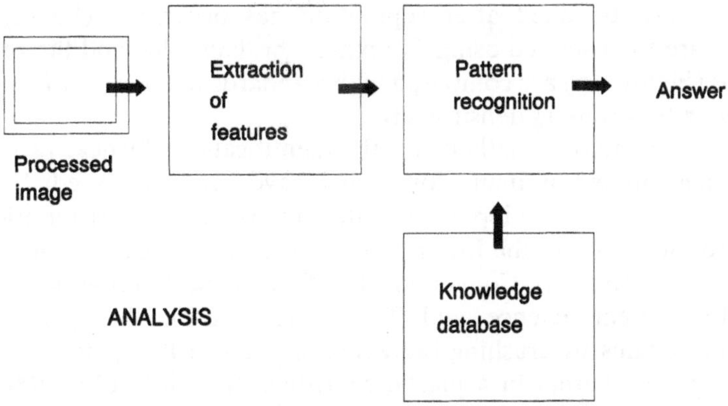

Figure 6.3 Sequence of data processing in image analysis.

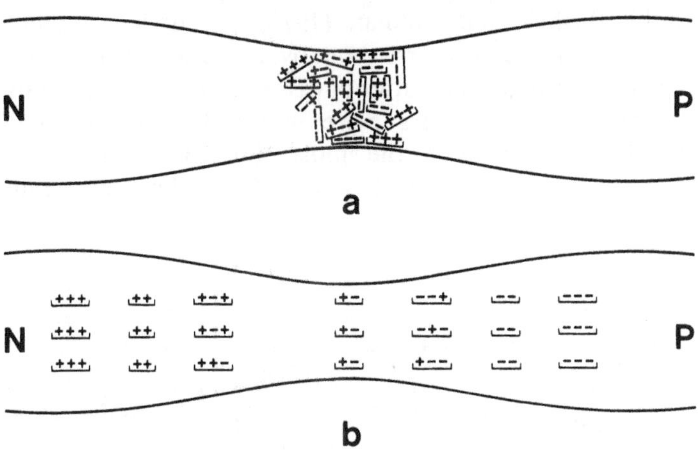

Figure 6.4 Principle of electrophoresis: (a) stylised depiction of charged molecules in a mixture; (b) electrophoretic separation of the mixture.

determination or confirmation of varietal identity and purity in case of trade disputes. Gliadin or hordein proteins are extracted from crushed single grains in 70% ethanol or urea solution. When these proteins are placed in an acid medium some of the amino acid residues become charged so that when placed in an electric field the proteins migrate towards the cathode. The greater the charge on the molecule, the faster the migration. Use of a porous gel to effect the separation results in the migration of the proteins being partially restricted by their size so that larger molecules will migrate more slowly than smaller ones. The overall separation is therefore effected on the basis of both charge and size of the individual protein molecules (Figure 6.4). The gel also hinders diffusion of the proteins so that they can be fixed after separation has occurred. The separated proteins are then stained using Coomassie brilliant blue and the resultant patterns (Figure 6.5) are compared with standards of known variety, either visually or by scanning densitometry.

At least 30 different methods for the identification of wheat varieties by PAGE and almost as many for barley have been published. Standard methods have been adopted by the International Organisation for Standardisation (ISO), the International Seed Testing Association (ISTA), the European Brewery Convention (EBC) and the International Association for Cereal Science and Technology (ICC). A typical PAGE procedure begins by crushing individual kernels of the grain sample and extracting each kernal in a plastic centrifuge tube with *circa* 100 μl 6% aqueous urea. The extracts are then centrifuged and 30 μl aliquots of the supernatants loaded onto polyacrylamide gels contained between two glass plates. The gels may be prepared immediately prior to use by the method described by Bushuk and Zillman (1978) or purchased precast. The electrophoresis separation is carried out over a period of *circa* 2 h in a tank containing acetic acid/glycine buffer of pH 3.1. The gels are removed from the tank and the upper glass plate removed prior to staining with Coomassie blue. Staining is the most time-consuming stage in the procedure. If several individual grains from the same sample are

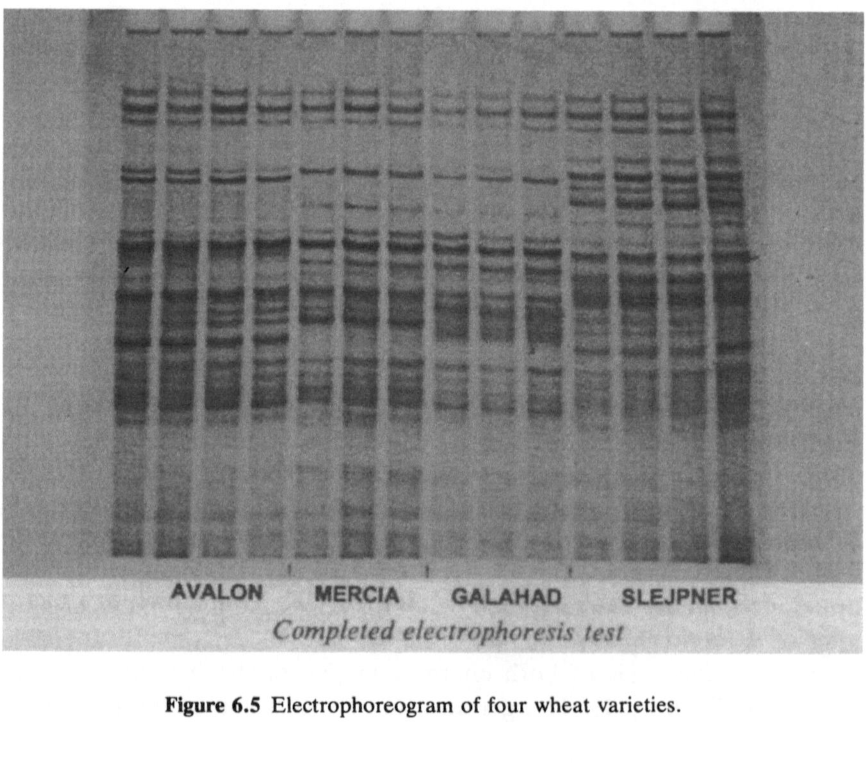

Figure 6.5 Electrophoreogram of four wheat varieties.

examined, it is possible to identify the variety of each grain then estimate any admixture on the basis of the numbers of grains identified of each variety present. Usually, either 25 or 50 grains are examined for this type of test, with an improvement in accuracy of the result as sample size is increased.

Simplified and speeded up versions of PAGE for cereal variety identification are available. One example is the Micrograd system (Wrigley *et al.*, 1991). The system (Figure 6.6) consists of an outer buffer tank with an inner chamber containing electrodes and gaskets for holding precast pocket gels. The inner chamber is immersed in the outer chamber during electrophoresis and a lid enclosing the power sockets encloses both chambers. The precast gels, buffer concentrate and Coomassie blue stain are supplied with the system. The Micrograd system enables PAGE separations to be achieved in as little as 10 min and the complete test in less than 45 min, with sufficient resolution to discriminate a limited number of the most relevant wheat varieties.

Samples of flour can also be tested by PAGE. However, these may be derived from a mixed grist rather than a pure variety. In such cases, protein bands that are characteristic of each of the major constituents can be seen and identified and used to assess their approximate proportions.

6.3.2.2 High-performance liquid chromatography (HPLC). HPLC is one of the most widely applicable techniques in analytical chemistry and

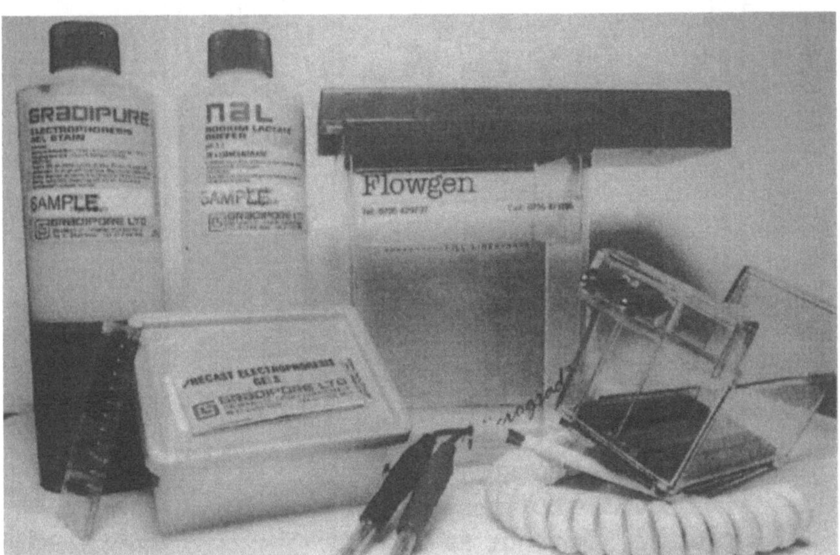

Figure 6.6 The Micrograd system.

has been shown to be capable of separating mixtures of all types of compounds. In 1983, the first successful separation of wheat endosperm storage proteins was achieved (Beitz, 1983). Methods are now available for producing HPLC varietal fingerprints that are analogous to those derived from PAGE, although the higher resolution of HPLC results in the separation of more components than PAGE. An example of the HPLC separation of hordeins from two barley varieties is shown in Figure 6.7. Thus, many of the advantages of PAGE apply equally to HPLC. However, the physical basis of the separation of the protein molecules is different in the two techniques (surface hydrophobicity in the case of HPLC and size/charge in the case of PAGE). Hence, the two techniques are complementary in that it may be possible to distinguish between varieties using HPLC that are indistinguishable by PAGE and *vice versa*. HPLC separations are more easily automated and the data more easily quantified so that computer-controlled variety identification is possible. On the negative

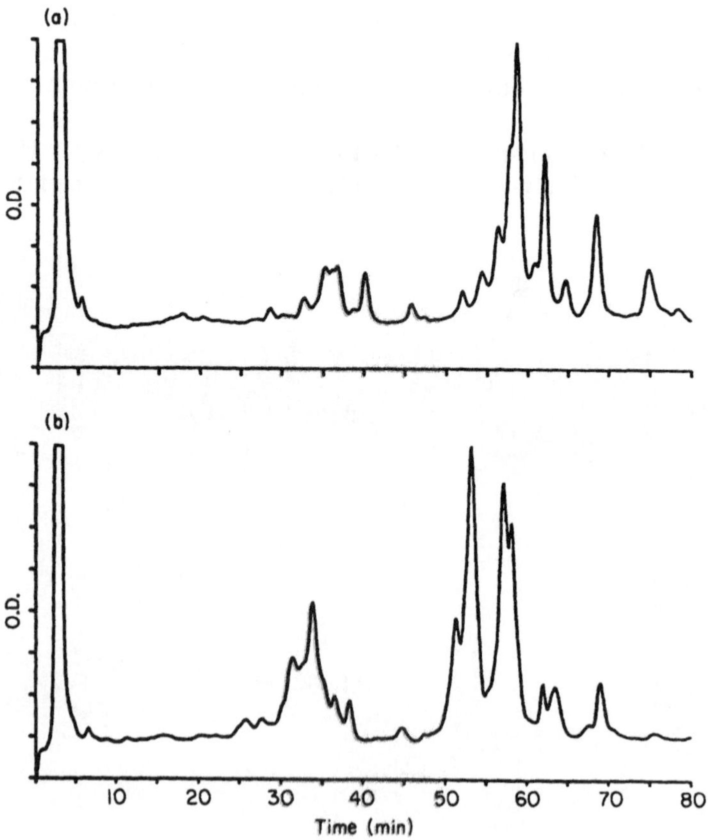

Figure 6.7 HPLC chromatogram of two barley varieties.

Morphological Characteristics of Wheat

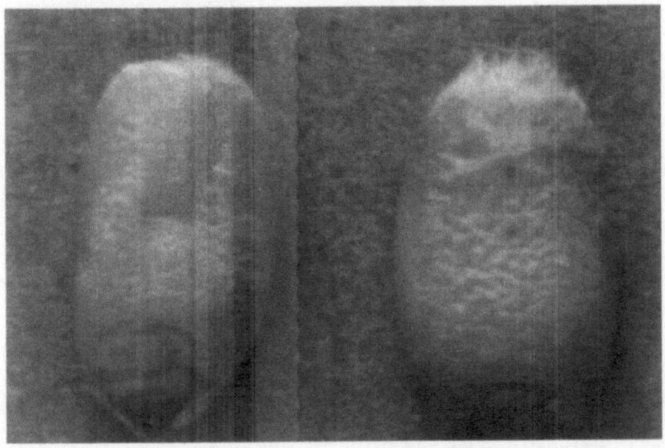

Short Long

Plate 1 Brush hair length.

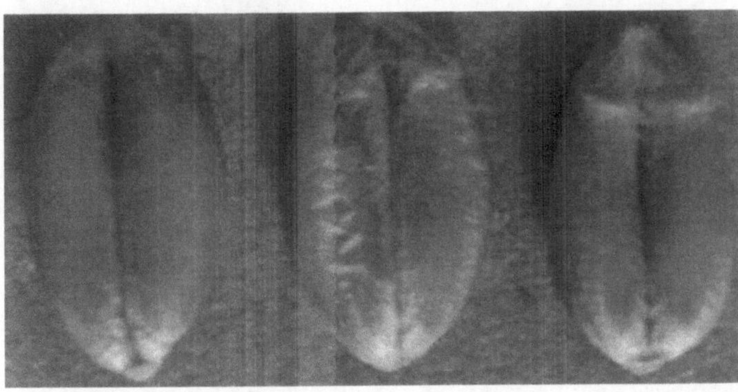

Absent Weak Strong

Plate 2 Flange.

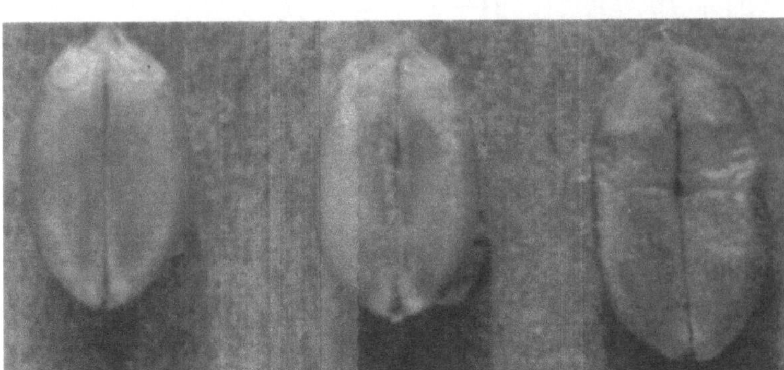

Absent Weak Strong

Plate 3 Pit.

(a) 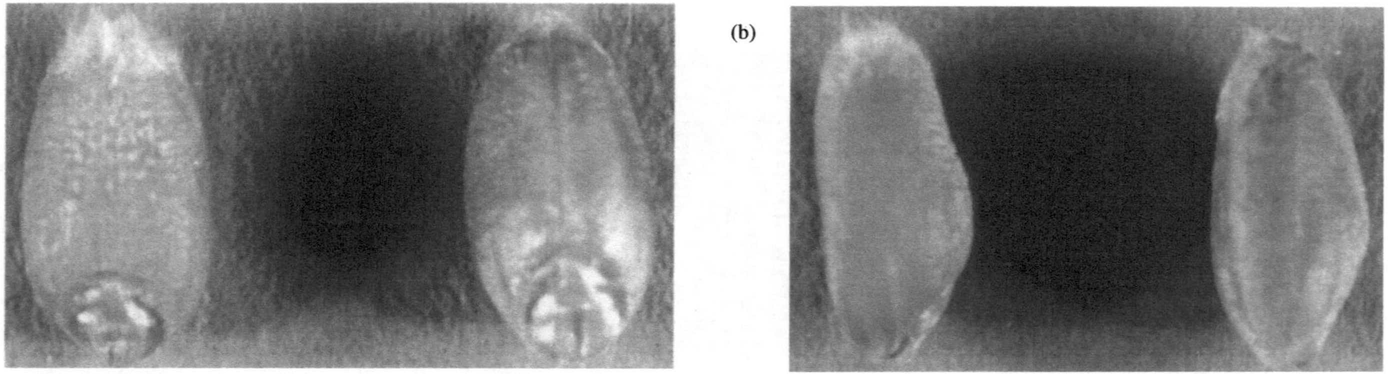 (b)

Absent Present Absent Present

Plate 4 Brush hair restriction (a) top view and (b) side view.

Thin Thick

Plate 5 Straw wall thickness.

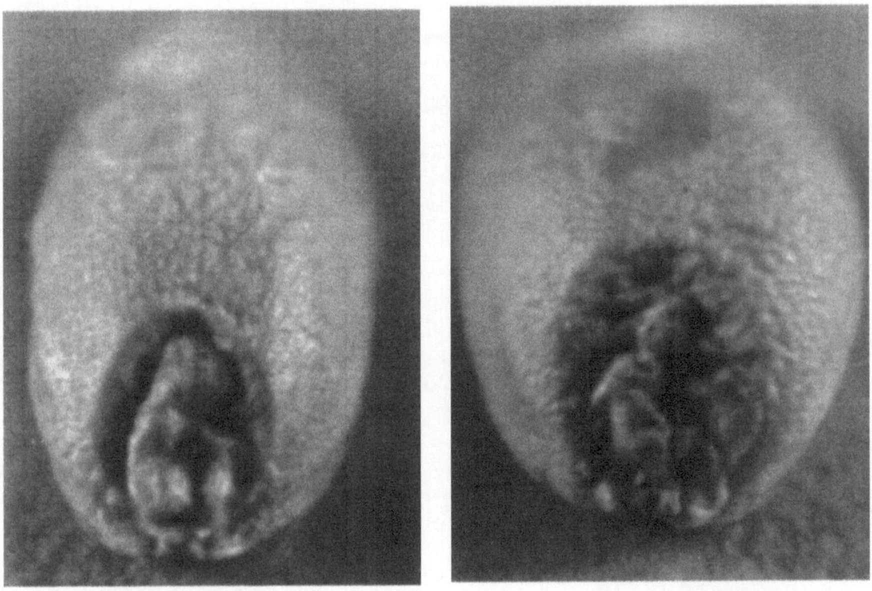

Oval

Round

Plate 6 Germ shape.

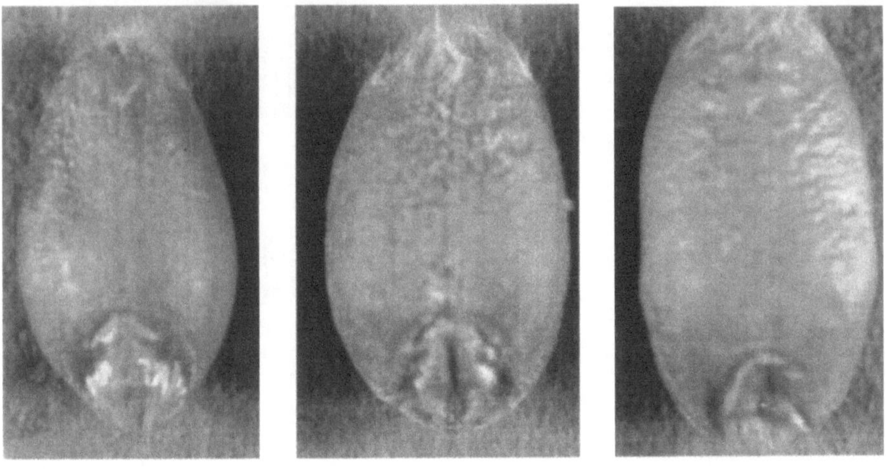

Ovate

Oval

Elliptical

Plate 7 Grain shape.

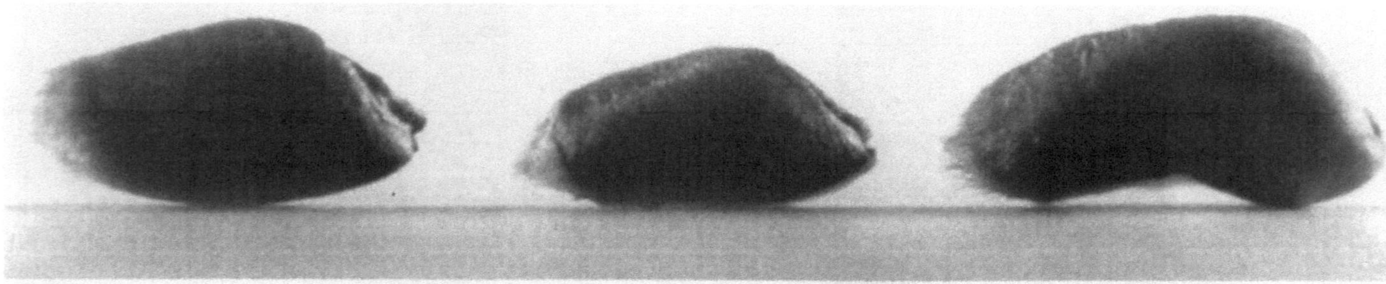

Convex Straight Concave

Plate 8 Ventral profile.

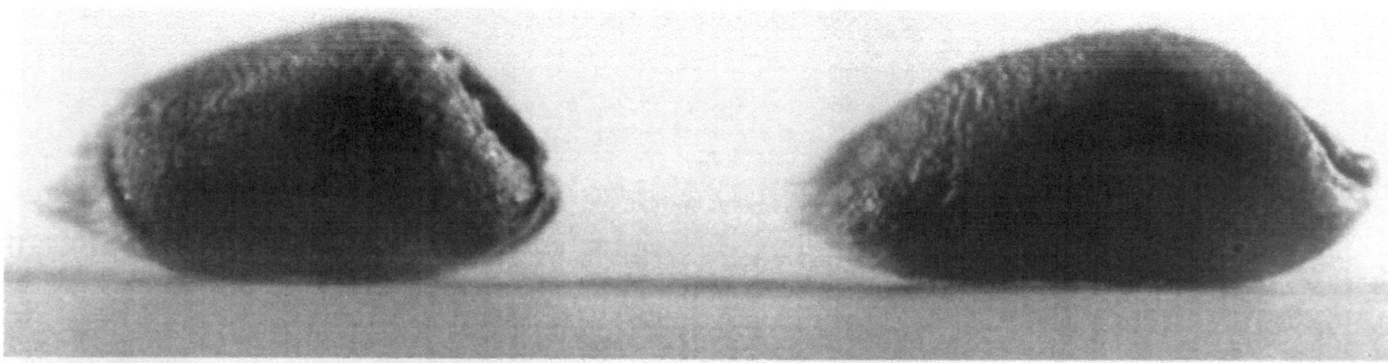

Absent Present

Plate 9 Germ beak.

side, HPLC is much slower than PAGE because it is not presently possible to analyse multiple samples concurrently. Furthermore, HPLC equipment is more expensive and sophisticated than that required for PAGE.

6.3.2.3 Capillary electrophoresis (CE). CE has only recently been applied to cereal variety identification and it has not yet reached the stage where it can be used routinely. Nevertheless, it offers several potential advantages compared with both PAGE and HPLC. In CE, instead of a gel, a very thin capillary is used as the separation medium. The charged proteins migrate as in PAGE but flow past a detector as in HPLC, thereby eliminating a lengthy staining procedure. The output of CE is analogous to that of HPLC and similar data processing techniques can be employed. CE separations of gliadins can be achieved in as little as 15 min but, like HPLC, only one sample can be run at a time.

6.3.3 Monoclonal antibody (Mab) methods

Immunological methods of analysis depend on a highly specific reaction between an antigen and an antibody. Proteins are antigens in that they cause an immunological reaction in the body of an animal, which results in the production of antibodies that recognise each antigen. Usually antibodies are produced in mixtures but single types (monoclonal) of antibody molecule (Mabs) can also be isolated. Being single antibodies, not mixtures, Mabs can form the basis for highly specific and sensitive immunological tests. Mabs that bind to certain hordeins and thus enable some varietal distinctions to be made have been isolated (Burbridge *et al.*, 1986). Thus, immunological tests for barley identification based on Mabs may be possible in the future.

6.3.4 DNA methods

The most direct method to determine variety is restriction fragment length polymorphism (RFLP), which is based on the DNA itself (Ainsworth and Sharp, 1989). In this process, restriction enzymes are used to cut the strands of DNA at specific points where a certain sequence of nucleotides is recognised (Figure 6.8). The resulting fragments are separated on the basis of size using an electrophoretic gel (Figure 6.9). This procedure may give rise to hundreds or even thousands of fragments. Simple dyestains are, therefore, inadequate and a more specific means of highlighting bands of interest must be used. This is achieved by using labelled (usually radioactively) pieces of DNA called gene probes. The power of RFLP lies in the large number of restriction enzymes available and the virtually unlimited number of probes; if two varieties cannot be discriminated by one particular combination of enzyme and probe, it is only necessary to

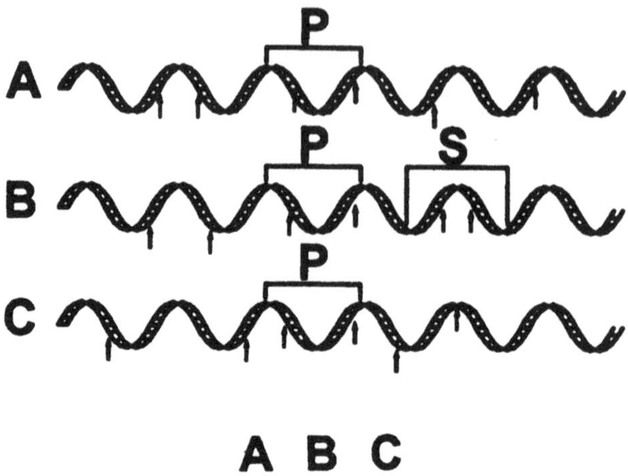

Figure 6.8 Three strands of DNA from different plant varieties being cut by restriction enzymes (represented by arrows).

find a different combination and this must exist for the varieties to be genetically distinct. The use of the RFLP for barley variety identification has already been proposed (Bunce *et al.*, 1986) but it has not yet reached the stage where it is suitable for general routine use.

6.3.5 Summary

Different sectors of the industry have varying authenticity requirements for wheat and barley, which no single technique is likely to satisfy. Visual examination will continue to be important for (DUS) testing purposes and serves as a rapid indicator of identity and purity in grain trading. This will increasingly be supplemented and ultimately superseded by IA. PAGE remains the method of choice for reference analysis in the medium term but Mab or DNA methods may eventually prove to be more suitable.

6.4 Common wheat in durum pasta

Durum wheat, *Triticum durum*, is preferred to common wheat, *Triticum aestivum*, in the manufacture of pasta because the latter may result in a product of lower quality. This is because the adhesion of the gluten protein matrix to the starch granules in pasta is less efficient when common wheat is present, leading to losses of starch into the boiling water during cooking and thus a cooked pasta with a sticky surface. However, since in the EU durum wheat is more expensive than common wheat by as much as £90 per

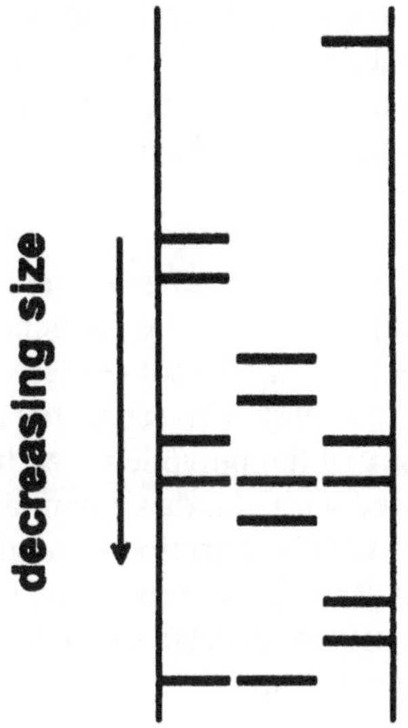

Figure 6.9 Separation by size of the fragments generated from Figure 6.8.

tonne, there is an incentive for adulteration. There are six main methods in use for the detection of common wheat in durum semolina or pasta.

6.4.1 *Gillies and Young method*

This method (Gillies and Young, 1964) is based on the detection of sitosterol palmitate, which is a component of the lipid fraction of wheat and is found in generally higher concentrations in common wheat than in durum wheat. It involves extraction of ground pasta with diethyl ether, concentration of the extract by evaporation and subsequent detection of the target analyte by thin-layer chromatography (TLC). The result is only qualitative and is limited by the fact that not all varieties of common wheat contain levels of sitosterol palmitate that are significantly higher than those of durum. However, the method is prescribed by the EU for use in customs laboratories and it is applicable to the analysis of pasta samples that have been dried at high temperatures.

6.4.2 *Genetically-based methods*

Durum wheat is more primitive in evolutionary terms than common wheat. It is derived from two wild grass ancestors, and is a tetraploid wheat with two similar but not identical sets of seven chromosome pairs, which are known as the A and B genomes. Common wheat, which evolved later, is hexaploid with a third set of chomosomes, the D genome. Most of the methods used to detect common wheat in durum are based on semolina components (e.g polyphenoloxidases, albumins, Ω-gliadins, friabilin, γ-gliadins) expressed by the D genome. Since the D genome is present only in common wheats and not in durum, these methods are all highly specific.

6.4.2.1 Electrophoresis of polyphenol oxidases. In the oldest method based on components expressed by the D genome of common wheat (Feillet and Kobrehel, 1974), polyphenol oxidases are extracted from the sample with water or aqueous urea then separated by PAGE and visualised with catechol. The intensity of the polyphenoloxidase band is measured by densitometry and compared with standards containing known amounts of common and durum wheats. The duration of analysis (10 h) is somewhat shorter than for the other PAGE methods. This method has been used in France for many years as the official procedure for the detection of adulteration of durum wheat pasta.

6.4.2.2 Electrophoresis of albumins. Albumins expressed by the common wheat D genome are not found in durum wheat. The official Italian method for detection of adulteration of durum wheat products (Resmini and De Barnardi, 1976) is based on extraction of albumins from finely ground pasta with magnesium and ammonium sulphates, followed by their precipitation and isolation by centrifugation. The albumins are then solubilised in urea and identified by isoelectric focusing. Bands are visualised using trichloroacetic acid and comparisons between samples and standards made using scanning densitometry. The Resmini method allows the detection of low levels of common wheat in pasta but it is not applicable to samples that have been dried at temperatures in excess of 60°C. The extraction procedure is complicated and specialised equipment is required for isoelectric focusing.

6.4.2.3 Electrophoresis of Ω-gliadins. This method devised by Autran involves the detection of Ω-gliadins expressed by the D genome. Following extraction of the sample with aqueous chloroethanol, the gliadins are separated using PAGE, stained with Coomassie brilliant blue and band intensities measured using scanning densitometry to obtain a quantitative result. To erase the thermal history of the samples, both samples and standards are boiled for 15 min, freeze-dried and ground to a fine powder prior to extraction. The Autran method is slower (24 h) than Durotest (see 6.4.2.4) and requires specialised electrophoretic equipment and expertise. However, it can be performed by a laboratory set up for wheat and barley variety identification.

6.4.2.4 Friabilin immunoassays. A major distinguishing feature of durum wheats is that they have very hard endosperm texture compared with common wheats. A 15k protein marker for hardness called friabilin (Greenwell and Schofield, 1986) is coded for by a gene on the short arm of chromosome pair 5D and is therefore absent from durum wheat. A monoclonal antibody has been raised against a component of friabilin known as friabilin (basic-1). This antibody has been used to develop an immunoassay for friabilin, and, by inference, common wheat. Figure 6.10 shows the result of a test for durum semolina purity based on friabilin. The right-hand side of Figure 6.10 shows an electrophoresis test where the friabilin (F) can be distinguished in the samples of hard (H) or soft (S) common wheat endosperm but not in samples of durum (D). The left side shows, for the same samples, the detection of friabilin using an immunological method based on a labelled probe to friabilin (basic-1) antibody. With the immunoassay technique, it is no longer necessary to separate the friabilin electrophoretically from other components in the sample because the antibody is specific to friabilin.

A commercially-available kit based on friabilin (basic-1) antibody and known as Durotest (Rhone-Poulenc Diagnostics, Glasgow) has been developed in ELISA format and in a rapid semi-quantitative dipstick format. In the Durotest procedure (Figure 6.11), samples of semolina or dried pasta (ground in a coffee grinder) are extracted, absorbed onto the membrane of a nitrocellulose strip and the untreated portion of the membrane is saturated with a blocking agent. The enzyme-labelled monoclonal antibody to friabilin (basic-1) is then added. If a common

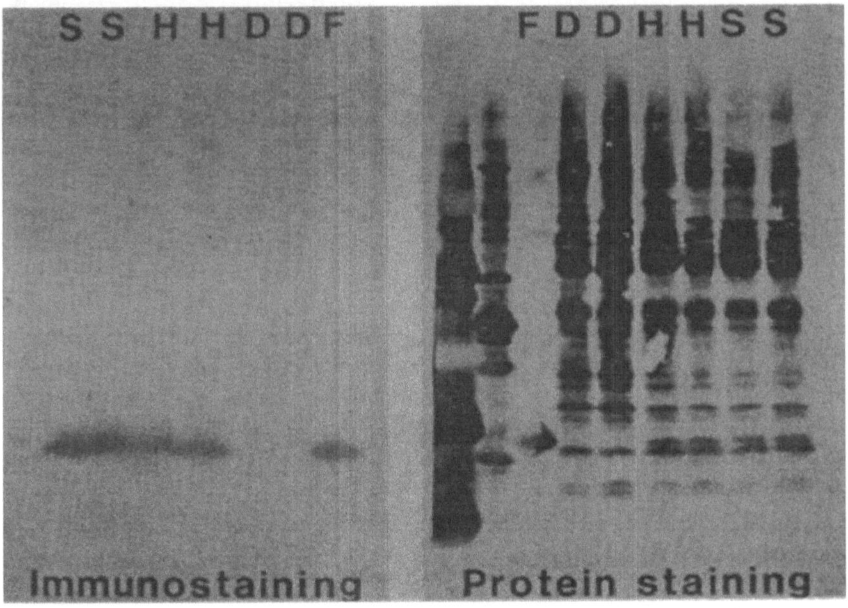

Figure 6.10 Test for durum semolina purity.

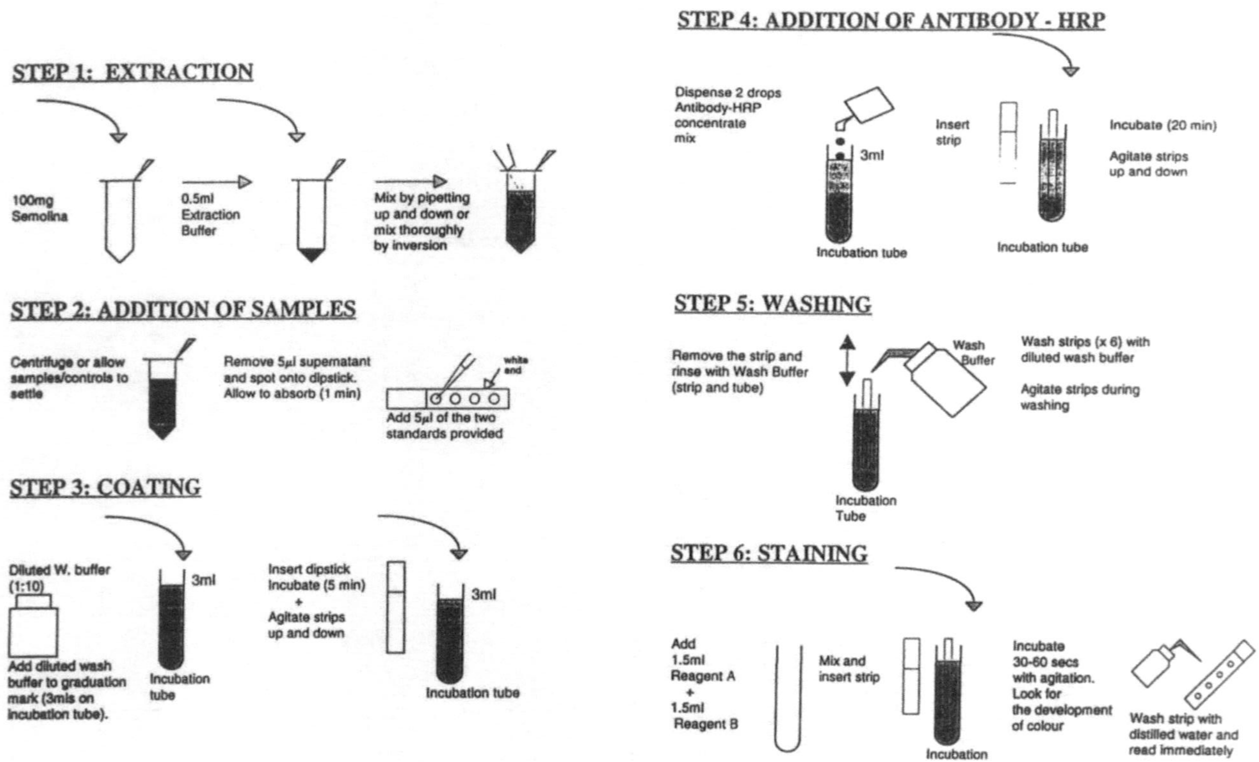

Figure 6.11 The Durotest procedure.

wheat is present in the sample, a complex is formed between its friabilin and the antibody. After washing the strip, a clear-coloured staining agent is added, which forms a blue/purple deposit in the presence of the fixed enzyme complex. The colour that has developed on the membrane is then compared with a negative control and a 3% adulterated standard, included in the kit. A colourless reaction indicates a pure durum, while a blue/purple result indicates the presence of common wheat, with the intensity of the colour proportional to the percentage of common wheat present. Durotest was designed for a yes/no decision about a 3% limit. It is simple, needs no special apparatus and is relatively rapid (producing eight samples in 2 h). The original Durotest has been found to work satisfactorily when detecting percentages of common wheat as low as 1% in semolina or pasta dried at less than 60°C; recently, a modification has been introduced to allow the test to work on samples of pasta that have been dried at higher temperatures.

6.4.2.5 HPLC of γ-gliadins. Another group of proteins encoded by the D genome is the γ-gliadins. The determination of γ-gliadins in semolina extracts has been carried out using HPLC on a large-pore reversed-phase column eluted with an acetonitrile-trifluoracetic acid gradient (McCarthy *et al.*, 1990). A γ-gliadin peak, which elutes after about 44 min is used as a quantitative marker for common wheat. The method has subsequently been extended to pastas where the drying temperature is known or can be estimated. HPLC of proteins requires specialised skill and equipment and the speed of analysis is about the same as for PAGE. This method has the advantage that Ω-gliadins can also be detected on the same chromatogram as the γ-gliadins for additional confirmation of the presence of common wheat.

6.4.3 Summary

The most widely used methods for testing the authenticity of durum semolina and dried pasta products are based on detection and quantitation of components encoded by the D genome, present only in common wheat and not in durum wheat. The available methods vary in their degree of technical complexity and sophistication of equipment required. Thus, a combination of a rapid screening method such as Durotest with a reference method such as PAGE or HPLC would provide the means to establish durum authenticity in most situations.

6.5 Basmati rice

Basmati is the name used for a class of rice comprising a few defined varieties grown in the Punjab region of India and Pakistan. Unusually for rice, Basmati yields only one crop per year but its highly favoured fragrance, flavour and slenderness apparently cannot be duplicated by growing the same seed in other regions. Consequently, Basmati rice can command at least twice the price of other long-grain rices. However, there are several varieties of ordinary, non-scented rice that resemble Basmati in one or several physical characteristics. The possibility of inclusion of such cheaper varieties in retail samples of Basmati is recognised by a GAFTA Code of Practice (Anon, 1992), which defines and describes Basmati using the export standards of the producing countries. It stipulates that the term 'Basmati' is applied only to long grain rice grown in India or Pakistan and requires that Basmati rice, where marked with its country of origin, must contain at least 97% of grains from the declared country of origin. Depending on grade and country of origin, tolerances of up to 20% contrasting or other varieties are permitted in a rice labelled as Basmati to allow for accidental contamination during harvesting. A 3% tolerance for non-Indian or Pakistani rice is permitted to allow for accidental contamination during, for example, milling at country of receipt. Indian Basmati is normally shipped as cargo rice and milled at its country of destination, whereas Pakistani is milled at the country of origin. Traditionally, the authenticity of Basmati (and other rices) has been assessed by an expert committee (Kirby-Johnson, 1993) on the basis of morphological and organoleptic evaluations. However, in response to a perception that more objective techniques were needed, three new methods (IA, RVA and NIR) were developed using a database of 98 samples (17 Basmati and 81 other) to be able, if required, to monitor the effectiveness of the GAFTA Code of Practice. Two further methods (analysis of fragrance components by GLC and RAPD) have also been studied.

6.5.1 Image analysis (IA)

The most obvious approach to an objective method of determining the authenticity of Basmati rice is to automate the assessment of morphological characteristics. IA (see 6.3.1) has been used as the basis of a device for market classification of milled rice in the USA (Mathewson *et al.*, 1990a). Basmati is prized partly for the slenderness of its grains and the tendency for elongation on cooking. However, although Basmati grains can elongate up to 100% on cooking compared with a maximum of 50% for non-fragrant varieties (Welsh *et al.*, 1990b), simply measuring elongation does not take account of the concomitant breadthwise swelling that occurs. Studies in India (Vaingankar and Kulkarni, 1986; 1989; Sood and Siddiq, 1980) suggest that the differential length to breadth ratio before and after cooking can be used to distinguish Basmati from other types of rice in mixtures. Work in the UK (M.B. Whitworth, unpublished results) has demonstrated the application of IA to cooked rice samples for establishing the authenticity of Basmati rice.

IA of cooked rice exploits the characteristic shape change that Basmati rice undergoes during cooking. At least 50 unbroken grains from the sample are cooked in a specially designed apparatus (Figure 6.12) then placed on a light box. A video camera is used to capture their silhouette images for computer processing. First, the computer makes a correction for background lighting, then it thresholds the image to identify dark regions and stores those measurements consistent with rice grains (Figure 6.13). The measurement of aspect ratio (the ratio of length to breadth) of cooked rices as a marker for Basmati is unsatisfactory because some Basmati grains curl during cooking and give rise to atypical values for aspect ratio. A combination called Rice Parameter 3 (RP3) of aspect ratio with circularity (4 × area/perimeter^2) and area has been shown to be a more reliable measure of Basmati authenticity. RP3 is greater for non-Basmati than for Basmati rice but there is a distribution of values for authentic Basmati rice. This is not surprising given the extent of variation in morphological characteristics in different Basmati grains (Sood and Siddiq, 1980). In the IA method, a weighted percentage of grains scoring above a threshold is given and is indicative of the degree to which the sample differs from the authentic Basmatis in the database.

This method detects directly grains that have a cooked appearance inconsistent with that observed for authentic Basmati grains. The appearance of individual adulterant grains is subject to a distribution, such that not all such grains are detectable. The proportion of adulterant grains likely to be detected depends on the type of adulterant: a few adulterants

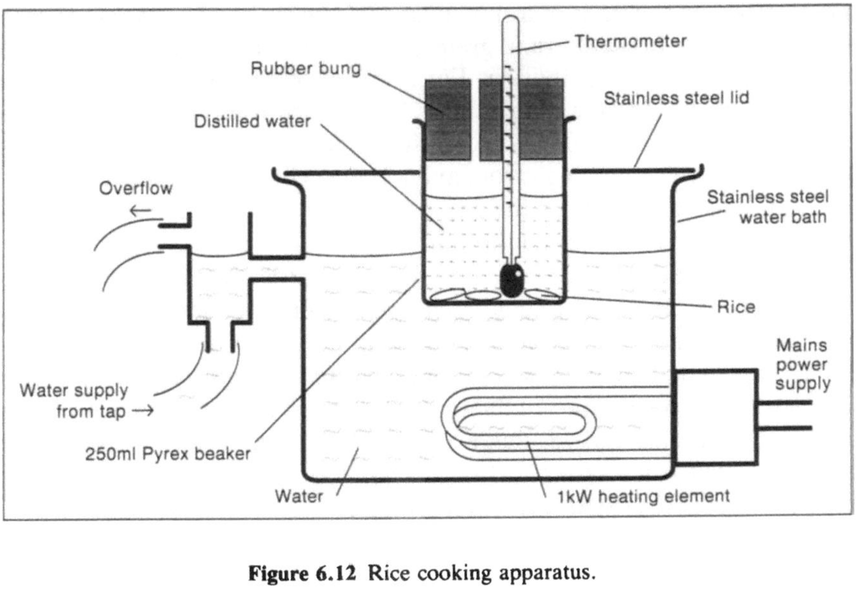

Figure 6.12 Rice cooking apparatus.

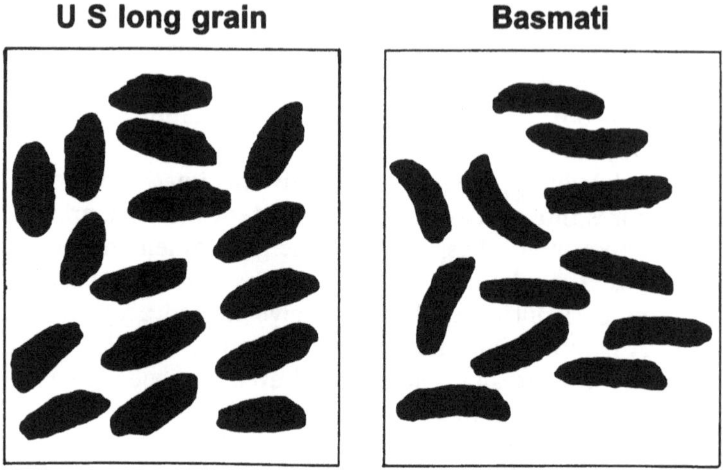

Figure 6.13 Raw images of US long grain and Basmati rice.

are completely undetectable but, for adulterants for which the method is applicable, the majority of adulterant grains would be detectable. The observed result for each grain is positive, negative, or uncertain. The stated result arises from a count of the proportion of positive results within a sample of many grains (uncertain results are also included but are given a weighting from 0 to 1 depending on the degree of uncertainty). Sampling error must thus be taken into account when drawing conclusions about a batch of rice on the basis of test results for a few grains. The sampling error in the number of adulterant grains, n is given approximately by $n^{1/2}$, assuming a Poisson distribution. For a test of 50 grains, as applied to the database of 98 samples, this may be a significant error for typical adulterant levels and affect the sensitivity accordingly. A 300 grain test would give lower sampling errors and a better chance of detecting smaller levels of addition (±1% at the 3% addition level). The sensitivity is also dependent on the type of adulterant.

IA of cooked grains has been largely successful in distinguishing populations of Basmati grains from populations of other types (Figure 6.14) although a few (7 out of 81) non-Basmati types were indistinguishable from Basmati on this basis. The samples that proved indistinguishable from Basmati by IA were three Italian-grown samples from varieties of Iranian origin, Australian grown Pelde and three Indian rices (Karnal and Kasturi). Authentic Basmati samples have been found to show a background level of typically 3–4% of grains above the threshold of RP3 (such grains including, in particular, red-striped grains); the maximum background level yet observed is about 7%. Adulterant grains are only

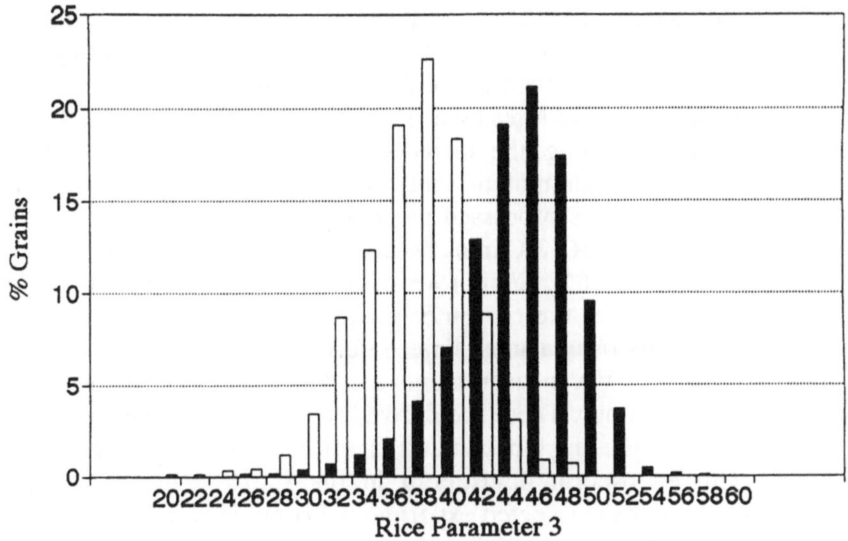

Figure 6.14 Distribution of RP3 for 17 Basmati and 81 other rice samples (solid bars = non-Basmati; open bars = Basmati).

detectable if present in sufficient quantities to raise the proportion of anomalous grains above the background level. This limit could be typically set at about 5% if a small proportion of false positives can be tolerated, or 10% if these are undesirable. It should be noted that, for many types of adulterant, such quantities of anomalous grains would correspond to slightly larger proportions (e.g. 10–20%) of total adulterant grains. IA is not currently applicable for testing broken grains.

For IA a very simple apparatus is required for sample preparation, no reagents are needed and little staff training is necessary. The duration of the 50-grain test is 30 min and would be approximately 1.5 h for a 300-grain sample. IA equipment is widely available but the algorithms are proprietary, and have currently only been implemented on one manufacturer's equipment.

6.5.2 *Rapid Visco-Analyser (RVA)*

The RVA (Newport Scientific Pty Ltd, Narrabeen, Australia) measures the electrical energy required to maintain a constant stirring speed of a starch–water dispersion as it is taken through a profile of temperature changes with time (Ross *et al.*, 1987). The RVA can be used to measure starch pasting properties in an analogous way to the Brabender Amylograph but using smaller samples and a more rapid testing cycle (Deffenbaugh and Walker, 1989). Such measurements on rice flours therefore exploit the differences in starch gelatinisation temperature between different types of rice (Sood and Siddiq, 1980). A modification has been made to bring the RVA under computer control with software that allows the heating and cooling cycles to be controlled precisely and readily varied over a broad range of temperatures and rates (Welsh *et al.*, 1990a). In the method used for rice authenticity testing, the stirred slurry (300 g rice flour in 25.0 ml water) in a disposable aluminium container is placed into the instrument (Figure 6.15) where it is processed using a programme of heating from 50°C to 95°C at a rate of 12°C/min, holding for 2.5 min and then cooling back to 50°C at a rate of 12°C/min (Welsh *et al.*, 1990a). At the end of the measurement sequence, the computer presents the measurements in the form of a plot of viscosity against time, referred to as a hot pasting curve (Figure 6.16). The RVA measurements are dominated by the raw starch granules within the flour. These are derived from the molecular structures of the amylose and the amylopectin and their interactions with each other, and also with endosperm lipids both in the starch granules and the hydrated gel. Under the selected experimental conditions, RVA curves of rice samples consist of an initial steep rise in viscosity to a peak as the temperature is raised to 95°C, followed by a thinning phase during stirring at 95°C. As cooling to 50°C commences, the viscosity reaches a minimum before rising again in the 'set-back' phase. In molecular terms, the increase in viscosity commenced at the temperature where the starch granules gelatinised and then continued rapidly in the subsequent pasting phase. The increase in viscosity during pasting is due to the granules becoming much larger particles as they hydrate and exude amylose molecules. The thinning at 95°C is largely a consequence of breakdown of the swollen granules by the shear forces imparted by the stirrer and the final increase in

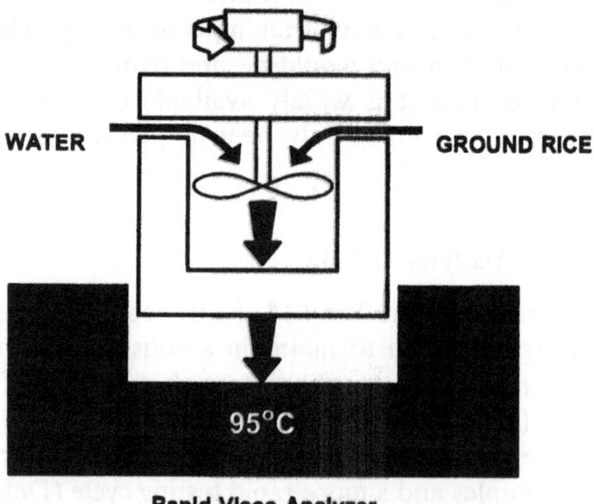

Figure 6.15 The Rapid Visco-Analyser.

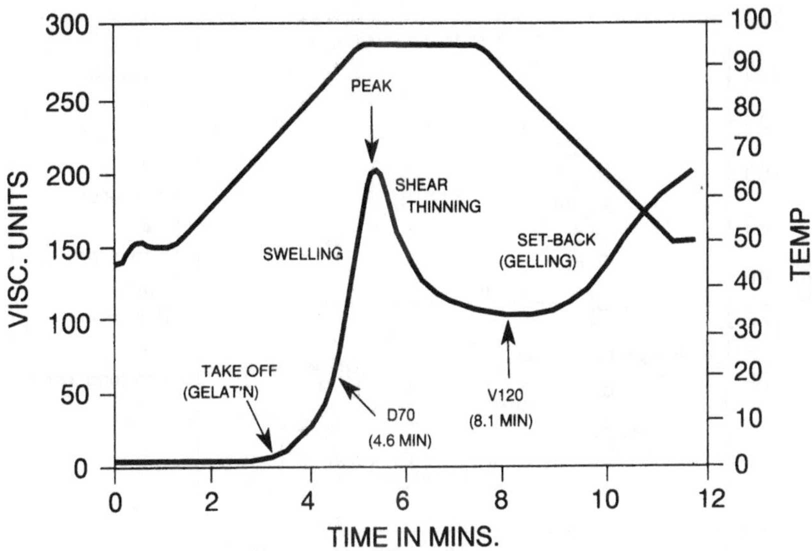

Figure 6.16 Pasting curve RVA parameters.

viscosity occurs as the cooling causes the solubilised amylose to aggregate into a fibrillar network that binds the swollen granules into a gel.

Two parameters from the RVA curves of rice flours (see Figure 6.16) have been identified as being useful for discriminating between different types of ground rice (T. Fearn and P. Greenwell, unpublished results). These are known as D70 (the rate of increase of viscosity at a particular time during the initial swelling of the starch granules) and V120 (the viscosity at a time point early in the cooling-induced gelling of the hot paste, resulting from disruption of the swollen starch granules by mechanical shear and/or enzymic hydrolysis). A two-dimensional plot of D70 *versus* V120 for the 98 samples in the database (Figure 6.17) may be interpreted in terms of the type and origin of the samples. Pure Basmati rice samples gave a tight cluster of points on the plot of D70 *versus* V120, which is distinct from most of the very wide scatter of points for the other long-grain rices of very diverse origin. Non-Basmati aromatic rice types tend to be confused with genuine Basmati by the RVA method but some of these are easily classified as non-Basmati using IA or NIR.

The RVA test takes approximately 30 min per sample, only a grinding mill is required for sample preparation, no reagents are needed other than distilled water and very little staff training is necessary. It is inapplicable to pregelatinised ('easy cook') rice but applicable to broken grain separated from whole grains by sieving, and could therefore detect adulteration by broken grains of different origin. The RVA method cannot be applied to single rice grains and is not quantitative but it can be used to compare the

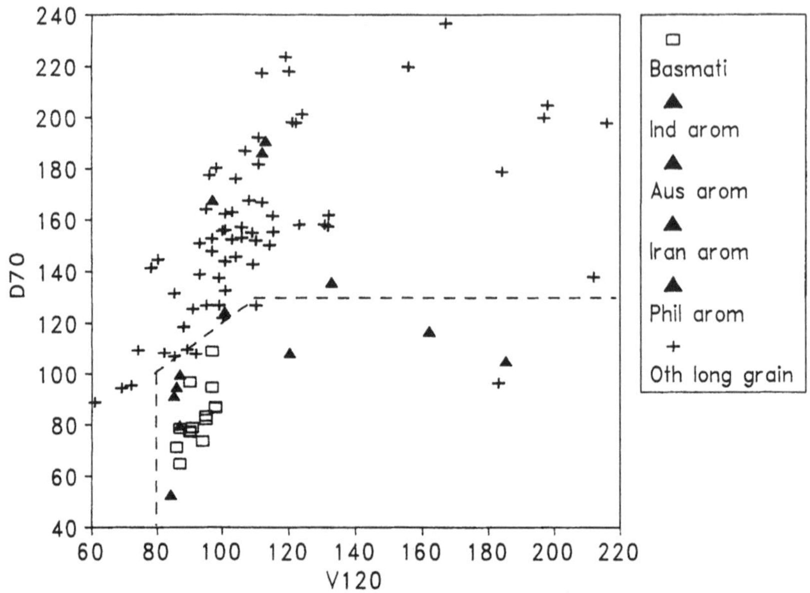

Figure 6.17 Plot of RVA parameters for 17 Basmati and 81 other rice samples.

intact and broken sub-fractions of a grain sample. No data are available on the ability of RVA to detect admixture but it is likely to be generally applicable only to gross adulteration, because the RVA parameters of potential adulterant varieties vary widely. The sensitivity in a given case would depend on the magnitude of the shift in RVA parameters between the two rices. It should be regarded as complementary to the other techniques.

6.5.3 Near-infrared spectroscopy (NIR)

NIR is widely used for the compositional analysis of food (Osborne *et al.*, 1993a) and has been applied to testing the authenticity of orange juice (see section 2.6.4 in chapter 2). The NIR method is empirical and is based on the absorption of near-infrared radiation by a sample as a result of changes in vibrational energy levels. Thus, different chemical groups (e.g. C–H, O–H and N–H) absorb energy at different wavelengths and the composite effect observed for a given sample is recorded in the form of a spectrum. In the case of granular materials such as cereal grains, the absorption is also strongly affected by physical characteristics such as the size and shape, hence the packing density, of the granules. Therefore, qualitative and quantitative analysis using NIR is accomplished by means of mathematical processing of the data, which may include both chemical and physical information (chemometrics). Foods may be in the form of solids, liquids or slurries and a range of measurement techniques and instrumentation has been developed to record their NIR spectra. In the 800–1100 nm region of the spectrum, transmittance through 1 cm of food is high enough to obtain analytically useful spectra. This magnitude of path length has the advantage that for cereal grains, such as milled rice, the need for grinding is eliminated.

For rice authenticity studies (Osborne *et al.*, 1993b), the Infratec 1225 Grain Analyzer (Perstorp Analytical, Hoganas, Sweden) has been used. This instrument (Figure 6.18) uses a monochromator based on a 50 W tungsten lamp and a diffraction grating to scan the NIR spectrum from 850–1050 nm by increments of 2 nm. The incident NIR radiation is collimated, and transmitted through the sample where a detector records the intensity of the transmitted radiation (T) relative to that of a standard (T_s). The measured absorbance is expressed at log 1/T. A single-seed adaptor is also available to measure NIR transmittance spectra of individual grains. Up to 23 grains are placed in the cassette, each grain being identified by its position, and either individual scans or only the average of the set may be stored. Whether for bulk samples or single grains, the spectra (Figure 6.19) are reduced to a few principal components and discriminant analysis is used to derive a rule for classification of rice types. With two groups, Basmati and other, this procedure results in a set

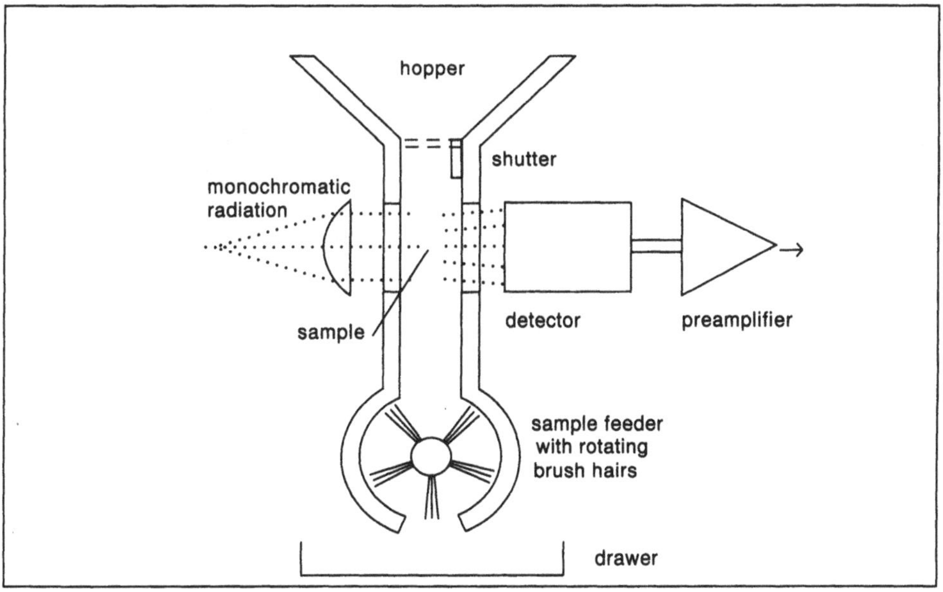

Figure 6.18 The Infratec 1225 optical system.

of weights that are applied to the spectral data to calculate a single number, the linear discriminant function (LDF) for each sample. The calculated LDF was then assessed using cross validation. In cross validation each individual result is removed from the dataset and tested against the residual data. This ensures that a sample does not influence its own classification. This evaluation found that nine Basmati and 53 other rice samples were classified correctly from NIR spectra measured on 200 g bulk samples but 8% of the Basmatis and 14% of the others were misclassified on the basis of spectra of individual grains. Using average spectra from 23 grains resulted in correct classification of 16 Basmatis but a 20% error rate on 100 other samples. The extent of overlap between the Basmati and other populations is shown in Figure 6.20.

NIR discriminates between Basmati and other rices via an empirical relationship between the spectra and size, shape and composition. The observed result for each grain is positive or negative. The sampling error in n samples is given approximately by $n^{1/2}$, assuming a Poisson distribution, and must be taken into account in drawing conclusions about a batch of rice on the basis of test results for a small sample such as the 23 grains used for NIR analysis of the database samples. NIR has been largely successful distinguishing populations of Basmati grains from populations of other

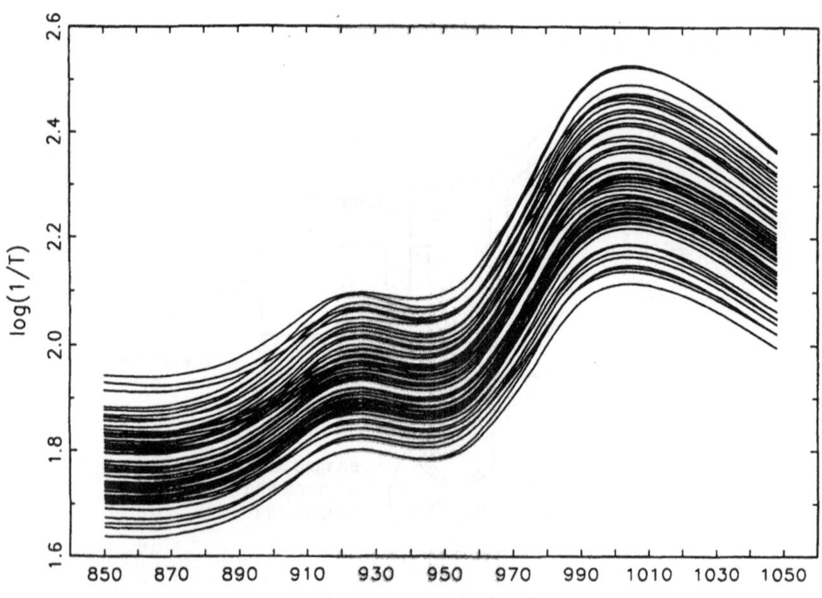

Figure 6.19 NIR transmittance spectra of 62 rice samples.

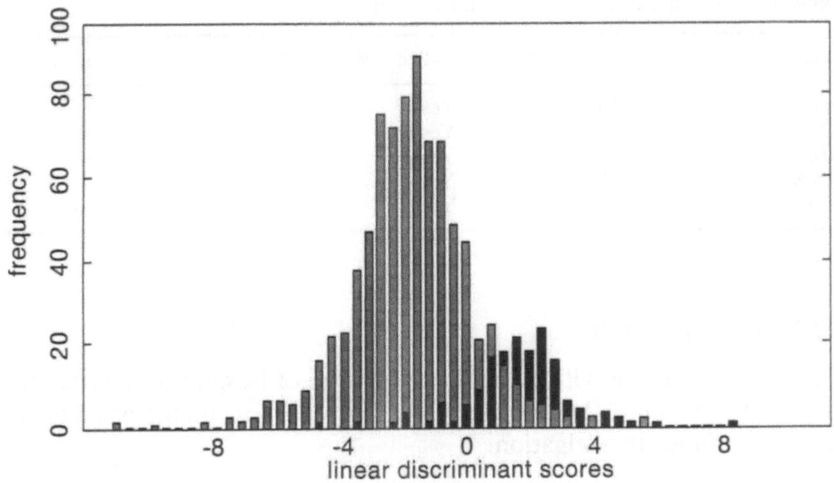

Figure 6.20 NIR discriminant analysis of individual grains of seven Basmati and 36 other rice samples (solid bars = Basmati; shaded bars = other single rice grains).

types and there is approximately the same degree of overlap between the populations as observed for IA (compare Figure 6.14 with Figure 6.20).

NIR is simple and rapid, no sample preparation or reagents are required, very little staff training is necessary and 23 individual grains may be tested in approximately 5 min. It provides at least as good a discrimination of Basmati from non-Basmati as IA. However, the equipment needed for analysis of single grains is expensive.

6.5.4 Measurement of fragrance

Since one of the valued characteristics of Basmati is its fragrance, this property has been investigated as a means of identification. Subjective assessments of aroma (Sood and Siddiq, 1978; Welsh *et al.*, 1990b) have been supplemented by direct determination of specific odoriferous compounds. One of these volatile components, which has a much higher odour potency than the others, has been identified as 2–acetyl–1–pyrroline. This compound could therefore potentially be used as a marker for fragrant rices, including Basmati. The presence of high levels of 2–acetyl–1–pyrroline in a sample of rice, while not conclusive on its own, could be used to corroborate its identification as Basmati. A method for the determination of 2–acetyl–1–pyrroline in rice has been developed (Buttery *et al.*, 1983) but its practical implementation has been limited because of the instability of this compound, the lack of a commercially available standard and the presence of other interfering compounds.

Table 6.1 Suitability of tests of Basmati rice authenticity

Technique	Cargo rice	Milled rice	Broken grains	Parboiled rice
Image analysis	No	Yes	No	Yes
RVA	No	Yes	Yes	No
NIR	Yes?	Yes	No?	No?
DNA	Yes	Yes	Yes	?
Aroma by GLC	Yes	Yes	Yes	Yes

6.5.5 *DNA methods*

The application of RAPD analysis to the issue of Basmati rice authenticity has been attempted (H.J. Rogers and H.C. Parkes, unpublished results) but is still under investigation.

6.5.6 *Summary*

Testing for Basmati rice authenticity is a complex problem because both genetic and environmental factors have to be considered. In addition, it is desirable that such testing be applicable to both cargo and milled rice, including broken grains and parboiled rice. Each of the five tests applied to the problem to date has its advantages and disadvantages and these are summarised in Table 6.1.

6.6 Conclusion

Cereals authenticity issues revolve around the variety and origin of grains and their processed products. In some cases, well-defined genetic-based markers are available, while in others the problem is not so clear-cut and measurement of indirect characteristics is employed.

References

Ainsworth, C.C. and Sharp, P.J. (1989) The potential role of DNA probes in plant variety identification. *Plant Var. Seeds* **2**, 27–34.

Anon (1992) GAFTA Rice Standards Section Code of Practice for Rice. The Grain & Feed Trade Association, London.

Beitz, J.A. (1983) Separation of cereal proteins by reversed-phase high-performance liquid chromatography. *J. Chromatogr.* **255**, 219–38.

Bertrand, D., Devaux, M.-F. and Robert, P. (1991) Basics of video image analysis. *Trends Anal. Chem.* **10**, 237–43.

Bunce, N.A.C., Forde, B.G., Kreis, M. *et al.* (1986) DNA restriction fragment length polymorphism at hordein loci: application to identifying and fingerprinting barley cultivars. *Seed Sci. Technol.* **14**, 419–29.

Burbridge, M.J., Batey, I.L., Campbell, W.P. *et al.* (1986) Distinction between barley varieties by grain characteristics, electrophoresis, chromatography and antibody reaction. *Seed Sci. Technol.* **14**, 619–29.

Bushuk, W. and Zillman, R.R. (1978) Wheat cultivar identification by gliadin electrophoreo grams. I. Apparatus, method and nomenclature. *Canad. J. Plant Sci.* **58**, 505–15.

Buttery, R.G., Ling, L.C., Juliano, B.O. (1983) Cooked rice aroma and 2-acetyl-1-pyrroline. *J. Agric. Food Chem.* **31**, 823–26.

Cooke, R.J. (1988) Modern methods for cereal variety identification. *HGCA Res. Rev. No. 9*, Home-Grown Cereals Authority, London.

Deffenbaugh, L.B. and Walker, C.E. (1989) Comparison of starch pasting properties in the Brabender Viscoamylograph and the Rapid Visco-Analyser. *Cereal Chem.* **66**, 493–99.

Draper, S.R. and Keefe, P.D. (1989) Machine vision for the characterisation and identification of cultivars. *Plant Var. Seeds* **2**, 53–62.

Feillet, P. and Koberhel, K. (1974) Determination of common wheat content in pasta products. *Cereal Chem.* **51**, 203–09.

Gillies, K.A. and Young, V.L. (1964) Evaluation of durum wheat and durum products. II. Separation and identification of the sitosterol esters of semolina. *Cereal Chem.* **41**, 502–13.

Greenwell, P. and Schofield, J.D. (1986) A starch granule protein associated with endosperm softness in wheat. *Cereal Chem.* **63**, 379–80.

Huebner, F.R., Bietz, J.A., Webb, B.D. *et al.* (1990) Rice cultivar identification by high-performance liquid chromatography of endosperm proteins. *Cereal Chem.* **67**, 129–35.

Hussain, F.R., Scanlon, M.G., Juliano, B.O. *et al.* (1989) Discrimination of rice cultivars by polyacrylamide gel electrophoresis and high-performance liquid chromatography. *Cereal Chem.* **66**, 353–56.

Kirby-Johnson, P. (1993) Pure and wholesome. *Internat. Milling Flour Feed* **187(9)**, 10.

Lockhart, G.L., Albers, L.D., Pomeranz, Y. *et al.* (1987) Identification of US rice cultivars by high-performance liquid chromatography. *Cereal Chem.* **64**, 199–206.

Mathewson, P.R., Zayas, I. and Rousser, R. (1990) A simple mechanical device for market classification of milled rice. *Cereal Foods World* **35**, 254–57.

McCarthy, P.K., Scanlon, B.F., Lumley, I.D. *et al.* (1990) Detection and quantification of adulteration of durum wheat flour by flour from common wheat using reverse phase HPLC. *J. Sci. Food Agric.* **50**, 211–26.

Morgan, A.G. (1989) Chromatographic applications in cultivar identification. *Plant Vari. Seeds* **2**, 35–44.

Osborne, B.G., Fearn, T. and Hindle, P.H. (1993a) *Practical NIR Spectroscopy*, 2nd edn., Longman, Harlow.

Osborne, B.G., Mertens, B., Thompson, M. *et al.* (1993b) The authentication of Basmati rice using near infrared spectroscopy. *J. Near Infrared Spectrosc.* **1**, 77–83.

Resmini and De Barnardi (1976) A rapid electrophoretic method to detect soft wheat (*Triticum aestivum*) in durum wheat (*Triticum durum*) in semolinas and in macaroni products. *Tecnica Molitoria* **27**, 97–109.

Ross, A.S., Walker, C.E., Booth, R.I. *et al.* (1987) The Rapid Visco-Analyzer: a new technique for the estimation of sprout damage. *Cereal Foods World* **32**, 827–29.

Sood, B.C. and Siddiq, E.A. (1978) A rapid technique for scent determination in rice. *Ind. J. Genet. Plant Breeding* **38**, 268–71.

Sood, B.C. and Siddiq, E.A. (1980) Studies on component quality attributes of Basmati rice, *Oryza sativa L. Zeitscrift fur Pflanzeneuchtung* **84**, 294–301.

Vaingankar, N.M. and Kulkarni, P.R. (1986) Preliminary observations on cooking quality parameters as indicators of adulteration of Basmati rice. *J. Sci. Food Agric.* **37**, 707–08.

Vaingankar, N.M. and Kulkarni, P.R. (1989) A cooking quality parameter as an indicator of adulteration of Basmati rice. *J. Sci. Food Agric.* **48**, 381–84.

Welsh, L.A., Blakeney, A.B. and Bannon, D.R. (1990a) Modified RVA for rice flour viscometry. *Chem. Austr.* **57**, A14–A15.

Welsh, L.A., Reinke, R.F., Reece, J.E. *et al.* (1990b) Quality selection procedures for the development of fragrant rice varieties. *Cereal Foods World* **35**, 820.

Wrigley, C.W., Gore, P.J. and Perry Manusu, H. (1991) A rapid (<10 min) electrophoresis method for identification of wheat varieties. *Electrophoresis* **12**, 384–85.

7 Authenticity of vegetable oils

D. FIRESTONE and R.J. REINA

7.1 Introduction

Vegetable oils used for food consist predominantly of triglycerides, that is, glycerol esters of fatty acids. Some significant fatty acids and their shorthand designations are shown in Table 7.1. Fatty acids contribute 94–96% of the total weight of the triglycerides. The fatty acids in vegetable oil triglycerides generally follow a pattern whereby those in the middle or 2-position of the glyceride molecule are unsaturated. Only small amounts of saturated acids are present in the 2-position, except when they are present at high concentrations in an oil or solid fat, such as coconut or palm kernal oil. Procedures for analysis of triglycerides were described by Litchfield (1972) and Christie (1986). Pancreatic lipase hydrolysis has been used to determine the fatty acid composition in the 2-position of oils and has been expanded to include stereospecific analyses (Litchfield, 1972; Breckenridge, 1978; Christie, 1986; Laakso and Christie, 1990; Christie *et al.*, 1991; Sempore and Bezard, 1991; Damiani *et al.*, 1994).

Vegetable oils have fewer component fatty acids than animal carcass fats and milk fats and fish oils, and they seldom contain odd-numbered, branched-chain, or unsaturated fatty acids with fewer than 16 carbon atoms. In addition, vegetable oils contain several minor components, including sterols, methyl sterols, and tocopherols, which are essentially absent in animal fats. Vegetable oils usually contain <25 mg/kg, and at most 100 mg/kg cholesterol (the principal zoosterol) in addition to higher levels of phytosterols, especially β-sitosterol (Rossell, 1991a). Refining of crude vegetable oils (e.g. degumming, alkali refining, bleaching with adsorption earths, and deodorization) removes free fatty acids, phosphatides, carbohydrates, protein fragments, and various resinous materials, while partially removing sterols, carotenoid pigments, chlorophylls, and tocopherols (Sonntag, 1979).

Vegetable oils are used as salad oils, cooking oils, liquid and solid shortenings, spreads, and ingredients in a variety of foods, including bakery products and fried foods. The relative prices of different oils may vary considerably from time to time, depending on costs of production and availability of stocks. With certain oils, particularly olive oil products, market price is the determining factor in a continuous problem of

Table 7.1 Significant fatty acids in food oils

Systematic name	Trivial name	Shorthand designation
Dodecanoic	Lauric	12:0
Tetradecanoic	Myristic	14:0
Hexadecanoic	Palmitic	16:0
Octadecanoic	Stearic	18:0
Eicosanoic	Arachidic	20:0
Docosanoic	Behenic	22:0
Tetracosanoic	Lignoceric	24:0
9-*cis*-Octadecenoic	Oleic	18:1 *n*–9
13-*cis*-Docosenoic	Erucie	22:1 *n*–9
9-*cis*, 12-*cis*-Octadecadienoic	Linoleic	18:2 *n*–6
9-*cis*, 12-*cis*, 15-*cis*-Octadecatrienoic	Linolenic	18:3 *n*–3

Table 7.2 International prices of some vegetable oils, US$/tonne, 9 December 1994[a]

Vegetable oil	Cost (US $/tonne)
Corn, USA fob Midwest	656
Cottonseed, USA PBSY, fob Gulf	698
Peanut (groundnut), cif Rotterdam	1055
Rapeseed, Dutch, fob ex-mill	692
Soybean, Dutch fob ex-mill	693
Sunflower, ex-tank, Rotterdam	690
Coconut, cif Rotterdam	700
Palm, RBD, cif USA West Coast	788
Palm kernel, Malay, cif Rotterdam	775
Linseed, ex-tank, Rotterdam	645
Rice bran, refined, drums tl[b]	2650
Sesame, USP, drums, lcl[b]	2293
Extra virgin olive oil, Spain, bulk at mill[c]	3120
Refined olive pomace, Spain, bulk at mill[c]	2620

[a]*Oil World* 9 December 1994, No. 49, Vol. 37, page 408, (cif = cost insurance freight; PBSY = prime bleachable/summer yellow; RBD = refined, bleached, deodorized).
[b]*Chemical Marketing Reporter*, 21 November 1994 (tl = truckload ton (short ton of 2000 pounds); lcl = less carload).
[c]*Information Bulletin* No. 770, 1 December 1994, International Olive Oil Council, Madrid.

adulteration and mislabeling. International prices of several vegetable oils are shown in Table 7.2. The price of sesame oil was recorded in *Chemical Marketing Reporter* (November, 1994); prices of other oils were reported by *Oil World* (December, 1994) and the International Olive Oil Council (IOOC, 1994a). The graphs in Figures 7.1 and 7.2, showing international prices for the 1992/93 and 1993/94 growing seasons, were prepared by the IOOC (1994a), based on international prices published in *Oil World*.

Availability as well as price can be a determining factor in incidences of fraud. Rossell (1991b) related an episode involving a Singapore dealer who

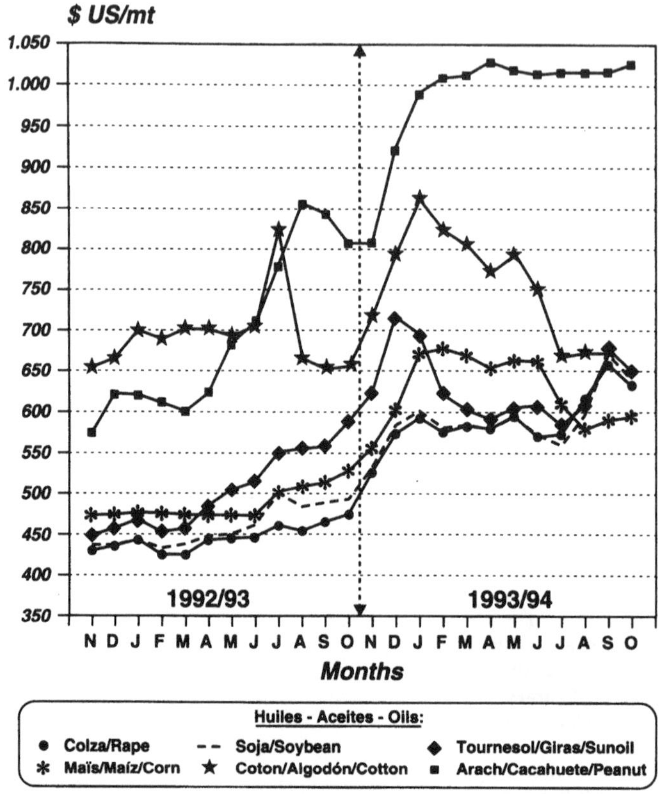

Figure 7.1 Trends of recent international prices of several vegetable oils based on data from *Oil World* (published by Ista Mielke & Co., Hamburg). (Redrawn from IOOC (1994a) by kind permission of the International Olive Oil Council.)

signed a contract to deliver a large quantity of cottonseed oil to the Egyptian government. The dealer had difficulty obtaining all the cottonseed oil required. The initial shipment was blended with about 15% palm olein, and delivered to Egypt without a problem. However, when the next shipment was prepared for delivery, a much larger amount of palm olein was added. The adulteration was detected and the dealer was sent to prison. The Malaysian Government, supplier of the palm olein, was unaware of the fraud before its exposure.

The disparity between the price of olive oils and other vegetable oils used for food has also generated fraud. An especially interesting case was the squalene olive oil racket in New York in the 1940s (Fitelson *et al.*, 1951). A method for estimating squalene in vegetable oils (Fitelson, 1943) showed that olive oil contained 136–706 mg squalene per 100 g oil, whereas other common edible vegetable oils contained 3–47 mg/100 g. After several successful legal actions against packers of adulterated olive

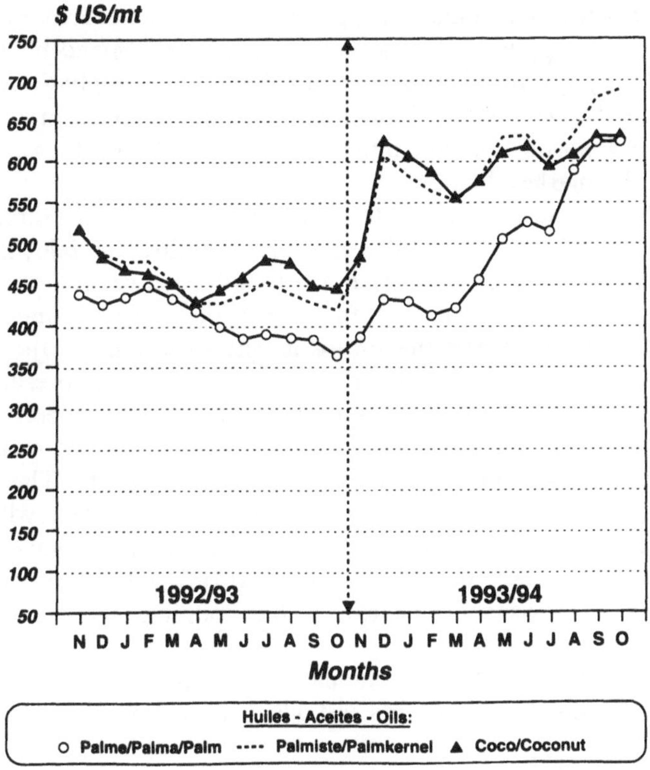

Figure 7.2 Trends of recent international prices of palm, palm kernel and coconut oils based on data from *Oil World* (published by Ista Mielke & Co., Hamburg). (Redrawn from IOOC (1994a) by kind permission of the International Olive Oil Council.)

oil blends in 1944 and 1945, mislabeling abated. Soon, olive oils with high squalene values commanded a premium price but their short supply spurred the search for a cheaper, commercially available source of squalene to be used as an adulterant. In early 1946, US Food and Drug Administration (FDA) inspectors heard rumors that squalene was being added to blends of olive and vegetable oils, making them appear to contain more than their actual content of olive oil. Laboratory examinations showed high squalene content but weak olive oil flavor. The source of the adulterant was traced to a vitamin manufacturer who had obtained the squalene as a by-product of shark liver oil fractionation. Two New York firms purchased the squalene, giving false statements about its intended use. Both firms were 'fronts' for a dealer who used it to make more than 1 million gallons of adulterated olive oil blended with cheaper vegetable oils.

Availability of the squalene was discontinued when the manufacturer was notified of its illegal use; 75 gallons were then fortified with a marker (anthranilic acid) to distinguish it from natural squalene. Experiments had shown that <0.1 ppm of the marker could be determined by extraction with dilute hydrochloric acid, diazotization, and coupling with an α-naphthol derivative, sodium 1–naphthol–2-sulfonate, to form a stable red product with absorption maximum at 508 nm. A court case followed seizure of the marked blends.

Consulting chemists for the defendants claimed that their examination of the seized oils failed to reveal the presence of anthranilic acid; that the β-naphthol derivative they used gave a color as sensitive as the α-naphthol color used by the government to extract the marker; and that one consultant had performed the entire government test on the oils with negative results. The judge ordered a demonstration in his presence. Dr J. Fitelson, director of FDA's New York District Laboratory, performed the test in court, weighing the chemicals and coupling known quantities of anthranilic acid with both the α- and β-naphthol agents. The expected results were obtained. Essentially no color was observed with the β-naphthol derivative but as little as 5 mg of anthranilic acid gave a distinct pink color in the government test. The court ordered the record to show the obvious superiority of the α-naphthol coupling agent and ruled in favor of the government.

A variety of physical and chemical tests have been developed to determine the authenticity of fats and oils. Classical physical and chemical tests have been supplemented with, or supplanted by, newer chemical and instrumental techniques, including silver ion chromatography, stable carbon isotope ratio analysis (SCIRA), chemometrics, gas liquid chromatography (GLC) with capillary or open-tubular columns, high-performance liquid chromatography (HPLC), supercritical fluid chromatography, Fourier-transform (FT) and attenuated total reflectance (ATR) infrared (IR) spectrophotometry, mass spectrometry (MS), and nuclear magnetic resonance (NMR) spectroscopy. Modern techniques for analysis of fats and oils are discussed in several recently published volumes (Hamilton and Rossell, 1986; Christie, 1987; 1989; McEwen and Larsen, 1990; Perkins, 1991; Christie, 1992a; 1993). Other recent texts cover the chemistry and technology of fats and oils (Hamilton and Bhati, 1987; Hoffman, 1989; Larsson, 1994; Lawson, 1994).

7.2 Specific tests for identity and estimation of impurities

Several good quality (virgin) vegetable oils are produced by mechanical extraction of raw materials without further processing (e.g. pressed cocoa butter, virgin olive oil, and several cold-pressed seed oils). However, most vegetable oils are obtained from oilseeds after a series of extraction and refining steps, including mechanical and heat pretreatment, mechanical expression (pressing), and solvent extraction (Krishnamurthy, 1982a); degumming, alkali refining, and bleaching (Krishnamurthy, 1982b); and deodorization (steam stripping) (Norris, 1985). More recent trends in oilseed and fats and oils processing are described by Carlson and Scott (1991). Other commercial processes used to produce fats and oils with special functionalities include fractionation and winterization (Thomas, 1985), hydrogenation (Allen, 1982; Haumann, 1994), and interestification (alkali-catalyzed randomization of the fatty acid (acyl) groups within the triglyceride molecules) (Haumann, 1994; Sonntag, 1982). Reduction, oxidation, polymerization, and other chemical/physical processes yield commercially important products. Tests have been developed to identify individual oils and oils resulting from hydrogenation or other processing. A greater challenge is identification and quantification of individual oils in mixtures.

The fat and oil industry uses specific tests to measure oil quality and chemical and physical characteristics, for example traditional (classical) quality control tests such as taste, iodine value (IV), acidity (free fatty acid content), free glycerol and 1–monoglyceride content, color, peroxide value (PV), anisidine value, thiobarbituric acid test, measurement of resistance to oxidative rancidity, and specific extinction in ultraviolet (UV) light. Traditional tests used to evaluate oil purity include refractive index, saponification value (SV), content of unsaponifiable matter, titer, Halphen test, Baudouin test, modified Villavecchia test, Crismer Value (critical temperature of dissolution of the oil in amyl alcohol–ethyl alcohol solvent), Kreis test (to indicate degree of rancidity), Renard (peanut oil) test, and carotene test. These and other quality control and purity tests were reviewed by Mehlenbacher (1960) and by Rossell (1986). Refractive index varies directly with IV and the levels of unsaturated fatty acids in the oil. IV can be calculated from the fatty acid composition of the oil (AOCS, 1987a). SV, which is related to the mean molecular weight of the fatty acids in the oil, can be calculated from the normalized fatty acid composition of the oil (AOCS, 1987b). Resistance to oxidation has been evaluated traditionally by the active oxygen method (AOCS, 1987c), which measures the time (in hours) required for an oil to reach a predetermined PV under empirical conditions at a temperature of 97.8°C. A newer method, the oil stability index (AOCS, 1987d), is an automated replacement for the active oxygen method. Two instruments are commercially available for measuring the oil-stability index. Tests measuring oxidative condition of oils and resistance to oxidation have been reviewed (Rossell, 1983). A series of traditional tests, including measurement of several physical and chemical properties are specified in the Codex Alimentarius Oils and Fats Standards (Codex, 1993a; 1993b) and are used by organizations such as the Federation of Oils, Seeds and Fats Associations, Ltd. (FOSFA, 1989) and the National Institute of Oilseed Products (NIOP, 1994). Typical physical

Table 7.3 Typical chemical and physical properties of some vegetable oils and fats

Oil or fat	Relative density (20%C/water at 20°C)	Refractive index (n_o40°C)	Saponification value (mg KOH/g oil)	Unsaponifiable matter (%)	Melting point (°C)	Iodine value (WIJS)
Almond kernel	0.910–0.916[b]	1.462–1.466	188–200	0.4–1.0	–	94–101
Apricot kernel	0.910–0.916[b]	1.462–1.466	185–199	–	–	97–110
Avocado	0.910–0.920[b]	1.458–1.465	177–198	1–12	–	85–90
Babassu[a]	0.914–0.917[c]	1.448–1.451	245–256	≤1.2	24–26	10–18
Blackcurrant	0.921–0.928	1.479–1.481	185–195	<1	–	173–182
Borage	–	–	189–192	1.2–1.9	–	141–160
Borneo tallow	–	1.456–1.457	189–200	0.4–2.0	37–39	29–38
Brazil nut	0.910–0.912[b]	1.458–1.462	192–202	<1.0	–	97–106
Cashew nut	0.911–0.918[e]	1.462–1.464	180–196	0.4–1.5	–	79–89
Castor bean	0.945–0.965[b]	1.466–1.473	176–187	<1.0	–	81–93
Chinese vegetable tallow	–	1.455–1.457	200–218	0.5–1.5	42–45	16–30
Cocoa butter	0.973–0.980[b]	1.456–1.458	192–200	0.2–1.0	31–35	32–40
Coconut[a]	0.908–0.980[d]	1.448–1.450	248–265	<1.5	23–26	6–11
Cohune nut	0.916–0.918[b]	1.449–1.450	251–260	0.2–0.5	18–24	9–14
Corn (maize)[a]	0.917–0.925	1.465–1.468	187–195	1.3	–	107–128
Cottonseed[a]	0.918–0.926	1.458–1.466	189–198	0.5–1.5	–	100–115
Cupuassu kernel	–	1.4563	189	–	–	44–45
Dhupa fat (Malabar tallow)	–	1.456–1.459	187–192	0.5–2.5	30–40	36–43
Evening primrose	–	1.4791[i]	192–198	1.5–2.0	–	147–155
Grapefruit seed	0.917–0.920[b]	1.469–1.470[b]	193–197	0.3–0.7	–	100–106
Grapeseed[a]	0.923–0.926	1.473–1.477	188–194	≤2	–	130–138
Hazelnut	0.908–0.915[b]	1.456–1.463	188–197	<0.5	–	83–90
Hempseed	0.923–0.925[b]	1.470–1.473	190–195	1.0–1.5	–	145–166
Illepé (Mowrah butter)	–	1.458–1.462	188–207	1–3	25–29	53–70
Jaboty tallow	–	1.449–1.452	228–236	0.3–1.6	–	4–23
Kapok seed	0.920–0.933[b]	1.460–1.466	189–197	0.5–1.8	≈30	86–110
Kokum butter	–	≈1.456	≈192	2.0–3.0	39–43	33–37
Limeseed	0.917–0.919[b]	1.462–1.469	191–198	0.4–0.8	–	93–111
Linseed	0.924–0.930[b]	1.472–1.475	188–196	0.1–2.0	–	170–203
Lupu fat	–	1.4565–1.4576	188–189	0.4–0.9	≈42	38–44
Mustard seed[a]	0.910–0.921	1.461–1.469	170–184	≤1.5	–	92–125
Neem	0.913–0.918[k]	1.4617–1.4627	195–204	1.0–2.0	–	68–75
Nutmeg butter	–	1.4659–1.4704	170–190	–	40–50	48–85

Oat	0.919–0.921[b]	1.464–1.470	190–199	1.0–3.0	–	105–116
Okra seed	0.916–0.919[b]	1.462–1.467	192–199	0.7–1.5	–	90–100
Olive	0.910–0.916	1.4677–1.4701[i]	184–196	0.7–2.0	−3.0–0	75–94
Orange seed	0.916–0.920[b]	1.460–1.465	192–197	0.4–1.0	–	98–104
Palm[a]	0.891–0.899[f]	1.449–1.455[j]	190–209	≤1.3	33–40	50–55
Palm kernel[a]	0.899–0.914[d]	1.452–1.488	230–254	≤1.0	24–26	14–21
Palm olein[a]	0.899–0.920[d]	1.4586–1.4592	194–202	≤1.3	–	≥56
Palm stearin[a]	0.881–0.891[g]	1.4472–1.4511	193–205	≤0.9	–	≤48
Peanut (groundnut)[a]	0.914–0.917	1.460–1.465	187–196	≤1.0	≈2.0	86–107
Poppy seed	0.918–0.920[b]	1.467–1.470	188–196	0.4–1.2	–	132–142
Rambutan tallow	–	1.458–1.459	193–195	0.5	38–42	39–44
Rapeseed[a]	0.910–0.920	1.465–1.469	168–181	≤2.0	–	94–120
Rapeseed[a]					–	
(low erucic [canola])	0.914–0.920	1.465–1.467	182–193	≤2.0	–	110–126
Rice bran	0.916–0.921[b]	1.465–1.468	181–189	3.0–5.0	–	99–108
Rye germ	0.932–0.941[e]	1.472–1.478[h]	172–192	1.0–11.0	–	111–142
Safflower seed[a]	0.922–0.927	1.467–1.470	186–198	≤1.5		136–148
Safflower seed						
(high oelic)	0.920–0.925	1.4680–1.4720[h]	–	≤2.0	–	91–95
Sal fat	–	1.456–1.457	175–192	0.6–1.3	30–36	31–95
Sesame seed[a]	0.915–0.923	1.456–1.469	187–195	≤2.0	–	104–120
Shea nut	0.916–0.918[e]	1.463–1.467	178–198	2.0–11.0	32–45	52–66
Soybean[a]	0.919–0.925	1.466–1.470	189–195	≤1.5	–	124–139
Sunflower seed[a]	0.918–0.923	1.467–1.469	188–194	≤1.5	–	118–141
Sunflower seed						
(high oleic)	0.912–0.913	1.4677[h]	–	0.8–1.0	–	84–87
Tea seed	–	1.462–1.464	188–195	0.1–1.0	–	80–92
Tobacco seed	0.918–0.9196[b]	1.4678–1.4717	187–200	≤3.0	–	129–142
Tomato seed	0.918–0.920[b]	1.466–1.468	186–198	0.4–1.4	–	112–125
Tonka bean	0.916[b]	1.457–1.468	183–198	≤1.0	–	72–79
Tucum kernel	–	1.449–1.451	240–250	≤1	30–33	10–14
Tucum pulp	–	–	189	–	–	64
Ucuhuba butter	–	1.4502–1.4525[j]	221–229	1.0–4.0	47.0	11–17
Walnut	0.923–0.925[b]	1.469–1.471	189–197	0.5	–	138–162
Watermelon seed	0.920–0.922[b]	1.4669–1.4748[i]	190–198	0.5–1.2	–	124–134
Wheat germ	0.925–0.933[b]	1.469–1.478	179–190	2.0–5.0	–	115–128

[a]Figures in *Codex Alimentarius* Commission (1993a) Proposed Draft Standard for Named Vegetable Oils, Appendix *1* (CX 1993/16); other values from references and general literature. [b]25°C/25°C. [c]25°C/20°C. [d]40°C/20°C. [e]15°C/15°C. [f]50°C/20°C. [g]60°C/20°C. [h]n_D25°C. [i]n_D20°C. [j]n_D50°C. [k]30°C/30°C.

and chemical properties of some refined vegetable oils and fats are presented in Table 7.3 along with the chemical and physical characteristics of important vegetable fats and oils in the Codex Alimentarius Commission Proposed Draft Standard for Named Vegetable Oils (Codex Alimentarius Commission, 1993a).

The Halphen test (AOCS, 1987e) identifies low levels of cottonseed, kapok, and other oils of the plant order *Malvales*, based on development of a characteristic pink-to-red color caused by the presence of cyclopropenoid fatty acids in the oils (Greenberg and Harris, 1982). Although well below 1%, crude cottonseed oil can be detected in blends with other oils; the test may be less sensitive with refined oils because cyclopropenoid fatty acids are destroyed by heat and hydrogenation. The test has also been modified for quantitative determination of cyclopropenoid fatty acids in oils and fats (Coleman and Firestone, 1972). The Baudouin (BSI, 1978) or modified Villavecchia (AOAC International, 1995a) test identifies <1% of sesame oil mixed with other oils and fats. It produces a characteristic pink-to-crimson color and is applicable to both hydrogenated and unhydrogenated sesame oil, although some Spanish and African olive oils also give pink-to-crimson colors (Jamieson, 1932). These colors disappear when water is added. Peanut oil can be identified qualitatively by the Bellier index (modified Bellier test) (AOAC International, 1995b), which is based on the temperature of crystallization of the fatty acids from 70% alcohol.

The Fitelson test (Fitelson, 1936; AOAC, 1995c) determines the presence of teaseed oil in olive oil; teaseed oil produces a deep red color within 1 min. Although the test is specific for teaseed oil, characteristic (but transient) colors are displayed by corn oil (deep blue) and soybean oil (red-brown).

The Bellier test (Bellier reaction), used in some Mediterranean countries to determine olive oil adulteration with other vegetable oils, was evaluated recently (Amr and Abu-Al-Rub, 1993). Soybean oil in olive oil yields a violet color (detection limit, 150 g/kg) as does cottonseed oil (detection limit, 80 g/kg) and corn oil (detection limit, 60 g/kg). Sesame oil gives a green color (detection limit, 10 g/kg), whereas linseed and sunflower oils do not respond. Use of this and other traditional tests for identifying specific oils has declined with the advent of modern analytical techniques to determine individual fatty acids and other components.

7.3 Fatty acid composition by GLC

The fatty acid distribution is used most frequently to characterize an oil and confirm authenticity. Oils are usually converted to fatty acid methyl esters (FAME) for GLC analysis using packed or capillary columns. Methyl esters are prepared by acid- or base-catalyzed transesterification.

The International Union of Pure and Applied Chemistry (IUPAC) method for preparing FAMEs includes a general procedure using boron trifluoride–methanol for methylation plus three alternative procedures (IUPAC, 1987a). GLC analyses are usually performed with polar stationary phases and packed columns (IUPAC, 1987b) or currently widely accepted capillary (wall-coated open tubular) columns (IUPAC, 1992a). Reviews have appeared regarding preparation (Christie, 1982; Bannon *et al.*, 1982; Christie, 1989; 1992b) and GLC (Christie, 1982; Slover, 1983; Ackman, 1986; Christie, 1989; Hammond, 1993) of FAMEs. A collaborative (interlaboratory) evaluation of errors in analysis of oils by GLC was made recently (Craske, 1993). AOCS official methods include several procedures for FAME analyses by capillary GLC (AOCS, 1987f). AOCS method Ce 1c-89 is designed to evaluate, with one capillary GLC procedure, three properties of a vegetable oil; (i) fatty acid composition; (ii) level of *trans* unsaturation in a partially hydrogenated vegetable oil; and (iii) *cis*, *cis* methylene-interrupted unsaturation caused by methyl linoleate and methyl linolenate in the FAMEs. A typical chromatogram of FAMEs for a partially hydrogenated soybean oil is shown in Figure 7.3. The fatty acid compositon of various vegetable fats and oils as determined by GLC is presented in Table 7.4. Soybean oil, mustard seed oil, rapeseed oil, and low erucic acid rapeseed oil (canola oil) are differentiated from other

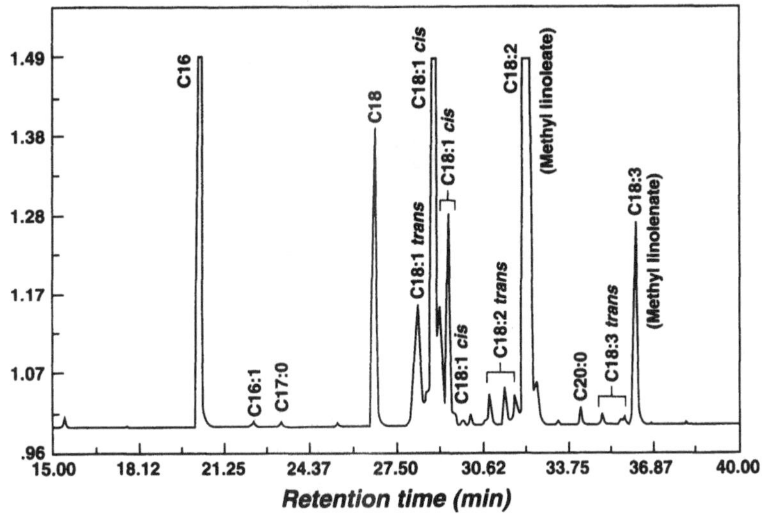

Figure 7.3 GLC chromatogram of fatty acid methyl esters from a collaborative study sample (partially hydrogenated soybean oil) using a 60 m × 0.25 mm internal diameter fused silica capillary column coated with SP-2340™. Oven temperature profile: initial temperature, 150°C, initial hold time, 0 min; program rate, 1.3°C/min; final temperature, 200°C; final hold time, 10 min (AOCS, 1987f). (Reproduced by kind permission of the American Oil Chemists' Society).

Table 7.4 Fatty acid composition of vegetable fats and oils as determined using GLC[a]

Oil/fat	6:0	8:0	10:0	12:0	14:0	16:0	16:1	18:0	18:1	18:2	18:3 (*n*–3)	18:3 (*n*–6)	20:0	22:0	22:1	24:0
Almond kernel						4–13	0.2–0.6	2–10	43–60	20–34			0.1–0.5			
Apricot kernel						5–6	1–2	0.5–1.0	58–59	29–33	0.5–1.0					
Avocado						9–18	3–9	0.4–1.0	56–74	10–17	0–2					
Babassu		2–7	1–8	40–55	11–27	5–11		2–7	9–20	1–7						
Blackcurrant[b]						6–8		1–2	9–13	45–50	12–15	14–20	0.2	0.1		0.1
Borage						9–12		2.5–5.0	15–21	36–43	0.2	17–25	0.2		2–3.0	
Borneo tallow						18–21		39–43	34–37	0.2			1.0			
Brazil nut					0–5.1	14–16		6–10	30–48	30–45						
Cashew nut						4–17	0.5–1.0	2–11	61–80	19–22	0.1–0.3					
Castor bean[c]						1–2		1–2	3–6	3–5	0.5 max					
Chinese vegetable tallow				0– 3	0.5–4	58–72		1–8	20–35	0–2						
Cocoa butter					0.1	25–27	0.1–0.3	33–37	34–35	3–4			0.2–1.0			
Coconut	0–0.6	4–9	5–8	45–50	17–21	7–10		2–4	5–8	1–2			0–0.2			
Cohune nut		7–9	6–8	44–48	16–17	7–10		3–4	8–10	1						
Corn				0– 0.3	0–0.3	10–17		1.6–3.3	25–42	39–61	0.7–1.3		0.3–0.6			
Cottonseed				0– 0.2	0.6–1.0	21–26	0–1.2	2–4	15–22	46–58	0–0.4		0.2–0.5			
Cupu assu kernel					0.6–1.0	6–12		22–35	39–47	3–9	0–1		10–12			
Dhupa fat					0–5.1	9–15		38–45	38–50	0.1–2.0			0.5–5.0			
Evening primrose					0.1	6–10		1.5–3.5	5–12	65–80	0.2	8–14	0.3			
Grapefruit seed					1.0	20–29	0 1.0	2–8	20–28	36–51	5–6		0.5–2.0			
Grapeseed				0–0.5	0–0.3	5–11	0–1.0	3–6	12–28	58–78	0–1		0–1.0	0–0.3		
Hazelnut						5–7	0.1–0.3	1.5–2.4	70–84	6–22	0–0.2					
Hempseed						6–12		1.0–2.0	11–16	45–65	15–30		2.0			0–0.1
Illipé																
(mowrah butter)		0.2		0.2	0.3	23	0.2	23	34	14	0.2		0.2			
Jaboty tallow			0.1	24	53	19			3							
Kapok seed						10–28		2–9	45–65	7–35						
Kokum butter					0–1.0	2–5		52–56	39–42	1–2						
Lime seed						25		5	21	35	12		1	1		
Linseed						7		4	20	17	52					
Lupu fat						5–10		34–50	39–51	3–5			2			
Mustard seed[d]					0–1	0.5–4.5	0–0.5	0.5–2.0	8–23	10–24	6–18		0–2	0.2–2.5	22–50	0–0.5
Neem						13–18		14–24	49–62	7–15			1–4			
Nutmeg butter				3–6	76–73	4–10			5–11	0–2						
Oat						14–23	0.1	1–3	29–53	24–48	1–3		0.2			

Olive					0–0.1	7.5–20	0.3–3.5	0.5–5.0	55–83	3.5–21	0–1.5	0–0.8	0–0.2		0–1
Orange seed					1.0	21–29	1	4–8	20–37	36–38	1–7	0.2			
Palm				0–0.2	0.7–1.3	40–46	0–0.3	4–7	36–41	9–12	0.1–0.4	0.1–0.7	0–0.1	0–0.2	0–0.2
Palm kernel	0–0.8	2.1–4.7	2.6–4.5	43–53	15–17	7–10		1–3	12–19	1–3	0–0.7	0–0.3			
Palm Olein				0.1–0.5	0.9–1.4	38–43	0.1–0.3	3.7–4.8	40–44	10–13	0.1–0.6	0.2–0.6			
Palm Stearin				0.1–0.4	1.1–1.8	48–74	0.05–0.2	4–7	15–36	3–10	0.1–0.6	0.3–0.6			
Peanut (groundnut)				0–0.1	0–0.1	8–14	0–0.2	1.9–4.4	36–67	14–43		1.1–1.7	2–5	0–0.3	1–2.2
Poppyseed					0.1–0.7	7–11	0.8–1.6	1–4	16–30	62–73					
Rambutan tallow						2.0		1–4	45			35			
								14							
Rapeseed[f]				0.1	0.2	1.5–6.0	0–3	0.5–3.1	8–60	11–23	5–13	0–3	0–2	5–60	0–2
Rapeseed low[g] eruocic (canola)					0–0.2	3.3–6.0	0.1–0.6	1–2.5	52–67	16–25	6–14	0.2–0.8	0–0.5	0–4.7	0–0.2
Rice bran					0.5–0.7	16–28	0.5	2–4	42–48	16–36	0.2–1.0	0.5–0.8	0.1–0.5		0–0.5
Rye germ					2	9–21		0.2	7–35	47–72	3–8				
Safflower seed					0–0.2	5–8	0–0.2	2–3	8–21	67–83		0.2–0.4	0.2–0.8	0–1.8	0–0.2
Safflower seed, high oleic					0.1 max	5–6	0–0.2	1.5–2	74–80	13–18	0.2 max	0.3 max	0.2 max		
Sal fat[h]						6–23		33–57	31–52	0.3–5.0		1.8			
Sesame seed					0–0.1	8–10	0.1–0.2	5–6	36–42	41–48	0.3–0.4	0.3–0.6	0–0.3		0–0.3
Shea nut				0.4	0.3	4–8		36–41	45–50	4–8	0–0.4	1–2			
Soybean				0–0.1	0–0.2	9.7–13.3	0–0.2	3.0–5.4	17–25	49–57	5.5–9.5	0.1–0.6	0.3–0.7	0–0.3	0–0.4
Sunflower seed				0–0.1	0–0.2	5.6–7.6	0–0.3	2.7–6.5	14–39	48–74	0–0.2	0.2–0.4	0.5–1.3	0–0.2	0.2–0.3
Sunflower seed (high oleic)						2.7–4.2		3–5	80–87	4–9					
Tea seed						5–9		0.3–4	79–37	7–10					
Tobacco seed						3–12		3–12	8–40	52–80	0.6–3.0	0.1–1.0			
Tonka bean						5–13		4–11	47–61	7–15		3–4	11–15		0.1
Tomato seed					0.2 max	12–16	0.6 max	4–7	16–25	50–60	1.5–2.5	0.6 max			
Tucum kernel		1	4	49	22	6		2	13	2					
Tucum pulp				0.2	0.5	30		2	60	3		4			
Ucuhuba butter			1	13–15	64–73	3–9		1	6–8	3–5					
Walnut						7–8	0.1–0.2	2	17–19	56–60	13–14	0.1	0.1		
Watermelon seed						9–11	0.5–1.0	5–10	13–19	62–71					
Wheat germ						12–19	0.5	0.5–3	14–23	52–56	3.5–7	0.3		3.0	1

[a]%m/m; figures in *Codex Alimentarius* Commission (1993a) Proposed Draft Standard for Named Vegetable Oils, Appendix 1 (CX 1993/16) and other sources in the general literature. [b]18:4 (*n*–3): 2–4%. [c]Ricinoleic: 81–91%; dihydroxystearic: 1%. [d]20:1: 5–13%; 20:2:0–1%; 22:2:0–1%; 24:1:0.5–2.5%. [e]20:1 0.7–1.7%. [f]20:1: 3–15%; 20:2:0–1%; 22:2:0–2%; 24:1:0–3%. [g]20:1: 0.1–3.4%; 24:1: 0–0.4%. [h]Hydroxystearic: 0.8%; epoxystearic: 0.4–1.2%.

major edible oils by the presence of significant levels (6–18%) of linolenic acid (18:3 *n*-3). Cottonseed oil contains 0.6–1.0% myristic acid (14:0), which is present only in trace amounts in corn, canola, peanut, safflower, sesame, and sunflower oils.

Assessment of identity and purity of vegetable oils in the food industry generally involves analysis of fatty acid composition (Ostermann, 1991). For example, in the production of sunflower oil margarine, the fatty acid composition of the sunflower oil ingredient can be used to demonstrate purity, as follows:

1. Lauric acid (12:0) content >0.1% suggests adulteration with lauric oils.
2. Palmitic acid (16:0) content should range from 5.6–7.6% – higher levels may be caused by adulteration with animal or vegetable oils such as palm oil and soybean oil.
3. Palmitoleic acid (16:1) content >0.1% may be caused by adulteration with animal fats.
4. Linolenic acid (18:3) content >0.4% suggests adulteration with rapeseed or soybean oil.
5. Erucic acid (22:1) content >0.1% points to adulteration with rapeseed oil or hydrogenated fish oil.

Adulteration is more difficult to determine when the composition of co-mixed oils approaches that of the oil under test. Additional tests recently developed include determination of sterols and tocopherols, and triglyceride composition.

7.4 Sterols and other minor components

Vegetable fats and oils contain a diverse group of minor components, including free fatty acids, mono- and diglycerides, glycolipids, phospholipids, saturated and unsaturated hydrocarbons, aliphatic alcohols, sterols, wax esters, tocopherols and other antioxidants, carotenoids, pigments such as chlorophyll, and trace metals.

7.4.1 Sterols

Sterols are probably the most important of the minor components and comprise a major portion of the unsaponifiable matter of most vegetable oils (Kochhar, 1983). They exist mainly as free sterols and sterol esters of fatty acids. Certain minor components may be unique to an oil, for instance the glycoside sesamolin in sesame oil. Vegetable sterols include a series of 4–desmethylsterols (e.g. brassicasterol, campesterol, stigmasterol, β-sitosterol), which are simply referred to as sterols, 4–monomethylsterols (e.g. citrostadienol, cycloeucalenol, obtusifoliol), and 4,4–dimethylsterols (e.g. cycloartenol, α-amryin, β-amyrin, lupeol), also known as triterpene alcohols. Although sterols are generally heat-stable, high temperatures and other treatment during refining processes affect the sterol content of vegetable oils (Kochhar, 1983).

Standard methods are available for separation and analysis of sterol mixtures (IUPAC, 1987c; ISO, 1983; AOCS, 1987g). These methods specify saponification of a 5 g portion of oil, extraction of the unsaponifiable matter, and separation of the sterols in the unsaponifiable matter by thin-layer chromatography (TLC), followed by analysis of the sterols (or their silyl ether or acetate derivatives) by GLC using packed or capillary columns. The related EEC/IOOC method (EEC, 1991; IOOC, 1992) specifies use of α-cholestanol (5α-cholestan–3β-ol) as an internal standard with capillary column GLC for determining individual sterol concentrations in an oil and the percentage of each sterol in the sterol fraction. The FDA's *Lipid Manual* (Sheppard, 1992) specifies preparation of sterol butyrates for GLC determination of sterol patterns and determination of oil adulteration. The Italian Technical Commission of the Oil and Fat Industry made a collaborative study of the determination of olive oil sterols by capillary column GLC (TCOFI, 1989). Many studies of the sterol fraction have been made to determine quantitative and qualitative differences among vegetable oils (Itoh *et al.*, 1973; Gutfinger and Letan, 1974; Mannino and Amelotti, 1975; Weihrauch and Gardner, 1978; Bielefeld, 1989; Rossell, 1991b; Plank and Lorbeer, 1993). A typical gas chromatogram of the sterol fraction (trimethylsilyl ether derivatives) of virgin olive oil (IOOC, 1992) is shown in Figure 7.4, peaks are identified in Table 7.5, and the sterol composition of vegetable oils (% of sterol fraction; total sterols in mg/kg) is shown in Table 7.6.

Oils from seeds of the Cruciferae family (e.g. rape and mustard) contain characteristically significant levels (5–20% of the sterol fraction) of brassicasterol (Itoh *et al.*, 1973). The high levels of 24–methylene cholesterol (15–20% of the sterol fraction) and Δ^5-avenasterol (15–28% of the sterol fraction) differentiate borage oils from other oils (e.g. blackcurrant, evening primrose) rich in γ-linolenic (all *cis*–6,9,12–octadecatrienoic) acid (Iatrides and Artaud, 1989) (Figure 7.5). Safflower and sunflower seed oils contain relatively high levels of Δ^7-stigmastenol (7–23%) in the sterol fraction compared with lower levels in other common edible oils (Codex Alimentarius Commission, 1993a). High levels of β-sitosterol and Δ^5-avenasterol plus low levels of campesterol and lower levels of stigmasterol are characteristic of olive oil sterols; a stigmasterol to campesterol ratio of 3:1 is characteristic of cocoa butter sterols (Homberg, 1991). The 4–methyl sterols and triterpenes in vegetable oils should be separated by TLC for exact determination of sterol content and compositon (Homberg and Bielefeld, 1990). Sterols are identified and their molecular structures confirmed by gas chromatography–mass spectrometry (GC–MS)

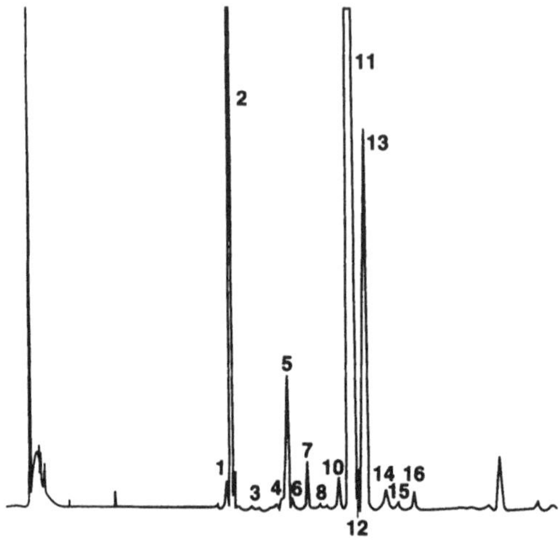

Figure 7.4 GLC chromatogram of the sterol fraction (trimethylsilylether derivatives) of unrefined (virgin) olive oil using a 30 m capillary column coated with SE-54 stationary phase and column temperature, 260°C; see Table 7.5 for identity of the GLC peaks (IOOC, 1992). (Redrawn by kind permission of the International Olive Oil Council).

(Knights, 1967; Lusby *et al.*, 1984; Dumazer *et al.*, 1986; Griffini *et al.*, 1987; Meneghetti *et al.*, 1987; Lognay *et al.*, 1993).

Several methods have been proposed to replace saponification for determining sterols and other minor components. One procedure involved separating sterol esters and free sterols by liquid chromatography (LC) on silica gel, TLC, and transesterification to liberate the sterols from their fatty acid esters (Worthington and Hitchcock, 1984). Another included separation of free sterols from the oil on silica gel cartridges or HPLC columns packed with silica gel (Horstmann and Montag, 1986). HPLC allowed separation of the triterpene (dimethyl), methyl and desmethyl sterols for subsequent GLC or GLC–MS analysis. HPLC was also used to separate total sterols from unsaponifiables. Grob *et al.* (1989; 1990) described an on-line (coupled) LC–GLC method for rapid determination of free sterols, esterified sterols, and wax esters (fatty acid esters of aliphatic alcohols). The free sterols were esterified with pivalic acid before fractionation of the oil on the LC column; this facilitated removal of the triglycerides and interfering components and elution of the free (derivatized) sterols, esterified sterols, and wax esters in one fraction for transfer to the GLC system. The method permitted improved identification of cold-pressed *versus* solvent-extracted oils. For example, the concentration of free sterols in solvent-extracted sunflower oil was more than double the level in cold-pressed sunflower oil. The LC–GLC method was modified (Artho *et al.*, 1993) using silylation under milder conditions, which eliminated undesirable esterification of free alcohols and sterols with free

Table 7.5 Relative retention times for sterols[a]

Peak	Identification	Relative retention time	
		SE54 column	SE52 column
1. Cholesterol	Δ–Cholesten–3β–ol	0.67	0.63
2. Cholestanol	5 α–Cholestan–3β–ol (internal standard)	0.68	0.64
3. Brassicasterol	[24S]–24–Methyl–Δ5,22–cholestadien–3β–ol	0.73	0.71
4. 24–Methylenecholesterol	24–Methylene–Δ5,24–cholestadien–3β–ol	0.82	0.80
5. Campesterol	[24R]–24–Methyl–Δ5–cholesten–3β–ol	0.83	0.81
6. Campestanol	[24R]–24–Methyl–cholestan–3β–ol	0.85	0.82
7. Stigmasterol	[24S]–24–Ethyl–Δ5,22–cholestadien–3β–ol	0.88	0.87
8. Δ7–Campesterol	[24R]–24–Methyl–Δ7–cholesten–3β–ol	0.93	0.92
9. Δ5,23–Stigmastadienol	[24R,S)–24–Ethyl–Δ5,23–cholestadien–3β–ol	0.95	0.95
10. Clerosterol	[24S]–24–Ethyl–Δ5,25–cholastadien–3β–ol	0.96	0.96
11. β–Sitosterol	[24R]–24–Ethyl–Δ5–cholesten–3β–ol	1.00	1.00
12. Sitostanol	24–Ethyl–cholestan–3β–ol	1.02	1.02
13. Δ5–Avenasterol	[24Z]–4(28)–Ethylidene–Δ5–cholesten–3β–ol	1.03	1.03
14. Δ5,24–Stigmastadienol	[24R,S]–24–Ethyl–Δ5,24–cholestadien–3β–ol	1.08	1.08
15. Δ7–Stigmastenol	[24R,S]–24–Ethyl–Δ7–cholesten–3β–ol	1.12	1.12
16. Δ7–Avenasterol	[24Z]–24(28)–Ethylidene–Δ7–cholesten–3β–ol	1.16	1.16

[a]Source: I00C (1992).

Table 7.6 Sterol (desmethylsterol) composition in vegetable oils as a percentage of total sterols[a]

Oil/fat	Chole-sterol	Brassica-sterol	Campe-sterol	Stigma-sterol	β-Stio-sterol	Δ^5-Avena-sterol	Δ^7-Stigma-stenol	Δ^7-Avena-sterol	Others	Total sterols (mg/kg)
Almond kernel			2–4	1–2	80	10–12	1–2	1–2		2660
Apricot kernel	<1		6	1–2	88–89	3–4				
Avocado	0–0.2	2	6–8	0–2	89–92	3		0.2		4040
Babassu	1–2	0–0.3	17–19	8–10	48–54	17–20		0.4–1		570–766
Blackcurrant	0.2–0.7		7–10	0.5–1	70–85	2–3	0.4–5	0.4–2		
Borage		0–2	25–30		22–42	15–28		1	15–20[b]	
Brazil nut	1		2	9	85			2		
Cashew nut			9	2	75	10	0.3	0.4	1[c]	1840
Castor bean			10	22	44–56	11–21	0–2	1		
Cocoa butter	1		8–11	24–31	58–63	3–5	1			
Coconut	0.6–3	0–0.9	7.5–11	11–16	32–51	20–41	0–3	0–3	0–4	470–1139
Corn (maize)	0.2–0.6	0–0.2	18–24	4–8	54–67	4–8	1–4	0.7–3	0–2	7950–22 150
Cottonseed	0.7–2.3	0–0.9	6–15	2–7	76–87	1.8–7.3	0–1.4	0.8–3	0–1.5	2690–6425
Cupu assu			4	9	80				7	245
Evening primrose			8–9		87–90	4	2			
Grapefruit seed			7	3	90					
Grapeseed	0.5 max	0.2 max	9–14	9–17	67–77	1–3	1–3	1 max	5 max	5826
Hazelnut			5	1	93	2				1200
Hempseed			17	15	44	2	2	1		3720
Illipé (Mowrah butter)			16	7	70	6	1			5500
Kapok seed			9	2	86	2	1			
Linseed	0–2		26–29	7–10	46–57	9–13	1–4			
Lupu			3	9	83				5	250
Mustard seed (black)		6	33		58	2				
Niger seed	0.2–0.8		12–13	13–14	38–43	6–7	4–5	4	14–22	
Olive	≤0.5	≤0.1	≤4	[d]	≥75	4–14	≤0.5	0–0.1		≥1000

Orange seed	0.2		8	3	88	0.2				
Palm	2.6–6.7		18–28	8–14	50–62	0–3	0.2–2	0–5		376–1627
Palm kernel	0.6–3.7	0–0.8	8–13	12–17	62–73	1–9	0–2	0–1.4	0–3	792–1406
Papaya seed	3		9–11	6–7	69–72	6–7	1–2	0.5–1	2–3[e]	
Peach kernel			3.5–5	0.4–1	90–93	2–3	0–0.5			
Peanut (groundnut)	0–4	0–0.2	12–20	5–13	47–68	8–19	0–5	0–6	0–1.4	901–2854
Pecan nut			4	2	81	12	1	1		1520
Pistachio nut			5	2	77	1				2010
Pumpkin seed				1			4	10	84[f]	5230
Rapeseed		12–13	30–33	0.4–0.6	49–55	1–2				88
Rapeseed										
low erucic (Canola)	0.5–1.3	5–13	25–39	0–1	45–58	3–7	0–1	0–1	0–4	4824–11 276
Rice bran			20–28	8–15	49–54	5–11	1–2	2–4		10 550
Safflower seed	0–0.5		9–13	6–10	40–50	2–4	16–23	3–5	0.5–3	2095–2647
Safflower seed										
(high oleic)	0–0.2		10–16	8–15	52–60	5–6	13–18	5–6		
Sesame	0.1–0.2	0.1–0.2	10–20	3–7	57–62	6–8	2–8	1–6	0.7–9	4501–18 957
Shea nut			38					11	51[g]	2470
Soybean	0.6–1.4	0–0.3	15–24	16–19	52–58	2–4	1–5	1–5	0–2	1837–4089
Sunflower seed	0.2–1.3	0–0.2	7–13	8–11	56–63	0–7	7–13	3–7	0–5	2437–4545
Sunflower seed										
(high oleic)			8–10	8–10	55–58	0–3	14–22	4–6		
Tea seed				34				2	64[h]	1020
Tobacco seed	16		7	13	60					1500
Tomato seed	7–27	0–3	4–7	7–17	54–73			1		
Walnut			5		89	5				1760
Wheat germ			19–29	1–4	60–67	6	3	2		5530

[a]Figures in *Codex Alimentarius* Commission (1993a) Proposed Draft Standard for Named Vegetable Oils, Appendix 1 (CX 1993/16) and other sources in the general literature. [b]24-Methylene cholesterol. [c]$\Delta^{7,25}$-Stigmastadienol. [d]Less than campesterol. [e]24-Methylene cholesterol, 1–2%; fucosterol, 1%. [f]24-Methyl-Δ^{7}-cholestenol, 6%; $\Delta^{7,22,25}$-stigmastatrienol, 29%; $\Delta^{7,22}$-stigmastadienol, 27%; $\Delta^{7,25}$-stigmastadienol, 22%. [g]24-Methyl-Δ^{7}-cholestenol, 6%; $\Delta^{7,22}$-stigmastadienol, 45%. [h]$\Delta^{7,22}$-Stigmastadienol, 60%; 24-methyl-Δ^{7}-cholestenol, 4%.

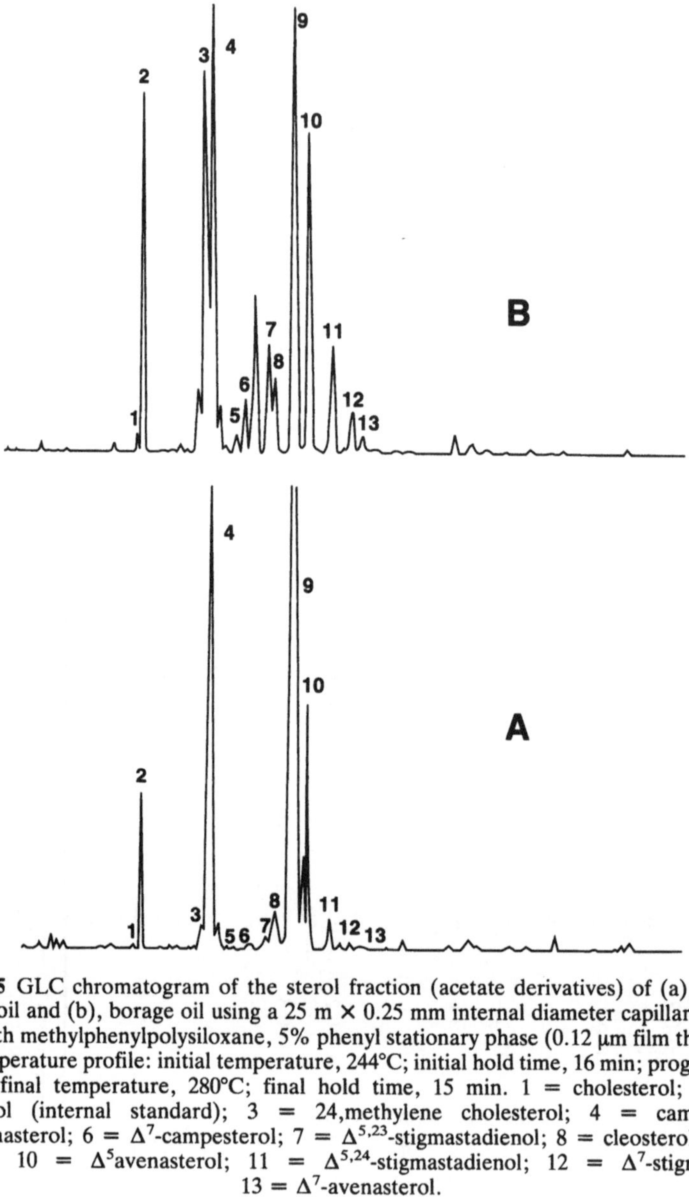

Figure 7.5 GLC chromatogram of the sterol fraction (acetate derivatives) of (a), evening primrose oil and (b), borage oil using a 25 m × 0.25 mm internal diameter capillary column coated with methylphenylpolysiloxane, 5% phenyl stationary phase (0.12 µm film thickness). Oven temperature profile: initial temperature, 244°C; initial hold time, 16 min; program rate, 5°C/min; final temperature, 280°C; final hold time, 15 min. 1 = cholesterol; 2 = 3*β*-cholestanol (internal standard); 3 = 24,methylene cholesterol; 4 = campesterol; 5 = stigmasterol; 6 = Δ^7-campesterol; 7 = $\Delta^{5,23}$-stigmastadienol; 8 = cleosterol; 9 = *β*-sitosterol; 10 = Δ^5avenasterol; 11 = $\Delta^{5,24}$-stigmastadienol; 12 = Δ^7-stigmastenol; 13 = Δ^7-avenasterol.

fatty acids present in the oil. Less than 10% of other oils admixed with olive oil could usually be determined with the coupled LC–GLC system unless harsh refining conditions removed the sterols and other minor components (Grob *et al.*, 1994a). Additional modification of the on-line LC–GLC technique (Biedermann *et al.*, 1993) involved transesterification of the oil with sodium methylate at room temperature followed by LC removal of the fatty acid methyl esters and separation of the triterpenes, methyl sterols, and sterols. An LC–GLC transfer procedure allowed analysis of individual classes (e.g. sterols and methyl sterols) from a single test portion of oil injected into the liquid chromatograph.

7.4.2 *Sterol dehydration products*

Instances of adulteration of expensive edible oils with cheaper oils from which sterols were removed to prevent their detection are increasing in the European market (Grob *et al.*, 1994b). Mariani *et al.* (1992) showed that under harsh conditions during refining (e.g. 150–160°C and 3–5% bleaching earth), sterols are dehydrated to steroidal hydrocarbons and adsorbed by the bleaching earth. Sterol esters are removed to a lesser extent. *Trans* fatty acids are also produced from the normal *cis* acids under these conditions. Lanzon *et al.* (1989) described a procedure for determining 3–100 ppm of stigmasta–3,5–diene (derived from β-sitosterol) in refined olive oils. The oil was saponified and extracted with hexane. The extract was cleaned up on a silica gel column and examined by GLC; 1% or less of refined oil in virgin olive oil could be detected. Cert *et al.* (1994) investigated the formation of stigmasta–3,5–diene during vegetable oil refining and demonstrated that the steroidal hydrocarbon was produced mainly during the bleaching step. Bleaching earth activity and temperature were the most important variables. Two methods were used to determine stigmasta–3,5–diene. The standard method involved saponification, solvent extraction, silica gel column chromatography (elution of a hexane fraction), and capillary GLC. An alternative method for oils with a high content of stigmasta–3,5–diene simply required silica gel column chromatography of the oil followed by capillary column GLC. The steroidal hydrocarbon, absent (<0.01 mg/kg) in virgin olive oil and crude seed oils, occurs at low concentrations (0.3–0.5 mg/kg) in crude olive pomace oils (caused by heating the pomace during a drying step).

Mariani *et al.* (1993) described a simple method for evaluating authenticity of olive oils with a single clean-up. Internal standards (3,5–cholestadiene, eicosanol, α-cholestanol, heptadecanoyl stearate, and cholesteryl heptadecanoate) were added to a test sample. The test portion was silylated and fractionated on a silicic acid column to sequentially obtain saturated hydrocarbon, steroidal hydrocarbon, squalene, and sterol–sterol ester–wax–tocopherol fractions, which were analyzed using capillary column GLC.

Although accurate determinations of steroidal hydrocarbons are difficult, Grob and Bronz (1994) noted that they are useful for; (i) determining stigmastadiene concentration to verify thermal treatment (refining) of an oil; and (ii) estimating steroidal hydrocarbon composition to check product identity and demonstrate admixture with desterolized (harshly refined) oil.

For the latter purpose, the concentration ratio of 3,5–stigmastadiene/3,5–campestadiene should be similar to that of β-sitosterol/campesterol, the parent sterols. However, accurate capillary column GLC determination of the 3,5–dienes following a preliminary LC–LC separation of test oil is hampered by the presence of interfering components, especially 2,4–dienes, which are not separated on apolar GLC stationary phases but are adequately separated on Carbowax-type phases (see Figure 7.6). The steroidal hydrocarbons present in refined edible vegetable oils were also determined by on-line LC–LC-GLC–MS (Grob *et al.*, 1994c). Schulte (1994) recommended an HPLC method for routine identification of refined oils involving preliminary separation of steroidal hydrocarbons by silica gel chromatography followed by HPLC analysis using an RP-18

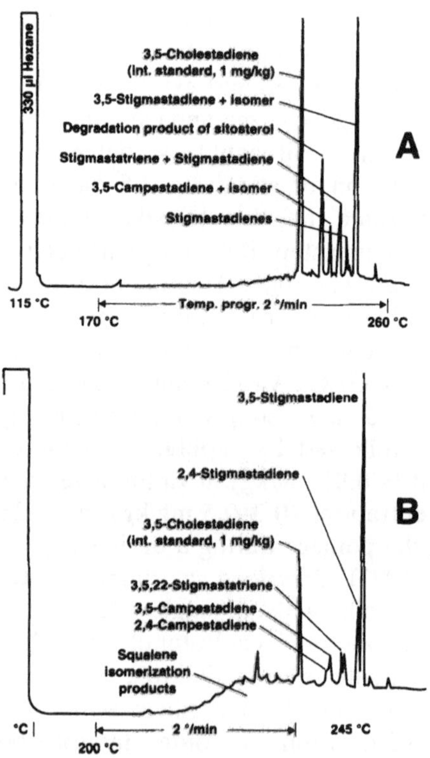

Figure 7.6 GLC chromatograms of (a) an HPLC fraction from refined olive oil containing 2,4- and 3,5-isomers of campestadiene and stigmastadiene using a 25 m × 0.25 mm internal diameter capillary column coated with a dimethylpolysiloxane stationary phase (2,4- and 3,5-dienes not resolved); (b) GLC chromatogram of the HPLC fraction from refined olive oil using a 30 m × 0.25 mm internal diameter capillary column coated with Superox 0.6 (polyethylene glycol) stationary phase (2,4- and 3,5-dienes fairly well separated). (Source: Grob and Bronz, 1994; redrawn by kind permission of the authors and La Rivista Italiana della Sostanze Grasse.

column with UV detection at 235 nm (wavelength typical of conjugated dienes).

7.4.3 *Tocopherols*

Vegetable oils contain up to 3000 mg/kg or more of tocopherols. These natural antioxidants are present as a mixture of isomers with one, two or three methyl groups on the phenol ring. A series of *unsaturated analogues, the tocotrienols*, are also present in some oils, especially palm oil. The series of α-, *β*-, γ- and δ-tocopherols and tocotrienols are collectively called tocols. The concentrations of various tocols vary from oil to oil and tocol composition is often a useful criterion of purity or evidence of the presence of one oil in another oil (Coors, 1991).

Measurements of individual tocols were initially carried out by colorimetric procedures (Emmerie and Engel, 1938, 1939; Harris *et al.*, 1950; Lange, 1950; Herting and Drury, 1963; Chow *et al.*, 1969; Muller-Mulot, 1976). More recently, GLC (Losi and Piretti, 1970; Carpenter *et al.*, 1976; Meijboom and Jongenotter, 1979; Lercker *et al.*, 1981; Mariani and Fedeli, 1982; Slover *et al.*, 1983) and normal phase HPLC (Thompson and Hatina, 1979; Cortesi and Fedeli 1980; Barnes and Taylor, 1980; Gertz and Hermann, 1982a, 1982b; Micali and Curro, 1984; Rammell and Hoogenboom, 1985; Syvaoja *et al.*, 1986; Indyk, 1988; Coors, 1991) have been used to separate and estimate tocopherol content. Govind Rao and Perkins (1972) isolated the tocols from vegetable oil unsaponifiable matter and identified and estimated the levels of individual tocols by means of GLC–MS of the TMS (trimethylsilyl ether) derivatives.

To prepare for HPLC separation, the oil is usually simply dissolved in hexane or mobile phase. Fluorescence is generally preferred over UV detection because of its greater selectivity and sensitivity. *β*- and γ-tocopherols are well separated on normal-phase (silica) columns but inadequately separated on reversed-phase columns. Richheimer *et al.* (1994) reported the separation of all four tocopherols using a commercially available pentafluorophenyl (PFP) column. A 25 cm × 4.6 mm Taxsil, PFP reversed-phase column with 5 μm spherical particles (Meta Chem Technologies, Torrance, CA) was preferred with a methanol:water mobile phase at a ratio of 92:8.

Evaporative light-scattering detection was recommended for HPLC of phytosterols and tocopherols in vegetable oils (Warner and Mounts, 1990). Tocopherols have also been examined by packed and capillary column supercritical fluid chromatography (Perrin and Prevot, 1988; Snyder *et al.*, 1993). HPLC has been applied to simultaneous determination of tocopherols, phenolic food antioxidants, and triglycerides (Andrikopoulos *et al.*, 1991); whereas capillary column GLC has been applied to simultaneous determination of tocopherols and sterols (Slover *et al.*, 1983).

Several standards methods are available for determining tocopherols (AOCS, 1987h; IUPAC, 1987d, 1992b; ISO, 1991). IUPAC method 2.411 (IUPAC, 1987d) describes a procedure for isolating tocopherols (extraction of unsaponifiables and TLC fractionation) plus three procedures for their determination (colorimetric determination with 2,2′-dipyridyl reagent, packed-column GLC, capillary-column GLC). The other methods (AOCS, 1987h; IUPAC, 1992b; ISO, 1991) are based on a validated HPLC procedure (Pocklington and Dieffenbacher, 1988) using a standard 5 μm particle size silica column with either UV or fluorescence detection. The order of elution of tocopherols and tocotrienols is shown in Figure 7.7; levels of tocopherols and tocotrienols in vegetable oils are shown in Table 7.7. The predominant tocol in olive oil is α-tocopherol; peanut oil, corn oil, cottonseed oil, and soybean oil, often used as olive oil adulterants, have significant levels of γ-tocopherol. Corn, palm kernel and palm oils contain tocotrienols, which are absent or present at very low levels in other common edible oils.

7.5 Triglycerides

Triglycerides composition is used increasingly in the food industry to confirm authenticity, detect adulteration, and to define the composition of mixtures. Analyses are carried out using high-temperature GLC (Karleskind *et al.*, 1974; Hamilton, 1975; Christie, 1989; Hammond, 1993); or reversed-phase HPLC (Pei *et al.*, 1975; Parris, 1978; Phillips *et al.*, 1984; Hamilton and Rossell, 1986; Christie, 1987; Shukla, 1988). Tallent *et al.* (1966) devised a procedure for triglyceride analysis involving pancreatic

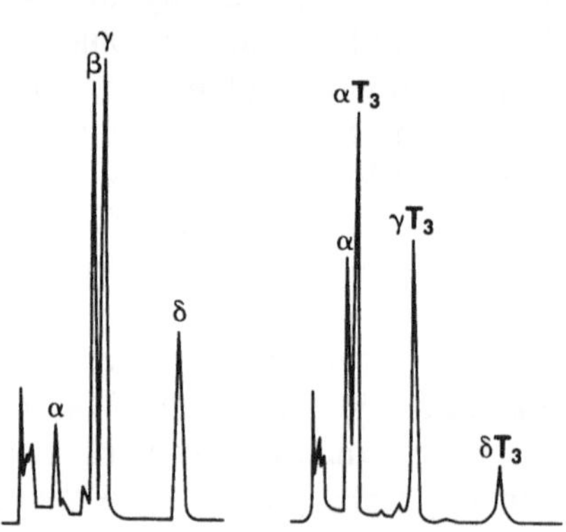

Figure 7.7 HPLC chromatogram showing the elution order of tocopherols and tocotrienols using a 250 mm × 4 mm internal diameter column packed with 5 μm diameter microparticulate silica (AOCS, 1987h). (Redrawn by kind permission of the American Oil Chemists' Society).

lipase hydrolysis followed by methylation, silylation, and temperature-programmed GLC of the hydrolysates. The method described by Brockerhoff (1965) for stereospecific analysis of triglycerides was used to determine the positional distribution of fatty acids in the triglycerides of several vegetable oils and fats (Brockerhoff and Yurkowski, 1966) (Table 7.8). Pancreatic lipase studies showed that saturated fatty acids were present almost entirely in the 1– and 3–positions, whereas linoleic acid tended to favor the 2–position.

El-Hamdy and Perkins (1981) studied the separation of critical pairs, that is, triglycerides exhibiting identical behavior on reversed-phase HPLC columns. They introduced the equivalent carbon number (ECN) to define critical pairs, where:

$$\mathrm{ECN} = \mathrm{CN} - 2n, \qquad (7.1)$$

where CN = actual carbon number, and n = number of double bonds per molecule. Thus, stearodiolein (SOO; 54:2) and stearooleopalmitin (SOP; 52:1) are critical pairs with stearodipalmitin (SPP; 50:0). They also developed the concept of theoretical carbon number (TCN) to define the separation of unsaturated triglycerides calculated from the carbon number, and experimentally determined factors for the fatty acid components. Determination of calculated and experimental TCN values was useful for identifying individual triglycerides, as demonstrated by Phillips *et al.* (1984), whose quantitative analyses of vegetable oil triglycerides involved lipase hydrolysis and reversed-phase HPLC.

Podlaha and Toregard (1982) devised a graphical system for tentative identification of triglyceride peaks in reversed-phase HPLC chromatograms based on equivalent carbon numbers. Plattner (1981) evaluated several different reversed-phase columns and suggested adding silver ion to the elution solvent to alter triglyceride selectivity. Shukla *et al.* (1983) reported use of a simple, direct HPLC method for separation of cocoa butter triglycerides, using UV detection at 220 nm. Perrin *et al.* (1987) determined the triglyceride composition of blackcurrant seed oil by reversed-phase HPLC with a light-scattering detector and gradient elution plus GLC to determine the fatty acid composition of the oil and certain HPLC fractions. Neff *et al.* (1994) determined the triglyceride composition of several vegetable oils by silver ion HPLC with flame ionization detection; Kapseu *et al.* (1992) used reversed-phase HPLC to determine the triglyceride composition of corn and peanut oils. Ouedraogo and Bezard (1982) determined the composition of cottonseed oil triglycerides by isolating the triglycerides by silicic acid column chomatography and separating them by silver ion TLC, pancreatic lipase hydrolysis of the triglyceride fractions, additional TLC fractionation of the mono- and diglycerides, and finally, GLC of isolated fatty acids as the butyl esters and diglycerides as the acetates. Singleton and Pattee (1987) characterized peanut oil triglycerides

Table 7.7 Levels of tocopherols (T) and tocotrienols (T_3) in vegetable oils (mg/kg)[a]

Oil/fat	α-T	β-T	γ-T	δ-T	α-T_3	β-T_3	γ-T_3	δ-T_3	Total
Apricot kernel	5–10		170–224	3–20					200–232
Avocado pulp	64–100		0–19						83–100
Babassu					25–46		32–80	9–10	67–128
Barley	13–23	2–3	1–3		440–590	80–150	70–90		800–1500
Blackcurrant	320	8	647	68					1043
Borage	0–46		33–272	690–1013					732–1111
Cocoa butter	1–19	0–10	18–196	0–17					25–220
Coconut	0–17	0–11	0–14		0–44		0–1		0–44
Corn	23–573	0–356	268–2468	23–75	0–239		0–450	0–20	331–3716
Cottonseed	136–674	0–29	138–746	0–21					389–1185
Cupu assu			122	6					128
Evening primrose	76–356		187–358	0–19					263–661
Grapeseed	16–38	9–89	0–73	0–4	18–107		115–205	0–3	240–405
Linseed	5		430–575	4–8					430–588
Lupu			78	8–31					86
Mustard seed	75–138		308–494	0–31					446–663
Niger seed	620–820	6–9	20–40						657–853
Oat	33	11			89	21	12	9	175
Okra seed	280–450		420–680						700–1130

Olive	63–135		7–15						70–150
Palm	4–193	0–234	0–526	0–123	4–336		14–710	0–377	141–1465
Palm kernel	0–44	0–248	0–257				0–60		0–257
Peach kernel	65	13	10						88
Peanut	49–373	0–41	88–389	0–22					176–1291
Pecan nut	210		210						420
Rapeseed	116	34	737	275					1165
Rapeseed (low erucic; Canola)	100–386	0–140	189–753	0–22					438–2680
Rice bran	576		324						900
Safflower seed	234–660	0–17	0–12				0–12		246–664
Safflower seed (high oleic)	480–600								480–600
Sesame seed	0–3	0–36	521–983	4–21			0–20		531–1003
Soybean	9–352	0–45	89–2307	154–932					601–3363
Sunflower seed	403–935		0–34	0–7					447–1514
Sunflower seed (high oleic)	94–96	3–4	0.5–2						98–102
Walnut	10–20		263–400	46–60					309–455
Wheat germ	880–1800	250–1150	175–700	0–100	<10–200	<10–200			1400–2400

[a]Figures in *Codex Alimentarius* Commission (1993a) Proposed Draft Standard for Named Vegetable oils, Appendix 1 (CX 1193/16) and other souces in the general literature.

Table 7.8 Positional distribution of fatty acids (mol%) in several vegetable oils and fats

Oil/fat	Position	16:0	18:0	18:1	18:2	18:3	20:0	20:1	22–24
Cocoa butter	1	34	50	12	1		1		
	2	2	2	87	9				
	3	37	53	9	0.4		2		
Corn	1	18	3	28	50	1			
	2	2	0.2	27	70	1			
	3	14	3	31	52	1			
Olive	1	13	3	72	10	0.6			
	2	1		82	14	1			
	3	17	4	74	5	1			
Peanut	1	14	5	59	19		1	1	2
	2	2	0.3	59	39			0.3	1
	3	11	5	57	10		4	3	9
Rapeseed	1	4	2	23	11	6		16	36[b]
	2	1		37	36	20		2	4[b]
	3	4	3	17	4	3		17	52[b]
Soybean	1	14	6	23	48	9			
	2	1	0.3	22	70	7			
	3	13	6	28	45	8			

[a]Source: Brockerhoff and Yurkowski (1966).
[b]Almost entirely 22:1 (erucic acid).

using reversed-phase HPLC, GLC, and electron-impact mass spectrometry. Each HPLC fraction was analyzed for carbon number, fatty acid composition, and mass fragmentation pattern.

The triglyceride compositions of several important vegetable oils are shown in Table 7.9. High-molecular-weight triglycerides are characteristic of peanut oil. Triolein (OOO) is a major triglyceride of olive oil and of an African peanut oil shown in the table. Sempore and Bezard (1991) observed that the major triglyceride of an African peanut oil from one locale was triolein; from another locale it was 2-oleolinoleoolein (OLO). Cocoa butter contains mainly three triglycerides, 2-oleodipalmitin (POP), 2-oleopalmitostearin (SOP) and 2-oleodistearin (SOS), whereas olive oil contains OOO, 1-palmitodiolein (POO) and 1-linoleodiolein (LOO). The very low level (≤0.5%) of trilinolein (LLL) in olive oil is substantiated by analysis of authentic olive oils from the major producing Mediterranean countries (Flor *et al.*, 1993). HPLC patterns of virgin olive oil and 5% and 25% soybean oil in olive oil (Kapoulas and Andrikopoulos, 1986) are shown in Figure 7.8. ECN 38 and 40 (1-linoleodilinolenin (LLnLn) and 1-linilenodilinolein (LnLL) increase as the level of soybean oil and ECN 42 (LLL) increase. Olive oil that is adulterated with 1–2% of linoleic acid-rich seed oil is readily identified using reversed-phase HPLC.

Triglyceride analysis, whether by GLC, HPLC, or MS (Hites, 1970), is useful for identifying esterified or interesterified oils and fats in which the fatty acid groups are chemically scrambled among the three glycerol positions. For example, esterified oils produced by reesterification of olive

Table 7.9 Triglyceride of several important vegetables oils (% of total triglycerides)

TG[a]	Cocoa butter	Corn	Cottonseed	Olive	Peanut	Soybean	Sunflower
PPS, PSS, PPP, SSS	6	2	8			4	1
LLO		22	10	4	8	21[b]	23
PLL, SLL		9	27	2	2	16	21
OLO		19	3		19		
PLP, SLP, SLS	6	2	8	2	1	4	1
OOO		9	1	34	27	3	1
POO, SOO	5	5	3	28	18	4	1
LOO			3	14		9	6
POP, SOP, SOS	85[c]	3	3	3	3	1	
LOP, LOS			8	11		12	8
LOL			3				
LLL		8	14	0.5	1	21[d]	36
LnLnL						1	
PLO		15	8		9		
PLS		5	1				
ALO, AOS	2				2		
BLO, BOP					4		
LgOO, GOO, BOO					3		
Reference	e	f	g	h	f	h	h

[a]TG = triglyceride; P = palmitic; S = stearic; O = oleic; L = linoleic; Ln = linolenic; A = arachidic (20:0); G = gondoic (20:1); B = behenic (22:0); and Lg = lignoceric (24:0) acid, respectively, attached to the glycerol molecule. [b]LLO, LnOO, LnOP. [c]POP, 15; POS, 40; SOS, 29. [d]LLL, LnLO, LnLP. [e]Shukla *et al.* (1983). [f]Kapseu *et al.* (1992). [g]Ouedraogo and Bezard (1982). [h]Neff *et al.* (1994).

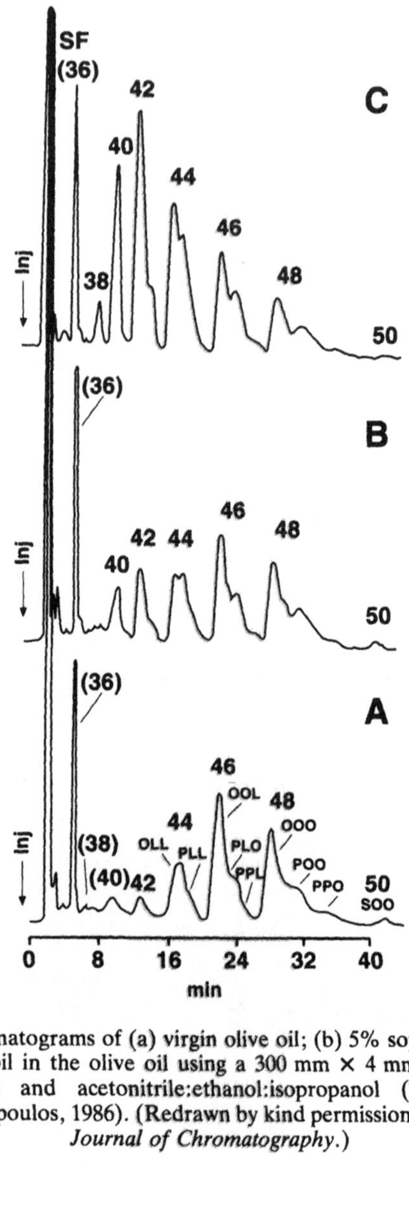

Figure 7.8 HPLC chromatograms of (a) virgin olive oil; (b) 5% soybean oil in the olive oil; and (c) 25% soybean oil in the olive oil using a 300 mm × 4 mm internal diameter. C18 bonded-phase packing and acetonitrile:ethanol:isopropanol (78:20:2) mobile phase (Kapoulas and Andrikopoulos, 1986). (Redrawn by kind permission of the authors and of the *Journal of Chromatography*.)

oil or olive-pomace oil containing high percentages of free fatty acids, or by esterification of glycerol with distilled fatty acids, can be identified by analysis of the triglycerides (Gegiou and Georgouli, 1983) or by determination of the fatty acids in the 2-position of the triglycerides (Gegiou and Georgouli, 1980). The specific distribution of fatty acids in several vegetable oil triglycerides (Tiscornia and Bertini, 1974) is presented in Table 7.10.

Standard methods have been validated for determining triglycerides by GLC (IUPAC, 1987e; AOCS, 1987i; Pocklington and Hautfenne, 1985) and by HPLC (IUPAC, 1992c; AOCS, 1987j; Wolff *et al.*, 1991). The IUPAC method 2.324 (AOCS method Ce5b-89) was developed for HPLC separation and determination of triglycerides in vegetable oils such as soybean, olive, palm and peanut oils, in terms of their molecular weight and degree of unsaturation as a function of their equivalent carbon number (Figure 7.9). Conditions are specified for use of a UV detector (220 nm) and a mass detector with 3 μm, C_{18}-bonded, reversed-phase column systems. IUPAC method 2.325 (AOCS method Ce5c-93) involves HPLC separation and determination of individual triglycerides with a differential refractometer or UV or mass detector. Figure 7.10 shows an HPLC chromatogram (AOCS, 1987j) of a blend of vegetable oils fortified with tripalmitin (PPP) and tristearin (SSS).

The standard method for determining fatty acids in the 2–position of oil and fat triglycerides (IUPAC, 1987f; AOCS, 1987k) consists of neutralization of acid oils (acidity above 3%) in a solvent, purification by passing through an alumina column, partial hydrolysis of the triglycerides by pancreatic lipase, separation of the formed 2-monoglycerides by TLC, methanolysis of the monoglycerides, and determination of the methyl esters by GLC. The total fatty acid composition of the same test sample is determined and compared with that of the fatty acids in the 2-position.

7.6 Data analysis techniques

Complex mixtures are calculated by focusing on one or two significant variables. Analysis of blends of complex mixtures such as vegetable oils is especially difficult when individual differences in oils are similar to or less than natural variations of each oil. The problem can be addressed by using multivariate statistics (Elliott *et al.*, 1971). Multivariate data analyses (chemometric methods), including discriminant, cluster, principal component, and canonical variate analyses used in food research, are discussed in a volume edited by Martens and Russwurm (1983). The use of chemometrics to validate food composition data was discussed by Feinberg and Bugner (1989).

The geographical origin of olive oils is determined by multivariate classification methods based on fatty acid profile (Tiscornia *et al.*, 1982; Forina and Tiscornia, 1982; Forina *et al.*, 1983; Eddib and Nickless, 1987; Forcadell *et al.*, 1988; Alonso and Aparicio, 1983; Tsimidou and Karakostas, 1993). Tsimidou *et al.* (1987a) analyzed Greek virgin olive oils by GLC for fatty acid profile and by HPLC for triglyceride composition. The data obtained were analyzed statistically; principal component analysis was used to classify the oils on the basis of olive variety, year of

Table 7.10 Fatty acid composition (%) in the triglycerides and in the 2-position of triglycerides of vegetable oils[a]

Oil	16:0	18:0	18:1	18:2	18:3	22:0	22:1
Rapeseed							
Triglycerides	3	1	13	13	9		50
2-Monoglycerides	1	0.3	36	38	21		2
Rapeseed low erucic							
Triglycerides	4	2	63	19	10		
2-Monoglycerides	0.5	0.2	57	30	12		
Peanut							
Triglycerides	10	3	55	23		3	
2-Monoglycerides	2	0.4	60	33		trace	
Soybean							
Triglycerides	11	4	23	53	8		
2-Monoglycerides	0.7	0.2	27	67	5		
Grapeseed							
Triglycerides	8	4	18	68	0.5		
2-Monoglycerides	0.6	0.2	26	73	0.3		
Olive							
Triglycerides	11	2.5	78	8	0.5		
2-Monoglycerides	1	0.1	90	7	0.4		
Olive pomace							
Triglycerides	10	2.5	76	9	0.5		
2-Monoglycerides	1	0.1	88	9	0.5		
Sesame							
Triglycerides	10	5	42	41	1		
2-Monoglycerides	1	0.4	50	48	0.2		
Safflower							
Triglycerides	8	4	18	69	0.3		
2-Monoglycerides	1	0.3	26	72	trace		
Safflower, high oleic							
Triglycerides	6	2	80	11			
2-Monoglycerides	1	0.2	86	13			
Sunflower							
Triglycerides	7	4	22	65			
2-Monoglycerides	0.6	0.3	24	76			
Corn							
Triglycerides	14	2	36	46	1		
2-Monoglycerides	2	0.3	46	51	0.5		

[a]Source: Tiscornia and Bertini (1974).

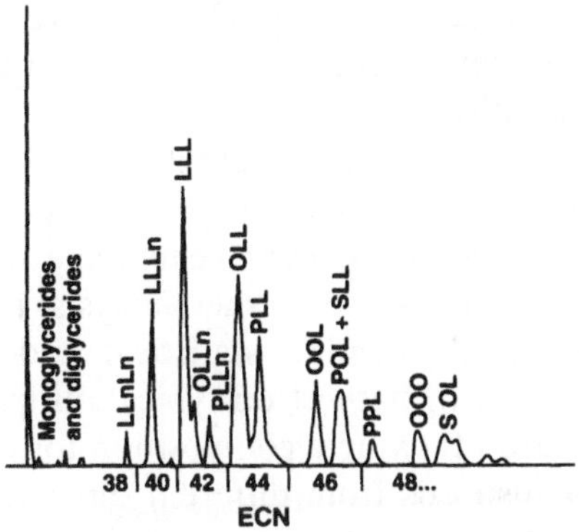

Figure 7.9 HPLC chromatogram of soybean oil triglycerides using a 250 mm × 4.5 mm internal diameter reversed-phase column packed with 5 μm diameter microparticulate silica coated with 22–23% octadecylsilane (AOCS, 1987j; Ce 5b-89). L = linoleic acid; Ln = linolenic acid; O = oleic acid; P = palmitic acid; S = stearic acid; thus OOO represents triolein. (Redrawn by kind permission of the American Oil Chemists' Society.)

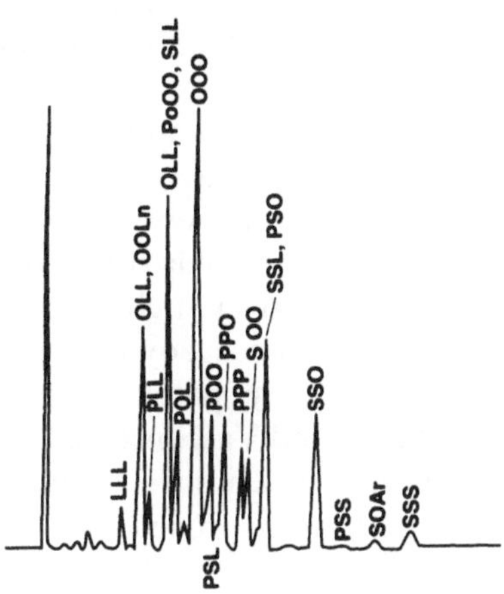

Figure 7.10 HPLC chromatogram of a vegetable oil blend fortified with tripalmitin and tristearin using a 250 mm × 4.5 mm reversed-phase column packed with a 5 μm diameter micro-particulate silica coated with 22–23% octadecylsilane (AOCS, 1987j; Ce 5c-93). Ar = arachidic acid; L = linoleic acid; Ln = linolenic acid; O = oleic acid; P = palmitic acid; Po = palmitoleic acid; S = stearic acid; thus POL represents palmitooleolinolein. (Redrawn by kind permission of the American Oil Chemists' Society.)

harvest, and geographical origin. The method relies on a 'reference set' of olive oils, which is upgraded continuously. HPLC-triglyceride profiles provided detailed fingerprints of the oils studied and gave a higher level of discrimination than that obtained by GLC of the fatty acid methyl esters. Greek virgin olive oils were evaluated further using principal component analysis of triglyceride and fatty acid profiles to study adulteration of the oils with seed oils (Tsimidou *et al.*, 1987b). Triglyceride profiles readily allowed identification of cottonseed, corn, rapeseed, soybean, or sunflower oil at 20% levels of addition.

Forina *et al.* (1983b) examined fatty acid data for Portuguese olive oil of a 4-year period. Two categories (South Portugal and Douro Valley, North Portugal) were characterized using several chemometric methods: principal component analysis; multivariate Bayesian analysis; and soft independent modeling of class analogy. A chemical–statistical model was constructed to measure the effect of year of harvest on typical oil quality.

Lopez *et al.* (1986) used fatty acid composition for discriminant analysis of olive oils to distinguish oils from different olive varieties. Rojas *et al.* (1994) demonstrated the utility of chemometric methods for identifying new sources of oleaginous plants and for classifying botanical species. Morales and Aparicio (1993) characterized various European virgin olive oils, using volatile components quantified by a dynamic headspace GLC method. The GLC data were subjected to discriminant and principal component analyses. Individual varieties were distinguished with a set of only six compounds from a set of 63 GLC peaks used initially.

Sterols were used by Paganuzzi (1986; 1987) to determine the area and country of origin of virgin olive oils by multivariate methods. Leardi and Paganuzzi (1987) used multivariate methods to characterize the origin (Greece, Spain or Tunisia) of virgin olive oils by analyzing sterol data. They concluded that multivariate analysis of sterol plus fatty acid data is a potent tool for determining the geographic origin of virgin olive oils.

Triglyceride profile data obtained by high-temperature capillary GLC for corn, grapeseed, olive, olive pomace, soybean, and sunflower oils were used to identify individual oils by pattern analysis and to confirm adulteration (Chaves Das Neves and Vasconcelos, 1989). POP, OOO, and POO were used to distinguish between olive oil and olive pomace oil. De Jong and de Jonge (1991) used two approaches to identify component oils in an unknown blend. Fatty acid data were used both for multiple regression analysis with a computer program ROBUST (Regression of Blends Using SAS Tools) and for mathematical programming with a technique known as CABLE (computer-assisted blend estimation) using linear programming techniques to produce a single optimal solution. The CABLE program, which is faster and more efficient than ROBUST, can consider up to 40 possible blend components. However, the best-fitting solution does not always match the true blend composition because of natural variations in fatty acid composition. Additional data from sterol and tocopherol analysis would significantly increase the percentage of accurate blend identifications.

Van Niekerk and Burger (1985) used the fatty acid, sterol, and tocopherol contents of vegetable oils to determine the composition of oil mixtures, using a weighted least-squares estimator with backwards elimination. They later refined the technique by applying a simplex optimization procedure to avoid the use of assumed values (Van Niekerk and Hasty, 1989). Abu-Hadeed and Kotb (1988) used a matrix inversion method based on data from GLC determination of fatty acids, sterols, methylsterols, triterpene alcohols, tocopherols, and squalene to identify and quantitate the component oils in vegetable oil blends. A computer program in BASIC was used to automate the calculations and print out results. Mixtures containing two, three, and six components of corn, soybean, cottonseed, palm, palm kernel, sesame, coconut, and olive oil were prepared. The system identified successfully the components of all blends tested and was more than 80% accurate in quantifying the components of the mixtures.

Bergquist and Kaufmann (1993) used multivariate optimization of a reversed-phase HPLC procedure to analyze triglycerides in several vegetable oils plus fish oil and tallow. Schwaiger and Vojir (1994) applied multivariate statistical methods to determine the authenticity of several edible plant oils based on the content of palmitic, stearic, oleic and linoleic acids. Sato (1994) applied principal component analysis to near-infrared (NIR) spectroscopic data of vegetable oils to classify correctly the oils and identify adulteration. Lai *et al.* (1994) used Fourier transform infrared (FTIR) spectroscopy with principal component analysis and discriminant analysis to investigate the potential of the FTIR technique for determining vegetable oil authenticity. Safer *et al.* (1994) used the combination of ATR and FTIR to characterize edible oils with principal component analysis of spectra in the mid-range of 4000–1560/cm. Bewig *et al.* (1994) carried out discriminant analysis of four vegetable oil types (cottonseed, peanut, soybean, and canola) using NIR reflectance spectroscopy to provide a qualitative method to classify oils and differentiate vegetable oil types. Second derivative spectra of 46 vegetable oils were subjected to discriminant analysis with Mahalanobis distance principles. A four-wavelength (1704, 1802, 1816 and 2110 nm) equation giving the best separation of the four oil types was derived. These four wavelengths were adequate for discriminating between oil groups. The Mahalanobis distances revealed adequate separation of the oil groups, with one exception. Observation of analyte locations on a log–log plot (Figure 7.11) showed absorption of one soybean oil in the region of the peanut oil group. The overall results, however, indicated that discriminant analysis could be used successfully to differentiate vegetable oil types.

Goodacre *et al.* (1992; 1993) used a combination of Curie-point (pyrolysis), mass spectrometric (Irwin, 1982), and multivariate data analysis with artificial neural networks (Hertz *et al.*, 1991) to identify virgin olive oils adulterated with seed oils. The method rapidly and easily identifies virgin olive oil contaminated with 5% corn, peanut, soyean, sunflower, or refined olive oil. The huge discriminating power and

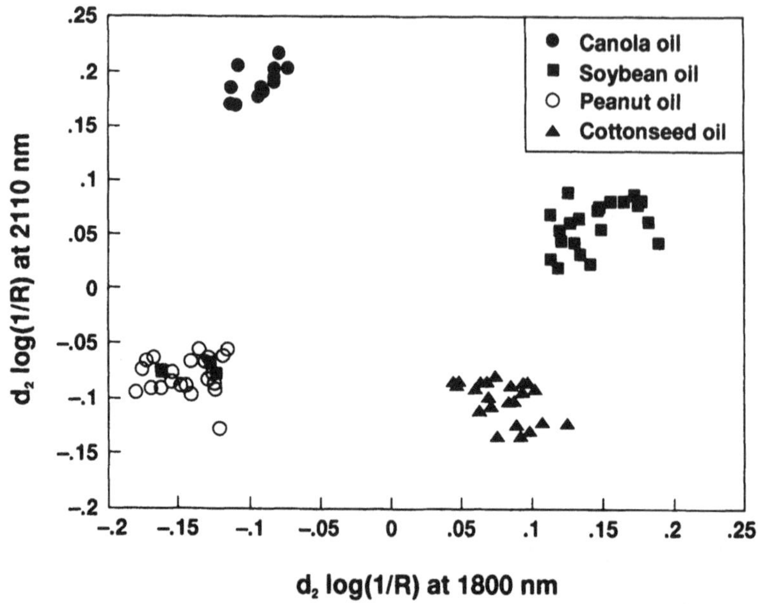

Figure 7.11 Second-derivative spectral data points of four vegetable oil groups (Canola, soybean, peanut and cottonseed) analysed in triplicate and plotted at wavelengths 2110 nm *versus* 1800 nm (R = reflectance). One soybean oil (in triplicate) is located in the peanut oil grouping (Bewig *et al.*, 1994). (Redrawn with kind permission of the authors and of the American Oil Chemists' Society.)

automation of pyrolysis mass spectrometry and the use of neural network analysis provide a powerful approach to assessment of food adulteration (Goodacre *et al.*, 1994).

7.7 Stable carbon isotope ratio analysis (SCIRA)

Measurement of relative abundance of the naturally occurring stable carbon isotopes ^{12}C and ^{13}C is useful for verifying the authenticity of apple juice (Doner *et al.*, 1980; Doner and Phillips, 1981; Brause and Raterman, 1982), orange juice (Brause *et al.*, 1984), and other foods (Gaffney *et al.*, 1979; Winkler and Schmidt, 1980). Winkler (1984) reviewed the general application of natural abundance stable isotope mass spectrometry in food control.

SCIRA is applicable because the ^{13}C:^{12}C ratio of plant materials depends upon the photosynthetic pathway used by individual plants (Winkler, 1984; Pollard, 1993). Almost all land plants use the C_3 pathway (normal or Calvin cycle) to incorporate CO_2 by carboxylation of ribulose diphosphate to phosphoglycerate (C_3 product). A few plants such as corn, millet, sorghum, sugarcane, and some tropical grasses, have evolved the alternative C_4 (Hatch-Slack) pathway involving carboxylation of phosphoenolpyruvate to oxalacetate (C_4 product). A much smaller group of plants (pineapple and cacti) use the crassulacean acid metabolic (CAM) pathway, which is capable of C_3 or C_4 CO_2 fixation (Winkler, 1984).

SCIRA is carried out by combustion of the test portion to CO_2 and measurement of ^{13}C:^{12}C ratio, with an isotope ratio mass spectrometer or other instrument able to determine the ^{13}C and ^{12}C content of CO_2 (Bricout *et al.*, 1974; Winkler and Schmidt, 1980; Doner and Phillips, 1981). Values of $\delta^{13}C$/ml (‰) are reported from comparisons with CO_2 generated from a reference standard (Pee Dee belemnite – PDB).

Gaffney *et al.* (1979) observed that $\delta^{13}C$ values for beef (-13.1 ± 1.6) and pork (-12.5 ± 0.4) were consistent with the heavy use of corn and sorghum (C_4 plants) as livestock feeds in the US ($\delta^{13}C$ value for corn oil was found to be -12.4 ± 0.6 *versus* coconut, palm, peanut, and soybean oils at −27 to −28). See Table 7.11 for the stable isotope ratios of various vegetable oils.

As the fatty acid compositions of several other vegetable oils are similar or equal to that of corn oil, and the large concentration of sterols in corn oil can swamp the sterol composition of blends, Rossell (1994) studied the SCIRA technique for determining corn oil authenticity. SCIRA values were obtained by combusting oils to CO_2, purifying the CO_2 using GLC, and determining the mass distribution in the CO_2 by isotope mass spectrometry. Stable isotope ratios of corn oil from various geographical

Table 7.11 Stable isotope ratios of vegetable oils

Type	Stable isotope ratio ($\delta^{13}C$ in ‰)	Reference[a]
Alm	−28.3	A
Avocado	−30.6	A
Cocoa butter	−31.9	A
Corn (maize)	−15.0	B
Cottonseed	−27.8	B
Linseed	−26.2	A
Olive	−28.7	A
Palm	−29.6	B
Palm olein	−29.7	B
Palm kernel	−29.5	B
Peanut	−27.9	B
Rapeseed	−28.6	B
Safflower	−28.9	B
Sesame	−27.9	B
Soybean	−30.1	B
Sunflower	−26.0	B
Wheat germ	−26.0	A

[a]A = Winkler and Schmidt (1980); B = Rossell (1994).

areas (Asia, Africa, Australia and New Zealand, Europe, North America, South America) ranged from −13.87 to −16.36, an average (of 42 samples) of −14.95. SCIRA identified the presence of as little as 10% of a foreign oil in corn oil. Rossell (1994) has continued this approach by applying statistical techniques to SCIRA data to improve the sensitivity for identifying other oils in corn oil. A combination of SCIRA and iodine value holds promise for determining about 5% of other liquid vegetable oils in corn oil.

7.8 Olive oil

Internationally recognized definitions and purity criteria have been promulgated by the International Olive Oil Council (IOOC, 1994b) and the Codex Alimentarius Commission (1993b). Virgin olive oil is defined as 'oil obtained from the fruit of the olive tree (*Olea europaea*) solely by mechanical or other physical means under conditions that do not cause alterations in the oil, and which has not undergone any treatment other than washing, decantation, centrifugation, and filtration'. Extra virgin olive oil has an organoleptic rating of 6.5 or more (on a scale of 1–10) and no more than 1% free acidity, expressed as oleic acid. The product known simply as olive oil is a blend of refined olive oil and virgin olive oil. Olive pomace oil is obtained by treating olive pomace (olive cake remaining after mechanical extraction of oil and containing residual oil) by refining methods (e.g. solvent extraction, degumming, alkali neutralization, bleaching, and deodorization) that do not lead to alterations in the initial glyceride structure. Identity and purity characteristics of olive oil products specified in the IOOC and Codex standards include fatty acid and sterol composition, maximum acceptable level of saturated fatty acids (16:0 and 18:0) in the 2–position of the triglycerides (≤1.5% for virgin olive oil; ≤1.8% for refined olive oil and olive oil; 2.2% for olive pomace oil), maximum acceptable level of *trans* fatty acids, and maximum permitted level of stigmastadiene (≤0.15 ppm stigmastadiene for edible virgin olive oil; ≤50 ppm for refined olive oil and olive oil). Criteria for sterol and *trans* fatty acid content are shown in Tables 7.12 and 7.13, respectively. Added α-tocopherol is permitted in refined olive and olive pomace oil to replace tocopherols lost during refining. A maximum of 200 mg/kg of total α-tocopherol is permitted in the final product. An AOCS Recommended Practice (AOCS, 1987l) lists international trade standards applicable to olive and olive pomace oils and includes a table listing methods of analysis applicable to these products.

Firestone *et al.* (1985; 1988) analyzed olive oil products from commercial sources for fatty acid and sterol composition to identify foreign vegetable oils, for triterpene diol (erythrodiol) content to identify undeclared olive

Table 7.12 Purity criteria for sterols in olive and olive pomace oils[a]

	Total sterols (%)	Total sterols (mg/100g)
Cholesterol	≤0.5	
Brassicasterol	≤0.1	
Campesterol	≤4.0	
Stigmasterol	<Campesterol in edible olive oils	
β-Sitosterol (true)	≥75.0	
Δ^7-Stigmastenol	≤0.5	
β-Sitosterol + Δ^5-avenasterol + $\Delta^{5,23}$ stigmastadienol + cholesterol + sitostanol + $\Delta^{5,24}$-stigmastadienol	≤93.0	
Virgin olive oil, olive oil, refined olive oil		≥100
Crude olive pomace oil		≥250
Refined olive pomace oil		≥180
Olive pomace oil		≥180

[a]Sources: *Codex Alimentarius* Commission (1993b); IOOC (1994b).

Table 7.13 Purity criteria for *trans* fatty acids in olive oil products[a]

	18:1 *trans* (%)	18:2 + 18:3 *trans* (%)
Edible virgin olive oils	<0.03	<0.03[b]
Refined olive oil	≤0.20	≤0.30
Olive oil	≤0.20	≤0.30
Crude oil pomace oil	≤0.20	≤0.10
Refined olive pomace oil	≤0.40	≤0.35
Olive pomace oil	≤0.40	≤0.35

[a]Sources: *Codex Alimentarius* Commission (1993b); [b]Provisional (tentative) limit. IOOC (1994b).

pomace oil, and for saturated fatty acids in the 2–position of the triglycerides to identify undeclared esterfied oil. GLC chromatograms of the sterol fraction from virgin olive oil, soybean oil, and a blend of 15% soybean oil and 85% virgin olive oil are shown in Figure 7.12. The presence of soybean oil in the blend is indicated by the presence of relatively high and approximately equal levels of campesterol and stigmasterol (peaks 4 and 5) in the sterol fraction. UV absorptivity at 232 and 268 nm and acidity were determined to confirm the identity of virgin olive oils. Undeclared artificial colors were identified by spectrophotometry at 425–750 nm.

HPLC is used increasingly for the analysis of olive oil (Cozzoli *et al.*, 1991; Cortesi *et al.*, 1992; Cortesi, 1993). Reversed-phase HPLC triglyceride analysis (Kapoulas and Andrikopoulos, 1986; Fellat-Zarrouck *et al.*, 1988) can determine 1–2% of linoleic acid-rich oils (e.g. corn, cottonseed, soybean, sunflower oils). However, Salivaras and McCurdy

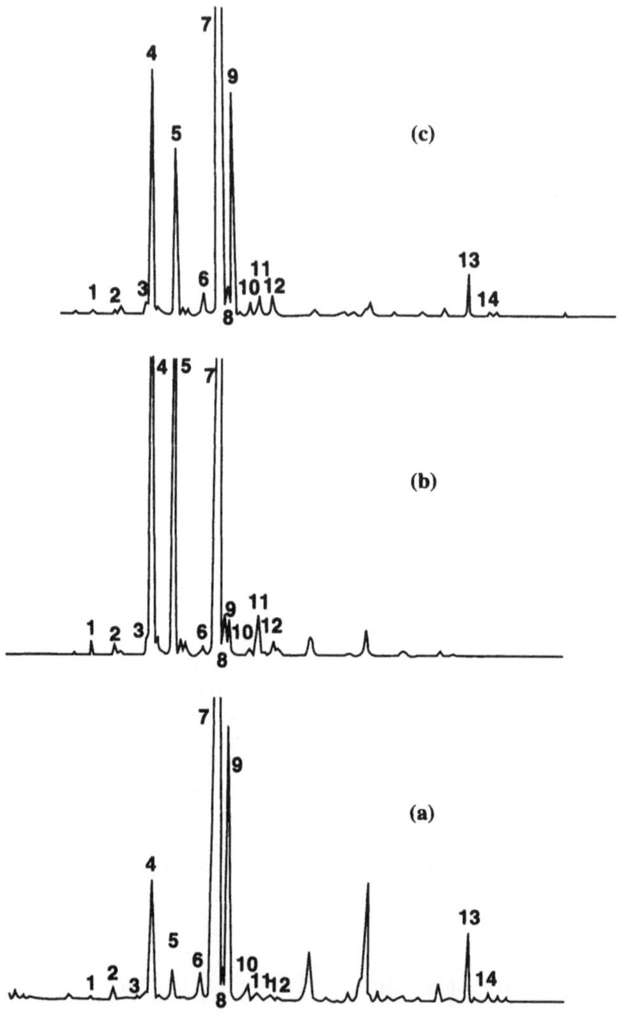

Figure 7.12 GLC chromatogram of the sterol fraction (acetate derivatives) of (a) extra virgin olive oil; (b) soybean oil; and (c) blend of 15% soybean oil and 85% extra virgin olive oil (GLC column and operating conditions as described for Figure 7.5). 1 = cholesterol; 2 = brassicasterol; 3 = 24–methylene cholesterol; 4 = campesterol; 5 = stigmasterol; 6 = clerosterol; 7 = β-sitosterol; 8 = sitostanol; 9 = Δ^5-avenasterol; 10 = $\Delta^{5,24}$-stigmastadienol; 11 = Δ^7-stigmastenol; 12 = Δ^7-avenasterol; 13 = erythrodiol; 14 = uvaol.

(1992) reported that olive oils that were adulterated with less than 7.5% canola oil could not be identified, and the trilinolein contents of some oils (e.g. hazelnut, safflower, sunflower oils) may occasionally be too low to permit identification without additional tests (Proto, 1992).

Since the rule governing distribution of fatty acids in the glycerol molecule is known and the triglyceride composition of vegetable oils including olive oil can be calculated from fatty acid composition, HPLC data may be compared with theoretical data from a computer to identify seed oils and mixtures on the basis of various triglycerides, including trilinolein (or ECN 42). A computer program, developed at the Oil and Fat Industry Experiment Station in Milan, that permits processing of triglyceride and fatty acid data to obtain ECN limits is available from the International Olive Oil Council (IOOC, 1990). The olive oil trade standard (IOOC, 1994b) specifies that the maximum difference between the real and theoretical ECN 42 triglycerides content of edible virgin oil is 0.3 (0.4 for refined olive oil and olive oil; 0.5 for low-grade (lampante) virgin olive oil and olive pomace oils).

Mariani *et al.* (1991a; 1991b) optimized GLC methodology for simultaneous determination of sterols, aliphatic alcohols, triterpene alcohols, squalene, tocopherols, and wax esters. Determinations of several of these minor components including wax esters are useful for identifying solvent-extracted and refined oils in virgin olive oil. Wax esters and other low-polarity components of vegetable oils are usually determined by fractionation on a hydrated silica gel column followed by capillary GLC (Mariani and Fedeli, 1986). Amelio *et al.* (1993) recommended replacing the hand-filled silica gel column used in initial fractionation of wax esters from olive oils with an HPLC column for greater reliability and repeatability and to automate the separation. Extra virgin olive oil is characterized by the virtual absence of waxes with 40–46 carbon atoms; increased levels of waxes and sterol esters are found in refined oil (Mariani and Fedeli, 1993). Virgin olive oil contains squalene, whereas refined oil also contains structural isomers of squalene formed largely during bleaching (decoloration). In addition, UV absorption at 232 nm (conjugated dienes) and 270 nm (conjugated trienes) is greater in refined oil.

Methods for identifying refined oils in virgin olive oil by determining sterol dehydration products are discussed in section 7.4.2. Cortesi and Rovellini (1994) proposed the separation of fatty acid benzyl esters using HPLC to identify refined oil in virgin olive oil, based on the UV detection of conjugated and nonconjugated fatty acids at 240 nm.

Gertz (1991) acnowledged an increasing demand by consumers for olive oil and other unrefined (natural) vegetable oils and the need for methods to authenticate unrefined oil and characterize refining operations. It was suggested that loss of polyphenols and polyphenolcarboxylic acids, which occurs during alkali refining (deacidification), could be identified using HPLC analysis. Bleaching can usually be identified by measuring UV absorption at 232 and 270 nm because of an increase in levels of conjugated dienes and trienes. Deodorization (high-temperature steam treatment under vacuum) can be identified by the presence of positional and stereoisomers of linoleic acid, which are determined by HPLC of the methyl esters. Deodorization of alkali-washing steps during oil refining can also be identified by HPLC determination of the diglycerides (Capella, 1993). The amount of diglycerides present in virgin olive oil as a result of natural lipase hydrolysis should be equivalent to the acidity of the oil. However, if fatty acids are removed during refining, the diglyceride level will be higher, indicating the presence of refined oil.

Galanos *et al.* (1968) studied the characteristic UV absorption spectra of olive oils and other vegetable oils in the 310–320 nm region where the conjugated tetraene systems of oxidation products in the oils display absorption maxima. Absorptivity values ($K^{4\%}{}_{\lambda}$) were used to determine 2–5% of refined oils in virgin oils. Kapoulas and Andrikopoulos (1987) found that experimental errors when measuring K at 310, 315, and 320 nm in low-absorbing oils and when determining the maximum in the 315 nm region could result in incorrect conclusions. The distance of two consecutive extremes (the maximum and minimum) of the second derivative spectra at 315 nm measured in units of absorptivity (ΔK_{315}) showed the most characteristic differences between virgin olive oil and refined olive, olive pomace, and seed oils. Ranges of ΔK_{315} values (1% oil solution) found for various oils are shown in Table 7.14. The 5% adulteration of virgin olive oil with refined oils was determined by this technique. Aued *et al.* (1993) determined ΔK for second-derivative spectra at 310–313 nm to distinguish virgin olive oil from other vegetable oils (refined olive oil, soybean oil, sunflower oil, and rice bran oil). ΔK values of 0.0005–0.0009 were determined for virgin olive oil *versus* 0.0023–0.0029 for refined olive oil. The ΔK value together with adsorptivity values at 270 nm were used to distinguish virgin olive oil from other vegetable oils. Chiricosta *et al.* (1993) applied UV second-derivative spectrophotometry of the diene, triene, and tetraene bands to three-dimensional representations of the ΔK values, which clearly distinguish between pressed and refined oils (virgin olive oil, refined olive oil, and refined seed oils).

Calapaj *et al.* (1993) found good agreement between second-derivative UV spectrophotometric values and sterol content (GLC analysis) for identifying refined seed oils in olive oil. Mariani *et al.* (1990) evaluated the use of spectrophotofluorimetric analysis for determining percentage virgin

Table 7.14 $K^{1\%}_{315}$ values of several vegetable oils[a]

Oil	Mean	Range
Virgin olive oil	0.015	0.008–0.030
Refined olive oil	0.725	0.450–1.020
Refined olive pomace oil	0.688	0.430–0.990
Refined cottonseed oil	0.574	0.250–0.870
Refined corn oil	0.929	0.630–1.420

[a]Source: Kapoulas and Andrikopoulos (1987).

olive oil in mixtures of virgin and refined oil. Good correlations were found between fluorimetric values and percentage virgin olive oil in mixtures with less than 20% virgin olive oil.

Wesley *et al.* (1995) presented an NIR method for predicting the level of adulteration of virgin olive oils with corn oil, sunflower oil, and crude olive pomace oil. The adulterant was identified using principal component analysis. Adulterant identification was correct for 75% of the test mixtures examined. Other data analysis techniques applied to olive oils are cited in section 7.5. Recent developments in methodology for the identification of olive oil adulteration have been reviewed by Li-Chan (1994). Reviews of the chemistry and technology of olive oil are also available (Kiritsakis and Markakis, 1987; Kiritsakis, 1991).

7.9 Other vegetable oils

Purity criteria recently developed for edible vegetable oils include fatty acid, triglyceride, sterol, and tocopherol composition and specific physical and chemical tests plus SCIRA measurement to distinguish corn oil from all other commercial vegetable oils. Rossell (1991b) pointed out that the Codex Alimentarius fatty acid compositions require periodic updating, because changes in agricultural practices result in variations of the composition of some vegetable oils. In addition to advances in biotechnology (Ratledge *et al.*, 1984) and genetic engineering (Wolter, 1993; Skillicorn, 1994; Anon., 1995), efforts to alter plant enzymes and vary fatty acid composition have resulted in high oleic corn oil, low linolenic plus high oleic soybean oil, high oleic safflower and sunflower oils, and high oleic and high lauric canola oils. Continuous updating of data is therefore needed on the composition of oils, including determination of various constituents in addition to fatty acids.

7.9.1 Cocoa butter and confectionery fats

Increases in the price of cocoa butter have led to a search for alternative fats for use in manufacturing chocolate. Alternatives include cocoa butter substitutes (e.g. lauric fats rich in C–36, C–38, and C–40 triglycerides), cocoa butter equivalents or CBE (fats with a fatty acid composition similar to that of cocoa butter), and non-lauric fats (blends or fractions of cottonseed, hazelnut, rapeseed, sunflower seed, and other oils) (Anon, 1994). Rossell (1992) described the attributes and uses of various vegetable fats needed to produce chocolate substitutes.

Biino and Carlisi (1971) suggested determining illipé butter in cocoa butter blends by measuring the stigmasterol/campesterol ratio and the POS/POP and SOS/POP triglyceride ratios. Iverson (1972) recommeded identification of palm kernel and coconut oils in cocoa butter by determining lauric acid content. Padley and Timms (1978; 1980) described a GLC method for analysis of triglycerides that determined low additions of CBE to cocoa butter. A graphical approach to identifying CBEs was also provided. Padley and Timms (1978b) also described the use of GLC to determine glyceride structure to distinguish between confectionery fat mixtures with the same proportions of individual fatty acids but different distributions of triglycerides. Young (1984) described an additional way of interpreting trigylceride data obtained by GLC – to determine CBE and other oils and fats in chocolate, based on the known linear relationship between the C–50 and C–54 triglycerides of cocoa butter.

Derbesy and Richert (1979) reported that HPLC of the unsaponifiables (sterol and triterpene alcohol fraction) revealed less than 1% shea butter in cocoa butter. With measurement at 280 nm, one specific peak area was linearly related to the percentage of shea butter. Homberg and Bielefeld (1982; 1990) suggested that sterols, methyl sterols, and triterpene alcohols be determined to identify cocoa butter adulteration.

Fincke (1980a; 1980b; 1980c; 1982) evaluated the use of GLC triglyceride analysis to determine foreign fats and cocoa butter substitutes in cocoa butter and chocolate products. Foreign fats were identified on the basis of carbon numbers of the triglycerides, and limiting values were recommended for the C–50 and C–52 triglycerides of authentic cocoa butters. It was suggested that, in individual cases, besides fatty acid composition, determination be made of *trans* fatty acids, unsaponifiables, sterols, methyl sterols, and triterpene alcohols, and that triglycerides be separated by number of double bonds followed by isolation and analysis of specific triglyceride fractions.

Geeraert and DeSchepper (1983) pointed out that, whereas triglyceride separations obtained by capillary GLC on apolar columns are based predominantly on molecular weight differences (carbon number (CN) determinations), reversed-phase HPLC fractionations (equivalent carbon number (ECN) separations) are based on the degree of unsaturation of the triglycerides. They also noted that optimization of HPLC conditions could refine the separations obtained, and proposed optimized reversed-phase HPLC analysis of cocoa butter and chocolate products (UV detection) to determine 5% or less of CBE in cocoa butter, based on monitoring the POP:POS:SOS ratio of the test sample. Podlaha *et al.* (1984) determined the triglyceride composition of 28 cocoa butters from five main geographical areas by reversed-phase HPLC. Ranges for the three major triglyceride types were 35–41% PSO, 24–29% SSO, and 14–18% PPO (see Table 7.9). The data were used to identify CBE fats in cocoa butters.

Fatty acids in the 2-position and the sterol and triterpene alcohol composition of cocoa butter and common CBE fats were used to develop a sensitive method for identifying CBE in chocolate (Gegiou and Staphylakis,

1985). The triterpene alcohol fraction, isolated from the unsaponifiables by preparative TLC and further fractionated on TLC plates impregnated with silver nitrate, was especially useful for identifying CBE in chocolate. As little as 5% CBE in cocoa butter was identified by GLC determination of α-amyrin in the triterpene alcohol fraction; this is because α-amyrin, absent in cocoa butter, is a major triterpene alcohol in CBE.

Geeraert and Sandra (1987) used a high-temperature phenylmethyl-silicone stationary phase for improved capillary GLC analysis of cocoa butters and other triglycerides, including mowrah (illipé) butter, Borneo tallow, butter oil, and almond, Brazil nut, hazelnut, and walnut oils. Levels of POP, POS, and SOS in 100 cocoa butters from different geographical areas were normalized, and calculated values were placed in a ternary diagram, to distinguish synthetic from natural cocoa butters. Palmer and Palmer (1989) used a 3 μm reversed-phase HPLC column with a light-scattering detector to separate various triglycerides, including CBE.

Jeffrey (1991) described a high-speed HPLC procedure for high-resolution separations of triglycerides in vegetable fats and cocoa butter equivalents. The HPLC system consisted of a ternary liquid chromatograph, flame ionization detector, and 100 × 4.6 mm internal diameter column packed with 3 μm spherical silica, loaded with 10% silver nitrate. Trisaturated and major isomeric unsaturated triglycerides were eluted within 12 min. A change of solvent composition allowed elution of triglyceride with up to nine double bonds within the same time frame. The author pointed out that silver-ion (silver-complexation) chromatography was currently the only direct means of analyzing triglycerides for both unsaturation and isomeric configuration.

7.9.2 Palm, palm kernel, and coconut oils

Rossell (1991b) described techniques for determining the purity of palm, palm kernel, and coconut oils. To identify palm oil fractions in palm oil the iodine value was plotted against the slip (melting) point. Carbon number analysis, an excellent technique for analyzing lauric oil, distinguished between coconut and palm kernel oils, which have similar fatty acid compositions. The composition of triglycerides with carbon numbers 32 to 42 is normalized to 100 to give 'K' values, and K34 + K40 is plotted against K36 + K38 to distinguish clearly between the two oils. Sassano and Jeffrey (1993) used capillary column GLC to characterize the triglycerides of palm oil and palm oil fractions.

Sen Gupta (1991) identified palm oil to be adulterated with palm stearin by checking the ratio of PPP (tripalmitin) to MOP (myristooleopalmitin) triglycerides. The ratio ranges from 3.5 to 4.5 in pure palm oil but is elevated in cases of adulteration. Palmitic acid and solid fat content provided additional indication of adulteration. Admixture of coconut and palm kernel oils with palm kernel olein was observed by measuring the iodine value ($\leqslant$11 for coconut oil and $\leqslant$19 for palm kernel oil) plus the content of stearic, oleic, and linoleic acids, the sum of which should be no greater than 11.5% for coconut oil and 22% for palm kernel oil. Triglyceride analysis was also helpful since palm kernel olein contains higher levels of C–46 to C–54 triglycerides than coconut and palm kernel oils.

Comparison of the sterols and sterol esters of coconut and palm kernel oils showed sufficient differences to form a basis for distinguishing between the two oils (Gordon and Griffith, 1992). Sterols were isolated as the digitonides and analyzed using packed-column GLC. Sterol esters, separated by preparative TLC, were analyzed by temperature-programmed capillary GLC (50% phenyl: 50% methlypolysiloxane stationary phase) and reversed-phase HPLC. Coconut oil had a much lower β-sitosterol content and higher Δ^5-avenasterol content than palm kernel oil. GLC profiles of the coconut oil and palm kernel oil sterol esters were quite similar (the triglycerides elute along with the sterol esters); however, differences were more evident in HPLC chromatograms. Palm kernel oil displayed two major peaks, apparently due to campesteryl myristate and unresolved sitosteryl myristate/avenasteryl palmitate, which were present at lower levels in coconut oil. In addition, variations in the concentrations of other components were observed.

7.9.3 *Other individual oils*

Imai *et al.* (1974) recommended GLC determination of sterols, fatty acids, and triglycerides to identify adulterated cottonseed oil at a time when soybean, rapeseed, and rice bran oils became less expensive and readily available in Japan. Brassicasterol was judged a primary indicator of rapeseed oil, whereas stigmasterol and linolenic acid content verified the presence of soybean and rice bran oils. Tests showed that 5% rapeseed oil in cottonseed oil and about 10% soybean or rice bran oil could be determined by analysis of test samples for sterol, fatty acid, and triglyceride content.

Manandhar *et al.* (1986) determined the fatty acid, sterol, and tocopherol compositions of soybean oils admixed with linseed oil. The monomethylsterol gramisterol was the most easily identifiable characteristic of linseed oil. As little as 5% linseed oil in soybean–linseed oil mixtures was found by GLC. Kornfeldt and Croon (1981) reported the content of monomethylsterols and other sterols in linseed, soybean, and other vegetable oils.

Nasirulla *et al.* (1992) developed tests for identifying rice bran, mustard, and karanja (*Pongamia glabra*) seed oils in other edible vegetable oils. Rice bran oil was identified by TLC analysis for oryzanol; mustard seed oil was identified by a color test for isothiocyanates; and karanja oil was identified by TLC analysis for karanjin, karanjone, pongaglabrone, and pongamol. Adhikari and Adhikari (1992) described techniques for routine identification of palmolein-adulterated peanut oil, using silver nitrate TLC, and ambadi and mustard oils in other vegetable oils, using spectrophotometry. Nasirulla and Nagaraja (1987) reviewed methods for identifying karanja seed oil and eight other inedible/toxic seed oils (e.g. kusum, ambadi, argemone, castor, tea, taramira, rubber, and jatropha) in edible vegetable oils.

Chakravorty (1978) reported a paper chromatographic method for identifying small amounts of rapeseed oil in mustard seed oil. A hydrochloric acid (HCl) extract of the oil was concentrated, chromatographed with HCl–water solvent, and visualized under UV light. Jha (1980) estimated rice bran oil in mustard seed oil by observing the shift in position of the 671.5 nm band in the visible spectrum. Han and Adams (1984) described a liquid chromatographic procedure for determining adulteration of sesame oil with soybean oil, based on observation of the C–42:C–46 triglyceride peak ratio. The fatty acid composition of sesame oil was analyzed mathematically to identify adulteration with soybean oil (Yamazaki *et al.*, 1993) and rice bran or rapeseed oil (Yamazaki *et al.*, 1994).

7.10 Castor oil in other oils

Sengupta *et al.* (1978) reported a TLC procedure for identifying castor oil in other oils. It is claimed that about 1% castor oil can be determined. Adhikari and Adhikari (1982) proposed TLC of the alcohol extract, using methanol:acetic acid as the developing solvent. Ahmed *et al.* (1986) investigated the infrared and UV spectra of vegetable oils. Castor oil showed a broad band at 725/cm caused by the hydroxyl group of ricinoleic acid. Husain *et al.* (1988) examined mixtures of coconut oil and castor oil using high-performance size-exclusion chromatrography. As little as 5% castor oil in coconut oil could be determined. Husain *et al.* (1993) applied ^{13}C NMR spectroscopy to identify as little as 2% castor oil in coconut, palm, and other edible vegetable oils. Krishnamurthy (1993) reviewed various methods for identifying castor oil and other adulterants in edible fats and oils.

7.11 Mineral oil contamination

Franzke and Kroll (1983) described a rapid procedure for identifying <1% mineral oil in edible fats and oils. The method involved TLC on silica gel G with hexane as the mobile phase and detection with ethanolic phosphomolybdic acid. A column chromatographic procedure for determining hydrocarbons by separating the unsaponifiables on an alumina column and weighing the hydrocarbon fraction was also described. Graciani *et al.* (1985) developed a method for determining mineral oil in vegetable oils by using TLC to isolate the hydrocarbon fraction without prior saponification and extraction of the unsaponifiables, followed by HPLC analysis with detection at 235 nm. Tan Kuntom (1993) described a method for determination of hydrocarbons in crude palm kernel oil by GLC. The hydrocarbons were isolated from the oil using silica gel column chromatography with hexane as the eluting solvent. The eluate was anlyzed by capillary column GLC on a 25 m crosslinked methyl silicone column with flame-ionization detection. The hydrocarbons in the elute were identified and confirmed using a gas chromatograph equipped with a 25 m Carbowax column and mass-selective detector. The presence of a contiguous envelope of evenly distributed hydrocarbons peaks indicated mineral oil contamination. Pristane and phytane were also found in a contaminated oil.

7.12 Animal fat in vegetable fat

Animal fats, in contrast to vegetable fats and oils, contain significant amounts of fatty acids having odd-numbered carbon chains and branched chains and a high concentration (up to 1000 mg/kg) or cholesterol (Rossell, 1991a; Enser, 1991; Krishnamurthy, 1993). In addition, animal fats are almost free of tocopherols, whereas vegetable oils contain up to 1000 mg/kg or more of tocopherols. Animal fats often contain appreciable levels of saturated fatty acids in the 2-position of the triglycerides in contrast to vegetable oils, which have very small amounts of saturated acids in the 2-position. Fish oils contain a wide range of fatty acids, including odd-numbered carbon chain fatty acids and mono-, di-, and polyunsaturated fatty acids with more than 18 carbon atoms. Accordingly, animal fats and fish oils can be identified in vegetable oils by determination of fatty acid, triglyceride and sterol composition and fatty acids at the triglyceride 2-position.

7.13 Toxic oil syndrome

In May 1981, an epidemic in Spain resulted from consumption of aniline-denatured rapeseed oil imported from France (Spanish law required addition of 2% aniline to rapeseed oil imported for industrial use to make the product unfit for human consumption). The epidemic affected more than 20 000 people and resulted in hundreds of deaths (Grandjean and

Tarkowski, 1984). Epidemiological investigation showed that the outbreak was caused by ingestion of the aniline-containing rapeseed oil, which was diverted from industrial use and sold as cooking oil or as 'pure olive oil' (Kilbourne *et al.*, 1988).

The disease progressed in three phases (Abaitua and Posada, 1992; Yoshida *et al.*, 1994). Symptoms during the initial acute phase (1–2 months) included cough, headache, fever, rash, myalgia, pulmonary edema, and eosinophilia. About 1% of the patients exhibited nervous system dysfunction. Approximately 60% of the patients advanced to an intermediate phase characterized by myalgia, weight loss, pulmonary hypertension, skin edema, alopecia, and thromboembolism. Some patients developed slowly progressing dysphagia (difficulty in swallowing), apparently from denervation and/or fibrosis of the oesophageal wall. Death at this stage occurred with symptoms of mesenteric thrombosis and ischemic colitis. Symptoms during the third (chronic) stage (beginning 4–6 months after onset of the disease) included pulmonary arterial hypertension, scleroderma, muscular wasting, chronic hepatitis, and weakness neuropathies. Death in this later stage was primarily caused by respiratory failure. Immunological mechanisms appear to have been involved in the toxic oil syndrome with symptoms similar to those in autoimmune diseases (Lahoz *et al.*, 1992).

The toxic oils contained several constituents related to the aniline denaturization, including a number of fatty acid anilides (fatty acid N-phenylamides) as well as aniline (Diachenko *et al.*, 1982; Guitart and Gelpi, 1992). Other compounds identified in the oils included *p*-benzoquinoniline and phenylisocyanate (Arribas, 1982), 3–hydroxymethylindole (Aramendia *et al.*, 1983), and chloropropanediol fatty acid esters (Gardner *et al.*, 1983). Although the specific toxic agents in the contaminated oils have not yet been identified, 3–(phenylamino)alanine was suggested as a possible etiologic agent, and *n*-phenyllinoleamide and *n*-phenyloleamide have become principal suspects among the various agents considered and subjected to experimental studies (Yoshida *et al.*, 1994; Bioque *et al.*, 1994; Das *et al.*, 1994; Mukherjee *et al.*, 1994). Kaphalia and Ansari (1991) developed a rapid method for isolating and determining fatty acid anilides from vegetable oils.

A trial began in Madrid on 30 March 1987 with 38 defendants (businessmen, entrepreneurs, chemists, and salesmen) facing a total of thousands of years in prison in connection with charges that the adulterated rapeseed oil sold in 1981 led to more than 500 deaths and as many as 24 000 injuries (Burns, 1987). In May 1989, 13 persons (including importers and retailers of the tainted oil) were ruled guilty on public health, negligence, and fraud charges in connection with the importation and distribution of the contaminated oil (Anon., 1989; Smith, 1994). Sentences were as long as 20 years in prison, and survivors of Spain's toxic oil scandal pressed for compensation and trial for criminal negligence of officials of the then-ruling government.

7.14 Conclusions

Many physical and chemical tests have been developed to determine the authenticity of vegetable oil. Classical tests have largely been replaced or supplemented in recent years with newer chemical and instrumental procedures. Determination of fatty acid composition by modern GLC techniques is used most frequently to characterize oils and fats and to verify purity. However, additional tests are frequently needed, including determination of sterol, tocopherol, and triglyceride composition and mass spectrometric analysis. Multivariate data analysis is often used to determine geographical origin and to estimate the composition of mixtures of oils and fats. Stable carbon isotope ratio analysis has proven useful for determining corn oil authenticity. Internationally recognized definitions and purity criteria for vegetable fats and oils have been promulgated by the Codex Alimentarius Commission.

References

Abaitua, I. and Posada, M. (1992) Clinical findings. In: *Toxic Oil Syndrome: Current Knowledge and Future Perspectives*, WHO Regional Publications, European Series No. 42, World Health Organization Regional Office for Europe, Copenhagen, pp. 27–38.

Abu-Hadeed, A.M. and Kotb, A.R. (1988) A method for the quantitative determination of individual oils in a blend. *J. Am. Oil Chem. Soc.* **65(12)**, 1922–26.

Ackman, R.G. (1986) WCOT (capillary) gas–liquid chromatography. In: *Analysis of Oils and Fats* (Ed. R.J. Hamilton and J.B. Rossell), Elsevier Applied Science Publishers, London, pp. 137–206.

Adhikari, S. and Adhikari, D.J. (1982) Detection of castor oil. *J. Oil Technol. Assoc. Ind.* **14(2)**, 57–58.

Adhikari, S. and Adhikari, J. (1992) Methods for the detection/estimation of admixed vegetable oils. *J. Oil Technol. Assoc. Ind.* **24(2)**, 39–43.

Ahmed, F.A., Osman, R.O., Nomany, H.M. *et al.* (1986) Infrared and ultraviolet spectra of some lipids of different structures. *Grasas Aceites* **37(5)**, 250–53.

Allen, R.R. (1982) Hydrogenation. In: *Bailey's Industrial Oil and Fat Products*, Vol. 2, 4th edn. (Ed. D. Swern). John Wiley & Sons, New York, pp. 1–95.

Alonso, M.V. and Aparicio, R. (1993) Characterization of European virgin olive oils using fatty acids. *Grassas Aceites* **44(1)**, 18–24.

Amelio, M., Rizzo, R. and Varazini, F. (1993) Separation of wax esters from olive oils by high-performance liquid chromatography. *J. Am. Oil Chem. Soc.* **70(8)**, 793–96.

Amr, A.S. and Abu-Al-Rub, A.I. (1993) Evaluation of the Bellier test in the detection of olive oil adulteration with vegetable oils. *J. Sci. Food Agricol. 61(4)*, 435–47.

Andrikopoulos, N.K., Bruschweiler, H., Felber, H. *et al.* (1991) HPLC analysis of phenolic antioxidants, tocopherols and triglycerides. *J. Am. Oil Chem. Soc.* **68(6)**, 359–64.

Anon. (1989) *International Newsletter on Fats and Oils*, Vol. 4, No. 5, May 1989, American Oil Chemists' Society, Champaign, Illinois, USA.

Anon. (1994) Fat is beautiful. *Kennedy's Confection* **1(5)**, 15, 17.
Anon. (1995) Transgenic oilseed harvests to begin in May. *INFORM* **6(2)**, 152–57.
AOAC International (1995a) *Official Methods of Analysis*, 16th edn. (Ed. P.A. Cunniff), AOAC, Arlington, VA. Method 893.01.
AOAC International (1995b) *Official Methods of Analysis*, 16th edn. (Ed. P.A. Cunniff). AOAC, Arlington, VA. Method 937.11.
AOAC International (1995c) *Official Methods of Analysis*, 16th edn. (Ed. P.A. Cunniff). AOAC, Arlington, VA. Method 936.12.
AOCS (1987a) *Official Methods and Recommended Practices*, 4th edn. (Ed. D. Firestone). American Oil Chemists' Society, Champaign, Illinois, USA. Method Cd 1c–85.
AOCS (1987b) *Official Methods and Recommended Practices*, 4th edn. (Ed. D. Firestone). American Oil Chemists' Society, Champaign, Illinois, USA. Method Cd 3a–94.
AOCS (1987c) *Official Methods and Recommended Practices*, 4th edn. (Ed. D. Firestone). American Oil Chemists' Society, Champaign, Illinois, USA. Method Cd 12–57.
AOCS (1987d) *Official Methods and Recommended Practices*, 4th edn. (Ed. D. Firestone). American Oil Chemists' Society, Champaign, Illinois, USA. Method Cd 12b–92.
AOCS (1987e) *Official Methods and Recommended Practices*, 4th edn. (Ed. D. Firestone), American Oil Chemists' Society, Champaign, Illinois, USA. Method Cb–1–25.
AOCS (1987f) *Official Methods and Recommended Practices*, 4th edn. (Ed. D. Firestone). American Oil Chemists' Society, Champaign, Illinois, USA. Method Ce 1e–91.
AOCS (1987g) *Official Methods and Recommended Practices*, 4th edn. (ed. D. Firestone), American Oil Chemists' Society, Champaign, Method Ch 6–91.
AOCS (1987h) *Official Methods and Recommended Practices*, 4th edn. (ed. D. Firestone), American Oil Chemists' Society, Champaign, Method Ce 8–89.
AOCS (1987i) *Official Methods and Recommended Practices*. 4th edn. (Ed. D. Firestone), American Oil Chemists' Society, Champaign, Method Ce 5–86.
AOCS (1987j) *Official Methods and Recommended Practices*. 4th edn. (Ed. D. Firestone), American Oil Chemists' Society, Champaign, Method Ce 5b–89 and Ce 5c-93.
AOCS (1987k) *Official Methods and Recommended Practices*. 4th edn. (Ed. D. Firestone), American Oil Chemists' Society, Champaign, Method Ch 3-91.
AOCS (1987l) *Official Methods and Recommended Practices*. 4th edn. (Ed. D. Firestone), American Oil Chemists' Society, Champaign, Method Ch 7-94.
Aramendia, M.A., Borau, C., Jimenez, C. *et al.* (1983) Characterization of two new toxic compounds in denatured rapeseed oil. *Annales Bromatologia* **34**(1), 1–8.
Arribas, S. (1982) The Spanish oil syndrome. *Trends in Anal. Chem.* **1(14)**, IV–VI.
Artho, A., Grob, K. and Mariani, C. (1993) On-line LC–GC for the analysis of the minor components in edible oils and fats – the direct method involving silylation. *Fat Sci. Technol.* **95(5)**, 176–80.
Aued, S., Almeida, M.I. and Mancini, J. (1993) Application of derivative spectrophotometry to evaluation of olive oil. *Cien. Tecnol. Aliment.* **13(2)**, 121–31.
Bannon, C.D., Breen, G.J., Craske, J.D. *et al.* (1982) Analyses of fatty acid methyl esters with high accuracy and reliability. III. *J. Chromatogr.* **247**, 71–89.
Barnes, P.J. and Taylor, P.W. (1980) The composition of acyl lipids and tocopherols in wheat germ oils from various sources. *J. Sci. Food Agricol.* **31(10)**, 997–1006.
Bergquist, M.H.J. and Kaufmann, P. (1993) A multivariate optimization of triacylglycerol analysis by high-performance liquid chromatography. *Lipids* **28(7)**, 667–75.
Bewig, K.M., Clarke, A.D., Roberts, C. *et al.* (1994) Discriminant analysis of vegetable oils by near-infrared reflectance spectroscopy. *J. Am. Oil Chem. Soc.* **71(2)**, 195–200.
Biedermann, M., Grob, K. and Mariani, C. (1993) Transesterification and on-line LC–GC for determining the sum of free and esterified sterols in edible oils and fats. *Fat Sci. Technol.* **95(4)**, 127–33.
Biino, L. and Carlisi, E. (1971) Research on illipé fat and cocoa butter. *Riv. Ital. Sostanze Grasse* **48(4)**, 170–76.
Bioque, G., Abian, J., Bulbena, O. *et al.* (1994) N-Phenyllinoleamide metabolism by human polymorphonuclear leukocytes. *Xenobiotica* **24(7)**, 613–21.
Brause, A.R. and Ratterman, J.M. (1982) Verfication of authenticity of apple juice. *J. Assoc. Off. Anal. Chem.* **64(4)**, 846–49.
Brause, A.R., Ratterman, J.M., Petrus, D.R. *et al.* (1984) Verification of authenticity of orange juice. *J. Assoc. Off. Anal. Chem.* **67(3)**, 535–39.

Breckenridge, W.C. (1978) Stereospecific analysis of triacyglycerols. In: *Handbook of Lipid Research 1 – Fatty Acids and Glycerides* (Ed. A. Kuksis). Plenum Press, New York, pp. 197–232.

Bricout, J., Fontes, J.-C. and Merlivat, L. (1974) Detection of synthetic vanillin in vanilla extract. *J. Assoc. Off. Anal. Chem.* **57(3)**, 713–15.

Brockerhoff, H. (1965) A stereospecific analysis of triglycerides. *J. Lipid Res.* **6(1)**, 10–14.

Brockerhoff, H. and Yurkowski, M. (1966) Stereospecific analysis of several vegetable fats. *J. Lipid Res.* **7(1)**, 62–64.

BSI (1978) *British Standard Methods of Analysis of Fats and Fatty Oils*. BS684, British Standards Institution, London. Method 2.30.

Burns. T. (1987) Mass poisoning trial begins in Madrid. *Washington Post*, 31 March 1987, p. A23.

Calapaj, R., Chiricosta, S. and Saija, G. (1993) Evaluation of gas chromatographic and spectrophotometric analytical results to check the presence of seed oils in olive oil samples. *Riv. Ital. Sostanze Grasse* **70(12)**, 585–94.

Capella, P. (1993) Olive oil analysis. *Olivae* **45**, 24–27.

Carlson, K.F. and Scott, J.D. (1991) Processing of oilseeds, fats and oils. *INFORM* **2(12)**, 1034–60.

Carpenter, D.L., Lehman, J., Mason, B.S. *et al.* (1976) Lipid composition of selected vegetable oils. *J. Am. Oil Chem. Soc.* **53(12)**, 713–18.

Cert, A., Lanzon, A., Carelli, A.A. *et al.* (1994) Formation of stigmasta-3,5-diene in vegetable oils. *Food Chem.* **49(3)**, 287–93.

Chakravorty, K.L. (1978) Paper chromatographic detection of rapeseed oil in mustard seed oil. *J. Assoc. Public Anal.* **16(3)**, 81–83.

Chaves Das Neves, H.J. and Vasconcelos, A.M.P. (1989) Characterization of fatty oils by pattern recognition of triglyceride profiles. *J. High Res. Chromatogr.* **12(4)**, 226–29.

Chemical Marketing Reporter November 1994.

Chiricosta, S., Saija, G., Calapaj, R. *et al.* (1993) Three-dimensional representation of second order derivative parameters of vegetable oils. *Riv. Ital. Sostanze Grasse* **70(11)**, 549–52.

Chow, C.K., Draper, H.H. and Csallany, A.S. (1969) Method for the assay of free and esterified tocopherols. *Anal. Biochem.* **32(1)**, 81–90.

Christie, W.W. (1982) *Lipid Analysis*, 2nd edn. Pergamon Press, Oxford.

Christie, W.W. (1986) The positional distribution of fatty acids in triglycerides. In: *Analysis of Oils and Fats* (Eds R.J. Hamilton and J.B. Rossell). Elsevier Applied Science Publishers, London, pp. 313–39.

Christie, W.W. (1987) *High-Performance Liquid Chromatography*, Pergamon Press, Oxford.

Christie, W.W. (1989) *Gas Chromatography and Lipids*, Oily Press, Ayr, Scotland.

Christie, W.W., Nikolova-Damyanova, B. and Herslof, B. (1991) Stereospecific analysis of triacyl-sn-glycerols via resolution of diastereomeric diacylglycerol derivatives by high-performance liquid chromatography on silica. *J. Am. Oil Chem. Soc.* **68(10)**, 695–701.

Christie, W.W. (1992a) *Advances in Lipid Methodology-One*, Oily Press, Ayr, Scotland.

Christie, W.W. (1992b) Preparation of fatty acid methyl esters. *INFORM* **3(9)**, 1031–34.

Christie, W.W. (1993) *Advances in Lipid Methodology – Two*. Oily Press, Ayr, Scotland.

Codex Alimentarius Commission (1993a) *Proposed Draft Standard for Named Vegetable Oils*, CX 1993/16, issued by the Joint FAO/WHO Food Standards Program, via delle Terme di Caracalla 00100, Rome.

Codex Alimentarius Commission (1993b) *Proposed Draft Standard for Olive Oils*, CL 1993/15-FO, issued by the Joint FAO/WHO Food Standards Program, via delle Terme di Caracalla 00100, Rome.

Coleman, E.C. and Firestone, D. (1972) A simplified Halphen procedure for cyclopropene fatty acids. *J. Assoc. Off. Anal. Chem.* **55(6)**, 1290–93.

Coors, U. (1991) Utilization of the tocopherol pattern for recognition of fat and oil adulterations. *Fat Sci. Technol.* **93(4)**, 519–26.

Cortesi, N. and Fedeli, E. (1980) Further research on the HPLC analysis of unsaponifiable of edible vegetable oils. Suggestion of a method of tocopherol determination. *Riv. Ital. Sostanze Grasse* **57(1)**, 16–19.

Cortesi, N., Rovellini, P.A. and Fedeli, E. (1992) Total quality evaluation of oils and fats. Note 1: Study of the triglyceridic fraction. *Riv. Ital. Sostanze Grasse* **69(1)**, 1–6.

Cortesi, N. (1993) High-performance liquid chromatography (HPLC) in the analysis of olive oil. *Olivae* **45**, 40–42.
Cortesi, N. and Rovellini, P. (1994) HPLC of benzylester derivatives of fatty acids and their isomers. *Riv. Ital. Sostanze Grasse* **71(12)**, 581–86.
Cozzoli, O., Cortesi, N., Gigliotti, C. *et al.* (1991) Use of high-performance liquid chromatography in the analysis of triglycerides in the oils. Results of a collaborative study. *Riv. Ital. Sostanze Grasse* **68(10)**, 543–48.
Craske, J.D. (1993) Separation of instrumental and chemical errors in the analysis of oils by gas chromatography – a collaborative evaluation. *J. Am. Chem. Soc.* **70(4)**, 325–34.
Damiani, P., Santinetti, F., Simonetti, M.S. *et al.* (1994) Comparison between two procedures for stereoscopic analysis of triacylglycerols from vegetable oils – 1: Olive oil. *J. Am. Oil Chem. Soc.* **71(10)**, 1157–62.
Das, S.K., Mukherjee, S. and Desai, U. (1994) Development of pancellular toxicity in guinea pig lung by ingestion of oleylanilide. *J. Biochem. Toxicol.* **9(1)**, 41–49.
De Jong, S. and de Jonge, J.R. (1991) Computer-assisted fat blend recognition using regression analysis and mathematical programming. *Fat Sci. Technol.* **93(4)**, 532–36.
Derbesy, M. and Richert, M.T. (1979) Detection of shea butter in cocoa butter. *Oleagineux* **34(8/9)**, 405–09.
Diachenki. G.W., Yurawecz, M.P., Puma, B.J. *et al.* (1982) Identification of fatty acid anilides in adulterated Spanish cooking oils. *Bull. Environ. Toxicol. Contam.* **29(2)**, 228–24.
Doner, L.W., Krueger, H.W. and Reesman, R.H. (1980) Isotope composition of carbon in apple juice. *J. Agric. Food Chem.* **28(2)**, 362–64.
Doner, L.W. and Phillips, J.G. (1981) Detection of high-fructose corn syrup in apple juice by mass spectrometric $^{13}C/^{12}C$ analysis: a collaborative study. *J. Assoc. Off. Anal. Chem.* **64(1)**, 85–90.
Dumazer, M., Farines, M. and Soulier, J. (1986) Identification of sterols by mass spectrometry. *Rev. Fr. Corps Gras* **33(4)**, 151–56.
Eddib, O. and Nickless, G. (1987) Elucidation of olive oil classification by chemometrics. *Analyst* **122(4)**, 391–95.
EEC (1991) European Economic Community. *Official Journal of the European Communities*, Luxembourg, 9 September 1991, No. L 248/15–22, Annex V.
El-Hamdy, A.H. and Perkins, E.G. (1981) High-performance reversed-phase chromatography of natural triglyceride mixtures: critical pair separation. *J. Am. Oil Chem. Soc.* **58(9)**, 867–72.
Elliott, S.C., Hartman, N.A. and Hawkes, S.J. (1971) Analysis of blends of mixtures using multivariate statistics. *Anal. Chem.* **43(14)**, 1938–39.
Emmerie, A. and Engel, C. (1938) Colorimetric determination of α-tocopherol (vitamin E). *Rec. Trav. Chim. Pays-Bas* **57**, 1351–55.
Emmerie, A. and Engel, C. (1939) Colorimetric determination of tocopherol (vitamin E); II. Adsorption experiments. *Rec. Trav. Chim. Pays-Bas* **58**, 283–89.
Enser, M. (1991) Animal carcass fats and fish oils. In: *Analysis of Oilseeds, Fats and Fatty Foods* (Eds J.B. Rossell and J.L.R. Pritchard). Elsevier Applied Science Publishers, London, pp. 329–94.
Feinberg, M. and Bugner, E. (1989) Chemometrics and food chemistry: data validation. *Anal. Chim. Acta.* **223(1)**, 223–35.
Fellat-Zarrouck, K., Boutellier, J.C. and Maurin, R. (1988) HPLC detection of adulteration of olive oil by seed oils. *Rev. Fr. Corps Gras.* **35(10)**, 383–85.
Fincke, A. (1980a) Potential and limitations of gas chromatographic triglyceride analysis for detection of foreign fats in cocoa butter and chocolate fats. I. Distribution of triglycerides according to atom number in cocoa butter. *Dtsch. Lebensm. Rundsch.* **76(5)**, 162–67.
Fincke, A. (1980b) Potential and limitations of gas chromatographic triglyceride analysis for detection of foreign fats in cocoa butter and chocolate fats. II. Distribution of triglycerides, classified according to C-number, in cocoa butter substitutes and other fats. *Dtsch. Lebensm. Rundsch.* **76(6)**, 187–92.
Fincke, A. (1980c) Potential and limitations of gas chromatographic triglyceride analysis for detection of foreign fats in cocoa butter and chocolate fats. III. Evaluation of gas chromatographic triglyceride analyses of cocoa butter. *Dtsch. Lebensm. Rundsch.* **76(11)**, 384–89.

Fincke, A. (1982) Potential and limitations of gas chromatographic triglyceride analysis for detection of foreign fats in cocoa butter and chocolate fats. IV. Evaluation of gas chromatographic triglyceride analyses of milk chocolate fats. *Dtsch. Lebensm. Rundsch.* **78(11)**, 389–96.

Firestone, D., Summers, J.L., Reina. R.J. *et al.* (1985) Detection of adulterated and misbranded olive oil products. *J. Am. Oil Chem. Soc.* **62(11)**, 1558–62.

Firestone, D., Carson, K.L. and Reina, R.J. (1988) Update on control of olive oil adulteration and misbranding in the United States. *J. Am. Oil Chem. Soc.* **65(5)**, 788–92.

Fitelson, J. (1936) A colorimetric method for the detection of tea seed oil in olive oil. *J. Assoc. Off. Agric. Chem.* **19(3)**, 493–97.

Fitelson, J. (1943) Detection of olive oil in edible oil mixtures. *J. Assoc. Off. Agric. Chem.* **26(4)**, 499–506.

Fitelson, J., Kirsten, G., Jablonski, C.F. and Offut, M.L. (1951) Chemical problems in the squalene olive oil racket. *Food Control Statement* No. 72, US Food and Drug Administration, Washington DC, pp. 51–55.

Flor, R.V., Hecking, L.T. and Martin, B.D. (1993) Development of high-performance liquid chromatography criteria for determination of grades of commercial olive oils. Part 1. The normal ranges for the triacylglycerols. *Olivae* **48**, 37–42.

Forcadell, M.L., Lopez, M.C. and de la Torre, M.C. (1988) Classification of virgin olive oils from different origin by discriminant analysis. *Riv. Ital. Sostanze Grasse* **65(3)**, 213–14.

Forina, M. and Tiscornia, E. (1982) Pattern recognition methods in the prediction of Italian olive oil origin by their fatty acid content. *Ann. Chim.* **72**, 143–55.

Forina, M., Armanino, C., Lanteri, S. and Tiscornia, E. (1983a) Classification of olive oils from their fatty acid composition. In: *Food Research and Data Analysis* (Eds H. Martins and H. Russwurm), Applied Science, London, pp. 189–214.

Forina, M., Armanino, C., Lanteri, S. *et al.* (1983b) Correlation of olive oil chemical features with the production year through multivariate methods of classification. *Riv. Ital. Sostanze Grasse* **60(10)**, 607–13.

FOSFA International (1989) *Standard Contractual Methods*, Federation of Oils, Oilseeds and Fats Associations Ltd., London.

Franzke, C. and Kroll, J. (1983) Detection of mineral oil in edible fats. *Lebensmittelindustrie* **30(5)** 223–24.

Gaffney, J., Irsa, A., Friedman, L. and Emken, E. (1979) ^{13}C-^{12}C analysis of vegetable oils, starches, proteins and soy-meat mixtures. *J. Agric. Food Chem.* **27(3)** 475–78.

Galanos, D.S., Kapoulas, V.M. and Voudouris, E. (1968) Application of UV spectrophotometry in the 315 nm region for oil control: Detection of olive oil adulteration with olive pomace oil. *Rev. Fr. Corps Gras* **15(5)**, 291–300.

Gardner, A.M., Yurawecz, M.P., Cunningham, W.C. *et al.* (1983) Isolation and identification of C16 and C18 fatty acid esters of chloropropanediol in adulterated Spanish cooking oils. *Bull. Environ. Contam. Toxicol.* **31(6)**, 625–30.

Geeraert, E. and DeSchepper, D. (1983) Structure elucidation of triglycerides by chromatographic techniques. Part 2: RP HPLC of triglycerides. *J. High Res. Chromatogr. Chromatogr. Comm.* **6(3)**, 123–32.

Geeraert, E. and Sandra, P. (1987) Capillary GC of triglycerides in fats and oils using a high-temperature phenylmethlysilicone stationary phase. Part II. The analysis of chocolate fats. *J. Am. Oil Chem. Soc.* **64(1)**, 100–05.

Gegiou D. and Georgouli, M. (1980) Detection of reesterified oils: determination of fatty acids at position 2 in the glycerides of oils. *J. Am. Oil Chem. Soc.* **57(9)**, 313–16.

Gegiou, D. and Georgouli, M. (1983) A rapid argentation TLC method for detection of reesterified oils in olive and olive-residue oils. *J. Am. Oil Chem. Soc.* **60(4)**, 833–35.

Gegiou, D. and Staphylakis, K. (1985), Detection of cocoa butter equivalents in chocolate. *J. Am. Oil. Chem. Soc.* **62(6)**, 1047–51.

Gertz, C. and Hermann, K. (1982a) High-performance liquid chromatographic separation of individual tocopherols and tocotrienols in food fats and food oils on short columns. *Lebensm. Gerichtl. Chem.* **36(1)**, 53–58.

Gertz, C. and Hermann, K.(1982b) Analysis of tocopherols and tocotrienols in foods. *Z. Lebensm. Unters. Forsch.* **174(5)**, 390–94.

Gertz, C. (1991) Native and non-refined vegetable oils and fats. *Fat Sci. Technol.* **93(4)**, 545–48.

Goodacre, R., Kell, D.B. and Bianchi, G. (1992) Neural networks and olive oil. *Nature* **359(6396)**, 594.

Goodacre, R., Kell, D.B. and Bianchi, G. (1993) Rapid assessment of the adulteration of virgin olive oils by other seed oils using pyrolysis mass spectrometry and artificial neural networks. *J. Sci. Food Agric.* **63(3)**, 297–307.

Goodacre, R., Kell, D. and Bianchi, G. (1994) Quick thinking. *Food Rev.* **21(3)**, 11–12.

Gordon, M.H. and Griffith, R.E. (1992) A comparison of the steryl esters of coconut and palm kernel oils. *Fat Sci. Technol.* **94(6)**, 218–21.

Govind Rao, M.K. and Perkins, E.G. (1972) Identification and estimation of tocopherols and tocotrienols in vegetable oils using gas chromatography–mass spectrometry. *J. Agric. Food Chem.* **20(2)**, 240–45.

Graciani, E., Gomez, R. and Vazquez, A. (1985) Determination of mineral oil in vegetable oil by high-performance liquid chromatography. Application to several rapeseed oils denatured with aniline. *Grasas Aceites* **36(4)**, 254–57.

Grandjean, P. and Tarkowski, S. (Eds) (1984) *Toxic Oil Syndrome: Mass Food Poisoning in Spain*. Report on a WHO meeting, Madrid, 21–15 March 1983, World Health Organization Regional Office for Europe, Copenhagen.

Greenberg, A. and Harris, J. (1982) Cyclopropenoid fatty acids. *J. Chem. Educ.* **59**, 539–43.

Griffini, A. Meneghetti, O. and Amelotti, G. (1987) Ammonia chemical inoization GC/MS of sterols from vegetable oils. *Riv. Ital. Sostanze Grasse* **64(12)**, 529–33.

Grob, K., Lanfranchi, M. and Mariani, C. (1989) Determination of free and esterified sterols and of wax esters in oils and fats by coupled liquid chromatography–gas chromatography. *J. Chromatogr.* **471**, 397–405.

Grob, Lanfranchi, M. and Mariani, C. (1990) Evaluation of olive oils through the fatty alcohols, the sterols and their esters by coupled LC–GC *J. Am. Oil. Chem. Soc.* **67(10)** 626–34.

Grob, K. and Bronz, M. (1994) Analytical problems in determining 3,5-stigmastadiene and campestadiene in edible oils. *Riv. Ital. Sostanze Grasse* **71(6)**, 291–95.

Grob, K., Giuffre, A.M., Leuzzi, U. *et al.* (1994a) Recognition of adulterated oils by direct analysis of the minor components. *Fat Sci. Technol.* **96(8)**, 286–90.

Grob, K., Giuffre, A.M., Biedermann, M. *et al.* (1994b) The detection of adulteration with desterolized oils. *Fat Sci. Technol.* **96(9)**, 341–44.

Grob. K., Biedermann, M., Artho, A. *et al.* (1994c) LC, GC, and MS of sterol dehydration products. *Riv. Ital. Sostanze Grasse* **71(11)**, 533–38.

Guitart, R. and Gelpi, E. (1992) Chemical composition of TOS-related oils. In: *Toxic Oil Syndrome: Current Knowledge and Future Perspectives*, WHO Regional Publications, European Series No. 42, World Health Organization Regional Office for Europe, Copenhagen, pp. 99–142.

Gutfinger, T. and Letan, A. (1974) Studies of unsaponifiables in several vegetable oils. *Lipids* **9(9)**, 658–63.

Hamilton, R.J (1975) Qualitative and quantitative gas liquid chromatography of triglycerides. *J. Chromatogr. Sci.* **13(10)**, 474–78.

Hamilton, R.J. and Rossell, J.B. (Eds) (1986) *Analysis of Oils and Fats*. Elsevier, London.

Hamilton, R.J. and Bhati, A. (Eds) (1987) Recent Advances in Chemistry and Technology of Fats and Oils, Chapman & Hall, London.

Hammond, E.W. (1993) *Chromatography for the Analysis of Lipids*, CRC Press, Boca Raton, Florida, USA.

Han, L.M. and Adams, J.M. (1984) Liquid chromatographic determination of adulteration of sesame oil. *J. Assoc. Off. Anal. Chem.* **67(5)**, 916–18.

Harris, P.L., Quaife, M.L. and Swanson, W.J. (1950) Vitamin E content of foods. *J. Nutr.* **40(3)**, 367–81.

Haumann, B.F. (1994) Tools: Hydrogenation and interesterification. *INFORM* **5(6)**, 668–78.

Herting, D.C. and Drury, E.-J.E. (1963) Vitamin E content of vegetable oils and fats. *J. Nutr.* **81(3)**, 335–42.

Hertz, J., Krogh, A. and Palmer, R.G. (1991) *Introduction to the Theory of Neural Computation*. Addison-Wesley, Redwood City, California, USA.

Hites, R.A. (1970) Quantitative analyses of triglyceride mixtures by mass spectrometry. *Anal. Chem.* **42(14)**, 1736–40.

Hoffman, G. (1989) *The Chemistry and Technology of Edible Oils and Fats and Their High Fat Products*. Academic Press, London.

Homberg, E. and Bielefeld, B. (1982) Sterols and methylsterols in cocoa butter and cocoa butter sutstitutes. *Dtsch. Lebensm. Rundsch.* **78(3)**, 73–77.

Homberg, E. and Bielefeld, B. (1989) Composition and content of sterols in 41 different vegetable and animal fats. *Fat Sci. Technol.* **91(1)**, 23–27.

Homberg, E. and Bielefeld, B. (1990) Main components of 4–methyl–sterol and triterpene fraction of twelve vegetable fats and their influence on sterol analysis. *Fat Sci. Technol.* **92(12)**, 478–80.

Homberg, E. (1991) Sterol analysis as means to detection of mixed and adulterated fats. *Fat Sci. Technol.* **93(4)**, 516–17.

Horstmann, P. and Montag, A. (1986) New methods for rapid isolation of sterols from lipids. *Fette, Seifen, Anstrichm.* **88(7)**, 262–64.

Husain, S., Kifayatullah, M., Sastry, G.S.R. *et al.* (1993) Quantitative determination of castor oil in edible and heat abused oils by ^{13}C nuclear magnetic resonance spectroscopy, *J. Am. Oil Chem. Soc.* **70(12)**, 1251–54.

Husain, S., Sastry, G.S.R., Raju, N.P. *et al.* (1988) High-performance size-exclusion chromatography of oils and fats. *J. Chromatogr.* **454**, 317–26.

Iatrides, M.C. and Artaud, J. (1989) Coexistence of two alkenyl-24 sterols with their alkyl-24 homologue in borage oils. *Rev. Fr. Corps Gras* **36(6)**, 279–82.

Imai, C., Watanabe, H., Haga, N. *et al.* (1974) Detection of adulteration of cottonseed oil by gas chromatrography. *J. Am. Oil Chem. Soc.* **51(7)**, 326–30.

Indyk, H.E. (1988) Simplified saponification procedure for the routine determination of total vitamin E in dairy products, foods and tissues by high-performance liquid chromatography. *Analyst* **113(8)**, 1217–21.

Information Bulletin (1994) No. 770, 1 December 1994. International Olive Oil Council, Madrid.

IOOC (1990) Report of meeting of chemists to study methods of analysis for olive oils and olive-pomace oils, IOOC/T.20/Doc. No. 18–3, 20 November 1990, International Olive Oil Council, Madrid.

IOOC (1992) Determination of the composition and content of sterols by capillary-column gas chromatography, IOOC/T.20/Doc. No. 10, 28 May 1992, International Olive Oil Council, Madrid.

IOOC (1994a) *Information Bulletin* No. 770, International Olive Oil Council, Madrid, 1 December 1994.

IOOC (1994b) Trade standard applying to olive oil and olive-pomace oil, COI/T.15/NC No. 2/Corr. 1, 17 November 1994, International Olive Oil Council, Madrid.

Irwin, W.J. (1982) *Analytical Pyrolysis: A Comprehensive Guide*, Marcel Dekker, New York.

ISO (1983) International Standards Organization, Geneva, ISO/DIS (Draft International Standard) 6799.

ISO (1991) International Standards Organization, Geneva, ISO/DIS (Draft International Standard) 9936.

Itoh, T., Tamura, T. and Matsumoto, T. (1973) Sterol composition of 19 vegetable oils. *J. Am. Oil. Chem. Soc.* **50(4)**, 122–25.

IUPAC (1987a) International Union of Pure and Applied Chemistry, *Standard Methods for the Analysis of Oils, Fats and Derivatives*, 7th edn. (Eds C. Paquot and A. Hautfenne), Blackwell Scientific Publications, Oxford. Method 2.301.

IUPAC (1987b) International Union of Pure and Applied Chemistry, *Standard Methods for the Analysis of Oils, Fats and Derivatives*, 7th edn. (Eds C. Paquot and A. Hautfenne), Blackwell Scientific Publications, Oxford. Method 2.302.

IUPAC (1987c) International Union of Pure and Applied Chemistry, *Standard Methods for the Analysis of Oils, Fats and Derivatives*, 7th edn. (Eds C. Paquot and A. Hautfenne), Blackwell Scientific Publications, Oxford. Method 2.403.

IUPAC (1987d) International Union of Pure and Applied Chemistry, *Standard Methods for the Analysis of Oils, Fats and Derivatives*, 7th edn. (Eds C. Paquot and A. Hautfenne), Blackwell Scientific Publications, Oxford. Method 2.411.

IUPAC (1987e) International Union of Pure and Applied Chemistry, *Standard Methods for the Analysis of Oils, Fats and Derivatives*, 7 edn. (Eds C. Paquot and A. Hautfenne), Blackwell Scientific Publications, Oxford. Method 2.323.

IUPAC (1987f) International Union of Pure and Applied Chemistry, *Standard Methods for the Analysis of Oils, Fats and Derivatives*, 7th edn. (Eds C. Paquot and A. Hautfenne), Blackwell Scientific Publications, Oxford. Method 2.210.

IUPAC (1992a) International Union of Pure and Applied Chemistry, *Standard Methods for the Analysis of Oils, Fats and Derivatives*, 1st supplement to the 7 edn. (Eds A. Dieffenbacher and W.D. Pocklington), Blackwell Scientific Publications, Oxford. Method 2.304.

IUPAC (1992b) International Union of Pure and Applied Chemistry, *Standard Methods for the Analysis of Oils, Fats and Derivatives*, 1st supplement to the 7th edn. (Eds A. Dieffenbacher and W.D. Pockington), Blackwell Scientific Publications, Oxford. Method 2.432.

IUPAC (1992c) International Union of Pure and Applied Chemistry, *Standard Methods for the Analysis of Oils, Fats and Derivatives*, 1st supplement to the 7edn. (Eds A. Dieffenbacher and W.D. Pocklington), Blackwell Scientific Publications, Oxford. Methods 2.324 and 2.325.

Iverson, J. (1972) Gas–liquid chromatographic detection of palm kernel and coconut oils in cocoa butter. *J. Assoc. Off. Anal. Chem.* **55(6)**, 1319–22.

Jamieson, G.S. (1932) *Vegetable Fats and Oils*, Chemical Catalog Co., New York, p. 213.

Jeffrey, B.S.J. (1991) Silver-complexation liquid chromatography for fast high-resolution separations of triacylglycerols. *J. Am. Oil, Chem. Soc.* **68(5)**, 289–93.

Jha, J.S. (1980) Spectrophotometric studies of rice bran oil in mustard oil mixtures: I. *J. Am. Oil Chem. Soc.* **57(2)**, 83–85.

Kaphalia, B.S. and Ansari, G.A.S. (1991) Rapid chromatographic analysis of fatty acid anilides suspected of causing toxic oil syndrome. *J. Anal. Toxocol.* **15(2)**, 90–94.

Kapoulas, V.M. and Andrikopoulos, N.K. (1986) Detection of olive oil adulteration with linoleic acid-rich oils by reversed-phase high-performance liquid chromatography. *J. Chromatogr.* **366**, 311–20.

Kapoulas, V.M. and Andrikopoulos, N.K. (1987) Detection of virgin olive oil adulteration with refined oils by second-derivative spectrophotometry. *Food Chem.* **23(3)**, 183–92.

Kapseu, C., Kamga, R., Tchatchueng, J.-B. *et al.* (1992) Comparative study of the composition of peanut and corn oil. *Rev. Fr. Corps Gras* **39(7/8)**, 215–18.

Karleskind, A., Valmalle, G. and Wolff, J.-P. (1974) Some examples of the application of analyses by high-temperature gas chromatography. 1–Triglyceride analysis. *Rev. Fr. Corps Gras* **21(11)**, 617–31.

Kilbourne, E.M., Bernert, J.T., Posada, M. *et al.* (1988) Chemical correlates of pathogenicity of oils related to the toxic oil syndrome epidemic in Spain. *Am. J. Epidemiol.* **127(6)**, 1210–27.

Kiritsakis, A.K. (1991) *Olive Oil*, American Oil Chemists' Society, Champaign, Illinois, USA.

Kiritsakis, A. and Markakis, P. (1987) Olive Oil: a review. *Adv. Food Res.* **31**, 453–82.

Knights, B.A. (1967) Identification of plant sterols using combined GLC/mass spectrometry. *J. Gas Chromatogr.* **5(6)**, 273–82.

Kochhar, S.P. (1983) Influence of processing on sterols of edible vegetable oils. In: *Progress in Lipid Research*, Vol. 22 (Ed. R.T. Holman), Pergamon Press, Oxford, pp. 161–88.

Kornfeldt, A. and Croon, L.-B. (1981) 4–Desmethyl–, 4–monomethyl– 4,4,–dimethylsterols in some vegetable oils. *Lipids* **16(5)**, 306–14.

Krishnamurthy, M.N. (1993) Updated methods for detection of adulterants and contaminants in edible oils and fats: a critical evaluation. *J. Food Sci. Technol. (India)* **30(4)**, 231–38.

Krishnamurthy, R.G. (1982a) Extraction of fats and oils. In: *Bailey's Industrial Oil and Fat Products*, Vol. 2, 4th edn. (Ed. D. Swern), John Wiley & Sons, New York, pp. 175–251.

Krishnamurthy, R.G. (1982b) Refining and bleaching. In: *Bailey's Industrial Oil and Fat Products*, Vol. 2, 4th edn. (Ed. D. Swern), John Wiley & Sons, New York, pp. 253–314.

Laakso, P. and Christie, W.W. (1990) Chemometric resolution of chiral diacylglycerol derivatives: Potential in the stereoscopic analysis of triacyl-sn-glycerols. *Lipids* **25(6)**, 349–53.

Lahoz, C., Rose, N.R. and Robinson, C.J.G. (1992) Immunology. In: *Toxic Oil Syndrome Knowledge and Future Perspectives*, WHO Regional Publications, European Series No. 42, World Health Organization Regional Office for Europe, Copenhagen, pp. 143–52.

Lai, Y.W., Kemsley, E.K. and Wilson, R.H. (1994) Potential of Fourier-transform infrared spectroscopy for the authentication of vegetable oils. *J. Agric. Food Chem.* **42(5)**, 1154–59.

Lange, W. (1950) Cholesterol, phytosterol, and tocopherol content of food products and animal tissues. *J. Am. Oil Chem. Soc.* **27(11)**, 414–22.

Lanzon, A., Cert, A. and Albi, T. (1989) Detection of refined olive oil in virgin olive oil. *Grasas Aceites* **40(6)**, 385–88.

Larsson, K. (Ed.) (1994) *Lipids-Molecular Organization, Physical Functions and Technical Applications*, Oily Press, Dundee, Scotland.

Lawson, H.W. (1994) *Food Oils and Fats: Technology, Utilization & Nutrition*, 2nd edn., Chapman & Hall, New York.

Leardi, R. and Paganuzzi, V. (1987) Characterization of the origin of extra virgin olive oils performed through the use of chemometrical methods on the sterolic fraction. *Riv. Ital. Sostanze Grasse* **64(4)**, 131–36.

Lercker, G., Frega, N., Conte, L.S. *et al.* (1981) High-resolution gas chromatography in the study of vegetable oil unsaponifiables. *Riv. Ital. Sostanze Grasse* **58(7)**, 324–30.

Li-Chan, E. (1994) Developments in the detection of adulteration of olive oil. *Trends Foods Sci. Technol.* **5(1)**, 3–11.

Litchfield, C. (1972) *Analysis of Triglycerides*, Academic Press, New York.

Lognay, G., Lacoste, F., Marlier, M. *et al.* (1993) The certification of the identity of individual sterols, in three BCR oil and fat reference materials by GM–MS. *Fat Sci. Technol.* **95(3)**, 98–104.

Lopez, M.C., Boatella, J., and Torre, M.C. (1986) Application of discriminant analysis to the differentiation of oils from different olive varieties. *Rev. Fr. Corps Gras* **33(2)**, 65–67.

Losi, G. and Piretti, M.V. (1970) Determination of tocopherols in vegetable oils and its application to genuineness control. *Riv. Ital. Sostanze Grasse* **42(10)**, 493–98.

Lusby, W.R., Thompson, M.J. and Kochansky, J. (1984) Analysis of sterol esters by capillary gas chromatography-electron impact and chemical ionization spectroscopy. *Lipids* **19(11)**, 888–901.

Manandhar, P.P., Nagao, A. and Yamazaki, M. (1986) Determination of content of linseed oil in edible soybean oil. *J. Jpn Oil Chem. Soc. (Yukagaku)* **35(9)**, 725–30.

Mannino, S. and Amelotti, G. (1975) Sterol composition of thirty vegetable oils determined by GLC using two phases of different polarity. *Riv. Ital. Sostanze Grasse* **52(3)**, 79–83.

Mariani, C. and Fedeli, E. (1982) Rapid methods for the analysis of vegetable oils. Note 1: Simultaneous determination of sterols and tocopherols. *Riv. Ital. Sostanze Grasse* **59(11)**, 557–65.

Mariani, C. and Fedeli, E. (1986) Detection of extraction oils in pressure ones, Note 1. *Riv. Ital. Sostanze Grasse* **63(1)**, 3–17.

Mariani, D., Grassi, L., Balestrieri, F. and Pascucci, E. (1990) Spectrophotofluorimetric analysis of olive oil. Possibility of application. *Riv. Ital. Sostanze Grasse* **67(2)**, 95–99.

Mariani, C., Fedeli, E., Grob, K. *et al.* (1991a) Variations in the free and esterified minor components of oils obtained from olives in connection with maturation and storage. *Riv. Ital. Sostanze Grasse* **68(4)**, 179–87.

Mariani, C., Fedeli, E. and Grob, K. (1991b) Evaluation of free and esterified minor components in fatty materials. *Riv. Ital. Sostanze Grasse* **68(5)** 233–42.

Mariani, C., Venturini, S., Bondioli, P. *et al.* (1992) Evaluation of the variations produced by the bleaching process on more meaningful minor components free and esterified in olive oil. *Riv. Ital. Sostanze Grasse* **69(8)**, 393–99.

Mariani, C., Venturini, S. and Fedeli, E. (1993) Evaluation of neo-formation hydrocarbons and free and esterified minor components in different classes of olive oils. *Riv. Ital. Sostanze Grasse* **70(7)**, 321–27.

Mariani, C. and Fedeli, E. (1993) Gas chromatrography in olive oil analysis. *Olivae* **45**, 34–39.

Martens, H. and Russwurm, H. (1983) *Food Research and Data Analysis*, Applied Science, London.

McEwen, C.N. and Larsen, B.S. (1990) *Mass Spectroscopy of Biological Materials*. Marcel Dekker, New York.

Mehlenbacher, V.C. (1960) *The Analysis of Fats and Oils*. Garrard Press, Champaign, Illinois, USA.

Meijboom, P.W. and Jongenotter, G.A. (1979) A quantitative determination of tocotrienols and tocopherols in palm oil by TLC–GLC. *J. Am. Oil Chem. Soc.* **56(1)**, 33–35.

Meneghetti, O., Amelotti, G. and Griffini, A. (1987) Identification of sterols in vegetable crude and refined oils by electron ionization GC–MS. *Riv. Ital. Sostanze Grasse* **64(11)**, 461–68.

Micali, G. and Curro, P. (1984) Determination of tocopherols in vegetable oils by HPLC. *Riv. Ital. Sostanze Grasse* **61(2)**, 95–98.

Morales, M.T. and Aparicio, R. (1993) Characterizing some European olive oil varieties by volatiles using statistical tools. *Grasas Aceites* **44(2)**, 113–15.

Muller-Mulot, W. (1976) Rapid method for the quantitative determination of individual tocopherols in oils and fats. *J. Am. Oil Chem. Soc.* **53(12)**, 732–36.

Mukherjee, S., Ghosh, S., Rodgers, L. *et al.* (1994) Toxic effects of fatty acid anilides on the oxygen defense systems of guinea pig lungs and erythrocytes. *J. Biochem. Toxicol.* **9(1)**, 1–7.

Nasirulla and Nagaraja, K.V. (1987) Detection of the presence of toxic oils in edible vegetable oils. *Oleagineux* **42(1)**, 35–38.

Nasirulla, Krishnamurthy, M.N. and Nagaraja, K.V. (1992) Methods for detection of rice-bran, mustard, karanja oils and rice-bran deoiled cake. *Fat Sci. Technol.* **94(12)**, 457–58.

Neff, W.E., Adolf, R.O., List, G.R. *et al.* (1994) Analysis of vegetable oil triacylglycerols by silver ion high performance liquid chromatography with flame ionization detection. *J. Liq. Chromatogr.* **17(18)**, 3951–68.

NIOP (1994), *Trading Rules*, National Institute of Oilseed Products, Washington DC, 20005.

Norris, F.A. (1985) Deodorization. In: *Bailey's Industrial Oil and Fat Products*, Vol. 3, 4th edn. (Ed. T.H. Applewhite), John Wiley & Sons, New York, pp. 127–65.

Oil World December 1994.

Ostermann, G. (1991) Assessment of identity and purity of vegetable fats and oils in the food industry. *Fat Sci. Technol.* **93(4)**, 554–56.

Ouedraogo, M.A. and Bezard, J.A. (1982) The triglycerides of cottonseed oil. *Rev. Fr. Corps Gras* **29(1)**, 11–16.

Padley, F.B. and Timms, R.E. (1978a) Determination of cocoa butter equivalents in chocolate. *Chem. Ind.* **23**, 918–19.

Padley, F.B. and Timms, R.E. (1978b) Analysis of confectionery fats. II. Gas-liquid chromatography of triglycerides. *Lebensm. Wiss. Technol.* **11(6)**, 319–22.

Padley, F.B. and Timms, R.E. (1980) The determination of cocoa butter equivalents in chocolate. *J. Am. Oil. Chem. Soc.* **57(9)**, 286–93.

Paganuzzi, V. (1986) Usefulness of sterols and erythrodiol determination for the detection of virgin olive oils origin through chemometric methods. *Riv. Ital. Sostanze Grasse* **63(10)**, 543–49.

Paganuzzi, V. (1987) Usefulness of the determination of sterols and erythrodiol for identifying the origin of virgin olive oils by chemometric methods. *Olivae* **16**, 19–22.

Palmer, A.J. and Palmer, F.J. (1989) Rapid analysis of triacylglycerols using high-performance liquid chromatography with light scattering detection. *J. Chromatogr.* **465(2)**, 369–77.

Parris, N.A. (1978) Non-aqueous reversed-phase chromatography of glycerides using infrared detection. *J. Chromatogr.* **149**, 615–24.

Pei, P.T.-S., Henley, R.S. and Ramachandran, S. (1975) New application of high-pressure reversed-phase liquid chromatography in lipids. *Lipids* **10(3)**, 152–56.

Perkins, E.G. (Ed.) (1991) *Analysis of Fats Oils and Lipoproteins*. American Oil Chemists' Society, Champaign, Illinois, USA.

Perrin, J.L., Prevot, A. Traitler, H. *et al.* (1987) Analysis of triglyceride species of blackcurrant seed oil by HPLC via a laser light scattering detector. *Rev. Fr. Corps Gras* **34(4)**, 221–23.

Perrin, J.L. and Prevot, A. (1988) Supercritical fluid chromatography: impact in the fat and oil field. *Rev. Fr. Corps Gras* **35(12)**, 485–94.

Phillips, F.C., Erdahl, W.L., Schmit, J.A. *et al.* (1984) Quantitative analysis of triglyceride species of vegetable oils by high-performance liquid chromatography via a flame-ionization detector. *Lipids* **19(11)**, 880–87.

Plank, C. and Lorbeer, E. (1993) Analysis of free and esterified sterols in vegetable oil methyl esters by capillary GC. *J. High Res. Chromatogr.* **16(8)**, 483–87.

Plattner, R.D. (1981) High-performance liquid chromatography of triglycerides: Controlling selectivity with reverse phase columns. *J. Am. Oil Chem. Soc.* **58(5)**, 638–42.

Pocklington, W.D. and Hautfenne, A. (1985) Determination of triglycerides in fats and oils: Results of a collaborative study and the standardized method. *Pure Appl. Chem.* **57(10)**, 1515–22.

Pocklington, W.D. and Dieffenbacher, A. (1988) Determination of tocopherols and tocotrienols in vegetable oils and fats by high performance liquid chromatography: Results of a collaborative study and the standardized method. *Pure Appl. Chem.* **60(6)**, 877–92.

Podlaha, O. and Toregard, B. (1982) A system for identification of triglycerides in reversed-phase HPLC chromatograms based on equivalent carbon numbers. *J. High Res. Chromatogr. Chromatogr. Comm.* **5**, 553–58.

Podlaha, O., Toregard, B. and Puschl, B. (1984) TG-type composition of 28 cocoa butters and correlation between some of the TG-type components. *Lebensm. Wiss. Technol.* **17(2)**, 77–81.

Pollard, M. (1993) Tales told by dry bones. *Chem. Ind.* **10**, 359–62.

Proto, M. (1992) Linoleic acid and triliolein contents of some seed oils and olive oil. *Ind. Aliment.* **31(300)**, 36–38.

Rammell, C.G. and Hoogenboom, J.J.L. (1985) Separation of tocols by HPLC on an amino-cyano polar phase column. *J. Liq. Chromatogr.* **8(4)**, 707–17.

Ratledge, C., Dawson, P. and Rattray, J. (Eds) (1984) *Biotechnology for the Oils and Fats Industry*, American Oil Chemists' Society, Champaign, Illinois, USA.

Richheimer, S.L., Kent, M.C. and Bernart, M.W. (1994) Reversed-phase high-performance liquid chromatographic method using a pentafluorophenyl bonded phase for analysis of tocopherols. *J. Chromatogr. A.* **667(1)**, 75–80.

Rojas, J.M., Serruya, H. and Bentes, M.H.S. (1994) Chemometric classification of two peach palm (*Bactris gasipaes* H.B.K.) landraces (*Jurua* and *Vaupes*), *J. Am. Oil Chem. Soc.* **71(2)**, 127–33.

Rossell, J.B. (1983) Measurement of rancidity. In: *Rancidity in Foods*, 2nd edn. (Eds J.C. Allen and R.J. Hamilton), Elsevier Applied Science Publishers, London, pp. 23–52.

Rossell, J.B. (1986) Classical analysis of oils and fats. In: *Analysis of Oils and Fats*. (Eds R.J. Hamilton and J.B. Rossell), Elsevier Applied Science Publishers, London, pp.1–90.

Rossell, J.B. (1991a) Vegetable oils and fats. In: *Analysis of Oilseeds, Fats and Fatty Foods*. (Eds J.B. Rossell and J.L.R. Pritchard). Elsevier Applied Science Publishers, London, pp. 261–327.

Rossell, J.B. (1991b) Purity criteria in edible oils and fats. *Fat Sci. Technol.* **93(4)**, 526–31.

Rossell, J.B. (1992) Vegetable fats for chocolate, couvertures and coatings. *Lipid Technol.* **4(5)**, 106–13.

Rossell, J.B. (1994) Stable carbon isotope ratios in establishing maize oil purity. *Fat Sci. Technol.* **96(8)**, 304–308.

Safar, M., Bertrand, D., Robert, P. *et al.* (1994) Characterization of edible oils, butters and margarines by Fourier transform infrared spectroscopy with attenuated total reflectance. *J. Am. Oil Chem. Soc.* **71(4)**, 371–77.

Salivaras, E. and McCurdy, A.R. (1992) Detection of olive oil adulteration with canola oil from triacylglycerol analysis by reversed-phase high-performance liquid chromatography. *J. Am. Oil Chem. Soc.* **69(9)**, 935–38.

Sassano, G.J. and Jeffrey, B.S.J. (1993) Gas chromatography of triacylglycerols in palm oil fractions with medium-polarity wide-bore columns. *J. Am. Oil Chem. Soc.* **70(11)**, 1111–14.

Sato, T. (1994) Application of principal component analysis on near-infrared spectroscopic data of vegetable oils for their classification. *J. Am. Oil Chem. Soc.* **71(3)**, 293–98.

Schulte, E. (1994) Determination of edible fat refining by HPLC of Δ3,5–steradienes. *Fat Sci. Technol.* **96(4)**, 124–28.

Schwaiger, I. and Vojir, F. (1994) Use of multivariate statistical methods for examination of authenticity of edible oils. *Dtsch Lebensm. Rundsch.* **90(5)** 143–46.

Sempore, G. and Bezard, J. (1991) Determination of molecular species of oil triacylglycerols by reversed-phase and chiral-phase high-performance liquid chromatography. *J. Am. Oil Chem. Soc.* **68(10)**, 702–09.

Sen Gupta, A.K. (1991) Palm oil and palm kernel oil. *Fat Sci. Technol.* **93(4)**, 548–54.

Sengupta, P., Sil, S. and Roy, B.R. (1978) Detection of castor oil in other oils by thin layer chromatography. *J. Inst. Chem. (India)* 50(3), 139–40.

Sheppard, A.J. (1992) *Lipid Manual*, William C. Brown, Dubuque, Indianapolis, USA.

Shukla, V.K.S., Nielsen, W.S. and Batsberg, W. (1983) A simple and direct procedure for the evaluation of triglyceride composition of cocoa butters by high-performance liquid chromatography – a comparison with the existing TLC-GC method. *Fette Seifen Anstrichm.* **85(7)**, 274–78.

Shukla, V.K.S. (1988) Recent advances in the high-performance liquid chromatography of lipids. *Prog. Lipid Res.* **27(1)**, 5–38.

Singleton, J.A. and Pattee, H.E. (1987) Characterization of peanut oil triacylglycerols by HPLC, GLC and EIMS. *J. Am. Oil Chem. Soc.* **64(4)**, 534–38.

Skillicorn, A. (1994) Oilseeds get a genetic makeover. *Food Proc.* **55(2)**, 48–52.

Slover, H.T. (1983) Gas chromatography-packed and capillary. In: *Dietary Fats and Health* (Eds E.G. Perkins and W.J. Visek). American Oil Chemists' Society, Champaign, Illinois, USA, pp. 90–109.

Slover, H.T., Thompson, R.H. and Merola, G.V. (1983) Determination of tocopherols and sterols by capillary gas chromatography. *J. Am. Oil Chem. Soc.* **60(8)**, 1524–28.

Smith, S. (1994) Toxic oil victims await justice day. *The European* (London), 9–15 September 1994, p. 3.

Snyder, J.M., Taylor, S.L. and King, J.W. (1993) Analysis of tocopherols by capillary column supercritical fluid chromatography and mass spectrometry. *J. Am. Oil Chem. Soc.* **70(4)**, 349–54.

Sonntag, N.O.V. (1979) Structure and composition of fats and oils. In: *Bailey's Industrial Oil and Fat Products*, Vol. 1, 4th edn. (Ed. D. Swern), John Wiley & Sons, New York, pp. 1–98.

Sonntag, N.O.V. (1982) Interesterification. In: *Bailey's Industrial Oil and Fat Products*, Vol. 2, 4th edn. (Ed. D. Swern), John Wiley & Sons, New York, pp. 127–73.

Syvaoja, E.-L., Piironen, V., Varo, P., *et al.* (1986) Tocopherols and tocotrienols in Finnish foods: oils and fats. *J. Am. Oil Chem. Soc.* **63(3)**, 328–29.

Tallent, W.H., Kleiman, R. and Cope, D.G. (1966) Gas–liquid chromatography of silylated mixtures from lipolysis of triglycerides containing unusual fatty acyl groups. *J. Lipid Res.* **7(4)**, 531–35.

Tan, Y.I. and Kuntom, A. (1993) Gas chromatographic determination of hydrocarbons in crude palm kernel oil. *J. AOAC Int.* **76(2)**, 371–76.

TCOFI (1989) Technical Commission of the Oil and Fat Industry, Sterol-Alcohol Study Group of the Subcommission on Vegetable Oils. *Riv. Ital. Sostanze Grasse* **66(9)**, 531–38.

Thomas, A.E. (1985) Fractionation and winterization: processes and products. In: *Bailey's Industrial Oil and Fat products*, Vol. 3, 4th edn. (Ed. T.H. Applewhite), John Wiley & Sons, New York, pp. 1–39.

Thompson, J.N. and Hatina, G. (1979) Determination of tocopherols and tocotrienols in foods and tissues by high-performance liquid chromatography. *J. Liq. Chromatogr.* **2(3)**, 327–44.

Tiscornia, E. and Bertini, G.C. (1974) Further investigations on the specific distribution of fatty acids in vegetable oil triglycerides. *Riv. Ital. Sostanze Grasse* **51(9)**, 333–47.

Tiscornia, E., Forina, M. and Evangelisti, F. (1982) Chemical composition of olive oil and its variations induced by refining. *Riv. Ital. Sostanze Grasse* **59(11)**, 519–55.

Tsimidou, M., Macrae, R. and Wilson, I. (1987a) Authentication of virgin olive oils using principal component analysis of triglyceride and fatty acid profiles: part 1 – classification of Greek olive oils. *Food Chem.* **25(3)**, 227–39.

Tsimidou, M., Macrae, R. and Wilson, I. (1987b) Authentication of virgin olive oils using principal component analysis of triglyceride and fatty acid profiles: part 2 – detection of adulteration with other vegetable oils. *Food Chem.* **25(4)**, 251–58.

Tsimidou, M. and Karakostas, K.X. (1993) Geographical classification of Greek virgin olive oil by non-parametric multivariate evaluation of fatty acid composition. *J. Sci. Food Agric.* **62(3)**, 253–57.

Van Niekerk, P.J. and Burger, A.E.C. (1985) The estimation of the composition of edible oil mixtures. *J. Am. Oil Chem. Soc.* **62(3)**, 531–38.

Van Niekerk, P.J. and Hasty, R.A. (1989) Optimization of a technique for the estimation of the composition of edible oil blends. *Anal. Chim. Acta* **223(1)**, 237–46.

Warner, K. and Mounts, T.L. (1990) Analysis of tocopherols and phytosterols in vegetable oils by HPLC with evaporative light-scattering detection. *J. Am. Oil Chem. Soc.* **67(11)**, 827–31.

Weihrauch, J.L. and Gardner, J.M. (1978) Sterol content of foods of plant origin. *J. Am. Diet. Assoc.* **73(7)**, 39–47.

Wesley, I.J., Barnes, R.J. and McGill, A.E.J. (1995) Measurement of adulteration of olive oils by near-infrared spectroscopy. *J. Am. Oil Chem. Soc.* **72(3)**, 289–92.

Winkler, F.J. and Schmidt, H.-L. (1980) Scope of the application of ^{13}C isotope mass spectrometry in food analysis. *Z. Lebensm. Unters. Forsch.* **171(2)**, 85–94.

Winkler, F.J. (1984) Application of natural abundance stable isotope mass spectrometry in food control. In: *Chromatography and Mass Spectroscopy in Nutrition Science and Food Safety* (Eds A. Frigerio and H. Milon). Elsevier, Amsterdam, pp. 173–90.

Wolff, J.P., Mordret, F.X. and Dieffenbacher, A. (1991) Determination of triglycerides in vegetable oils in terms of their partition numbers by high performance liquid chromatography: results of a collaborative study and the standardized method. *Pure Appl. Chem.* **63(8)**, 1173–82.

Wolter, F.P. (1993) Altering plant lipids by generic engineering. *INFORM* **4(1)**, 93–98.

Worthington, R.E. and Hitchock, H.L. (1984) A method for the separation of seed oil steryl esters and free sterols: application to peanut and corn oils. *J. Am. Oil Chem. Soc.* **61(6)**, 1085–88.

Yamazaki, M., Nagao, A. and Yamamoto, H. (1993) Estimation of the mixing ratio of a foreign oil in adulterated sesame oil. I. Soybean oil. *Jpn Oil Chem. Soc. (Yukagaku)* **42(8)**, 606–11.

Yamazaki, M., Nagao, A. and Yamamoto, H. (1994) Estimation of the mixing ratio of a foreign oil in adulterated sesame oils. II. Rice bran oil or rapeseed oil. *Jpn Oil Chem. Soc. (Yukagaku)* **43(1)**, 10–17.

Yoshida, S.H., German, J.B., Fletcher, M.P. *et al.* (1994) The toxic oil syndrome: a perspective on immunotoxicological mechanisms. *Regul. Toxicol. Pharmacol.* **19(1)**, 60–79.

Young, C.C. (1984) The interpretation of GLC triglyceride data for the determination of cocoa butter equivalents in chocolate: a new approach. *J. Am. Oil Chem. Soc.* **61(3)**, 576–81.

8 Authenticity of honey

P.C. MOLAN

8.1 Introduction

Honey is a completely natural product. Although most commercially distributed honey has been subjected to some degree of physical processing, raw honey can be used straight from the comb as taken from the beehive. The distinction of honey as a pure natural product is maintained by food regulations worldwide: anything sold as honey has to have been produced by bees from the secretions of plants, and be free from any additives and contaminants. Nevertheless, the fraudulent practice of selling adulterated or simulated honey is widespread and has gone on since ancient times. To understand the challenge presented in detecting such fraud it is necessary to be aware of the complexity of honey, a natural product with a very varied nature resulting from the wide variability of its sources.

A single section of honeycomb may have been produced from the nectar of a single type of plant abundantly in flower at the time of collection by the bees, or from the blended nectar of several different plants: alternatively it could have been produced from the phloem sap of other types of plants, in the form of honeydew. Even honey from the same floral source can vary if it comes from different locations, or because of seasonal climatic variations. Extracted honey can be even more variable. The honeycombs from a single hive may have been filled over a period of time in which several species of flowers came into bloom. On a larger scale, the honey may have have been extracted from hives in different locations with quite different local flora, before being processed.

Honeys from some of these different sources have organoleptic qualities that are in consumer demand which gives rise to another form of fraud that is fairly common and also is a challenge to detect. This is the misrepresentation of the source of honey on sale.

8.1.1 The nature of honey

8.1.1.1 Definition of honey. Honey is defined as 'the natural sweet substance produced by honey bees from nectar of blossoms or from secretions of living parts of plants or excretions of plant sucking insects on the living parts of plants, which honey bees collect, transform and combine with specific substances of their own, store and leave in the honeycomb to ripen and mature' in the *Codex Alimentarius* produced by the Food and Agriculture Organisation of the United Nations. This definition of honey sets the standard for use in trade in honey between all countries that adopt the *Codex*. The food regulations governing the sale of honey in various individual countries around the world have been compared by Fasler (1975). These regulations, which apply where the *Codex* has not been adopted, vary a lot in the standards set for such aspects as heat treatment, mineral and water contents, but show a remarkable uniformity in the definition of what honey is.

All of the definitions make it clear that honey is a completely natural product. The *Codex* specifies that no food additives are permitted in honey. Even the product stored in the cells of the honeycomb when bees are fed sugar or syrups cannot be considered to be honey. Although some countries permit the sale of manufactured substitutes for honey so long as these are not deceptive, other countries specifically prohibit the sale of products described as 'artificial honey' or 'imitation honey' (Fasler, 1975).

There are further definitions of honey in the *Codex* and the regulations of various countries. Blossom honey or nectar honey is distinguished as coming from the nectaries in flowers; honeydew honey is distinguished as coming from the excretions of insects that suck the sap of plants. Other definitions are of the various physical forms of honey: the types of honey resulting from the different methods of extraction are defined (i.e. pressed from the comb or centrifuged or drained from the comb after removing the caps from the cells), and the styles of honey presented are defined. Comb honey, sold as whole combs or sections of combs, is in its natural state. Chunk honey is pieces of comb contained in honey removed from the comb. Crystallised or granulated honey is honey removed from the comb that has undergone a natural process of solidification resulting from the crystallisation of the glucose in the honey. Creamed honey is the product of a physical process in which honey glucose crystals are crushed to achieve a fine crystalline structure to make the honey easy to spread. Some countries also define a separate category for honey that has been heated excessively during processing (Fasler, 1975).

A common feature of the various food regulations is the specification of standards that set upper or lower limits for the levels of water, reducing sugar, sucrose, minerals and hydroxymethylfurfural in honey. These limits serve to some degree to exclude honeys that have been subjected to some of the common practices of adulteration.

8.1.1.2 Composition of honey. Honey is primarily a highly concentrated solution of a complex mixture of sugars, often containing crystals of glucose monohydrate suspended in the syrupy solution. Depending on the degree to which it has been strained or filtered during processing, honey may also have suspended in it other particulate matter. In extracted honey, there may be particles of wax from the honeycomb and debris picked up on the comb from the beehive. Honey that has not been removed from the sealed cells of the honeycomb usually still contains pollen grains, and microorganisms such as yeasts, bacteria, bacterial and fungal spores and fungal hyphae. It also has components other than sugars in solution – organic acids, enzymes, vitamins, acetylcholine, flavonoids, minerals and a wide variety of organic chemicals in trace quantities – which give honey its colour, flavour and aroma. Even when added together, however, the very large number of minor components of honey contribute very little to the total mass of honey: the average compositon of 490 samples of honey from the USA was found to be 79.6% sugars and 17.2% water (White *et al.*, 1962).

The mixture of sugars in honey is composed of mainly glucose and fructose, which together account for 85–95% of the total carbohydrate content (White, 1975a). There is usually more fructose than glucose present, the average ratio being 1.2:1 (White, 1975a; 1980b). On average, sucrose comprises about 1.5% of honey, reducing disaccharides about 6.5% and higher oligosaccharides about 1.5% (White, 1975a). In his review on the sugars of honey, Siddiqui (1970) describes the structures of 27 oligosaccharides found in honey. White (1992) cites reports of identification of 22 oligosaccharides found in honey. White also refers to reports of other oligosaccharides being found in honey from honey-dew sources: melezitose (*O*–α–D–glucopyranosyl–[1→3]–*O*–*ß*–D– fructofuranosyl–[2→1]–α–D–glucopyranoside), and a series of higher sugars resulting from the stepwise addition of glucose to erlose (*O*–α–D–glucopyranosyl–[1–4]–α–D–glucopyranosyl–*ß*–D–fructofuranoside).

Another carbohydrate in honey is the gluconic acid, which is reponsible for most of the acidity. Honey typically has a pH of 3.2–4.5, averaging about 3.9 (White, 1975a). Although there have been many other carboxylic acids identified in honey, these are minor contributors to the total acidity (White, 1975a). Part of the gluconic acid is present in a non-acidic form as its lactone, which is in equilibrium with the free acid through a slow, spontaneous hydrolysis reaction. The acidity is also reduced by the presence of various mineral ions in honey.

Although honey contains a wide range of minerals, including the trace elements of nutrition (see tables in White, 1975a), the quantities present in the amounts of honey that is likely to be eaten are too low to be of nutritional significance when compared with the recommended daily intake (Haydak *et al.*, 1975). The same applies to the wide range of amino acids and vitamins found in honey (White, 1975a; Haydak *et al.*, 1975).

The Maillard reaction between the amino groups and sugar in honey is though to account for part of the colour of honey, this being in addition to the colour arising from the content of carotenoids and phenolics, and the caramelisation of the fructose in acidic solution in honey (White, 1975a). Honey, maple syrup and caramel all have a similar absorption spectrum (White, 1975a).

The turbidity of liquid honey that has not been clarified by filtration is the result of colloidal material that is primarily protein (White, 1975b). The protein content of honey is also very low: by calculation from the average nitrogen content of 490 samples of honey from the USA (White *et al.*, 1962), and taking into account the high proportion of non-protein nitrogen in honey (White 1975a), the protein content is seen to be less than 0.2%. Several different protein molecules have been separated, some being found to come from the bee and some from particular floral sources. This is also true for some of the enzymes found in honey (White, 1992a).

Reports of the source of the diastase (starch-digesting enzyme) in honey have been reviewed by White (1975a). This enzyme activity, with both α-amylase and β-amylase components, is concluded to be added to honey by the bees. The major invertase (sucrase) activity in honey is caused by a transglucosylase that also is added by the bees and is distinctly different in its mode of action from the invertase of plants and yeasts but some may be the result of a cell-wall transfructosylase from pollen grains (White, 1992a). The third major enzyme activity present in honey is glucose oxidase derived entirely from the bees (White *et al.*, 1963). It is the activity of this enzyme that gives rise to the gluconic acid in honey.

Catalase is present in some honeys, and is derived from the pollen and nectar of certain plants (Dustmann, 1971). Acid phosphatase is also found in some honeys: this too is apparently of plant origin (White, 1975a).

8.1.1.3 Characteristics of honey. The physical characteristics of honey result from the quantities of major components that are present, fructose, glucose and water. Fructose is much more soluble than glucose, and is primarily responsible for the hygroscopic nature of honey. Glucose is present in many honeys at levels that make the syrup super-saturated. When its crystallisation from a super-saturated solution is 'seeded' with fine crystals it gives a 'creamed' spreadable honey, which consists of small crystals suspended in syrupy liquid. If crystallisation occurs spontaneously, the crystals that grow are usually coarser and a 'granulated' honey results. If the level of glucose in a honey is not high enough to give a super-saturated solution, or if crystallisation is prevented by removal of particles that could nucleate crystal growth, the honey remains as a viscous liquid.

'Creamed' honey may be very plastic if the amount of glucose crystallised is low, or may be very solid if there is a lot of glucose crystallised or the liquid honey suspending the crystals is very viscous. The viscosity of 'creamed', granulated and liquid honey depends on the amount of water present, and on the temperature. Warming to as low a temperature as 35°C or adding quite small amounts of water bring about very marked reductions in the viscosity of honey. If placed in a refrigerator, honey increases in viscosity so much that the glucose molecules cannot diffuse to crystal surfaces for growth of crystals to occur.

Since the proportions of fructose and glucose are determined by the compositon of the plant secretion, the crystallisation characteristics of honeys can serve as an indicator of the authenticity of the claimed floral source. For instance, honey from rape (*Brassica napus*) and dandelion (*Taraxacum officinale*) has an unusually high proportion of glucose and granulates very rapidly, while honey from black locust 'acacia' (*Robinia pseudoacacia*) has a very high proportion of fructose and does not granulate for several years; honey from tupelo (*Nyssa* spp.) has such a high proportion of fructose that it does not granulate at all. *Salvia* and sweet chestnut (*Castanea sativa*) honeys also contain a very high proportion of fructose.

Another physical characteristic useful as an indicator of the authenticity of the claimed floral source is the unusual property of thixotropy in a few types of honey, particularly honey from ling heather (*Calluna vulgaris*) and manuka (*Leptospermum scoparium*). This is a particularly useful indicator as these two types of honey both fetch a premium price and thus are most likely to have their identity falsely claimed.

The flavour, aroma and colour of honeys are also characteristics that serve to indicate the plant source. The number of species of plants around the world that are significant sources of honey is very large, and many of the honeys derived from these have distinctive features: the characteristics of the honey from 150 of these species are described by Crane (1975). Honeys from different plant sources can vary from almost colourless to a dark brown colour, or may even have a greenish colour. They may be very delicately flavoured or very strongly flavoured. Their identification by apiarists and honey packers is often based just on organoleptic characteristics, and this is quite adequate for the consumer who is seeking a particular type of honey because of a preference for particular characteristics, even if the plant source has been misidentified. More likely to be aggrieved are regulatory authorities or the producers whose business is adversely affected by misrepresentation. However, where a particular plant source is desired because of features other than organoleptic characteristics associated with that source, then other means of identification are obviously required. Most sensibly this would involve testing for the desired features.

One characteristic of certain types of honey that has been in demand in recent times is the antibacterial activity. All honey will prevent microbial growth if its water content is not too high: the very high sugar content sequesters the water present leaving insufficient for the microorganisms to function. In some honeys there is a potent antibacterial activity, which arises from the enzymic production of hydrogen peroxide by the oxidation of glucose when glucose oxidase is activated by dilution of the honey. Certain floral sources are associated with high or low levels of hydrogen peroxide accumulation in the honey derived from them. Other floral sources also give honeys with additional antibacterial factors of phytochemical origin. The antibacterial activity of honey has been reviewed by Molan (1992a; 1992b).

Other biological activities are characteristic of honey. Acetylcholine is present in honey at up to 6 mg/100 g (White, 1975a). There is also an unidentified anti-inflammatory factor present (Church, 1954; Kandil *et al.*, 1987; Ali *et al.*, 1990).

8.1.2 Production of honey

Production of honey can be considered at two different levels, the collection and processing of plant fluids by the bees, and the extraction and processing of honey by apiarists and honey packers.

8.1.2.1 Sources of honey. The principal source of honey is nectar from flowers, although in some forest areas honeydew can be a major source. In the conifer forests of Europe, honeydew is produced prolifically and is the main source for the production of honey (Crane, 1975), which is sold as 'forest honey'. Nectar from the extrafloral nectaries is an important source for production of honey from a few types of plants (Crane, 1975; Crane *et al.*, 1984). The phloem sap that exudes for a few days from the stumps of sugar cane (*Saccharum* spp.) after cutting is also collected by bees for the production of honey (Crane, 1975). All of these secretions are derived from the phloem sap of plants but, other than the exudate from cut sugar cane, all have their composition modified from that of the sap.

The major sugar in phloem sap is sucrose but the metabolic activity of the nectary may change the composition and the concentration of sugars and other components in the sap that is secreted as nectar (Durkee, 1983). The sugars found in nectar are principally sucrose, fructose and glucose; less commonly minor amounts of the disaccharides maltose and melibiose (6–*O*–α–galactopyranosyl–D–glucose) are found; uncommonly the trisaccharides melezitose and raffinose (6–*O*–α–D–galactopyranosyl–[1→6]–α–D–glucopyranosyl–*β*–D–fructofuranoside) are found in small amounts (Baker and Baker, 1983). Nectars can be 5–80% sugar, with sucrose the major or only sugar in some, and a mixture of fructose and glucose (in proportions ranging from 28:1 to 0.7:1) the major sugar content in others (Maurizio, 1975a). An unusual case is the nectar of the sunflower (*Helianthus annuus*), which has raffinose as a major sugar (Shuel, 1992). Fructose and glucose are formed by hydrolysis of sucrose by invertases in the nectaries. It is the transglucosylase and transfructosylase activities of these invertases that give rise to the oligosaccharides (Maurizio, 1975a). The results of more recent analyses using GC and HPLC (reviewed by Shuel, 1992) show maltose to be a rare constituent, and galactose (tentatively identified) to be present in some nectars. It is possible that the occurrence in honey of some of these rarer sugars could be used as an indicator of the plant source, assuming that the sugars are not produced by metabolism in the ripening honey as well as in the nectary.

The occurrence of the different sugars in nectar is characteristic of families of plants (Wykes, 1952) and the proportions are generally constant for each species of plant; a constant sugar composition was found in 828 out of 889 species examined (Percival, 1961). The constancy of the sugar composition of nectar, and of the amino acid composition, has been demonstrated despite environmentally induced variation in the volume and concentration of the secretion (Baker and Baker, 1983).

Although nectar is primarily a solution of sugars in water, it may also contain small amounts of amino acids, minor amounts of protein, lipids, vitamins, antioxidants, organic acids, alkaloids, phenolics, saponins and minerals (Baker and Baker, 1983). Nectar also contains some of the essential oils responsible for the aroma of flowers (Shuel, 1992). The difference in composition of nectar from floral and extrafloral nectaries of the same plant (Baker *et al.*, 1978) indicates the importance of the metabolism of the nectaries in modifying the phloem sap that is secreted.

Major modification of phloem sap by metabolism occurs also in the secretion of honeydew. In the case of this secretion, the sap leaves the plant via the gut of various sucking insects such as aphids, scale insects and leaf-hoppers. A very large amount of sap is ingested by these insects, it being forced out of the plant by the pressure of the phloem sieve tubes. The gut of these insects is constructed to allow excess fluid to bypass the mid-gut. The excess fluid is secreted prolifically, some insects secreting more than their body weight per hour (Maurizio, 1975a).

In passing through the insects, the sap has added to it various enzymes, which modify its composition depending on the species of insect. Endosymbionts in the insects also add metabolites to the sap. Thus the honeydew has a composition that is quite different from that of the original sap (Maurizio, 1975a). Some honeydews have been found to contain amino acids and sugars that are not in the plant sap. For instance, invertase from the gut of many sap-sucking insects has a transglucosylase activity that produces the characteristic trisaccharides melezitose and erlose (Maurizio, 1975a) from the sucrose in the sap. Some insects produce honeydews containing one or more of: maltose, trehalose (α–D–glucopyranosyl α–D–glucopyranoside), melibiose, mannose, rhamnose, stachyose (α–D–galactopyranosyl–[1→6]–α–D–galactopyranosyl–[1–6]–α–D–glucopyranosyl–*β*–D–fructofuranoside) and raffinose. In the manna found on some species of *Eucalyptus* the raffinose content may be up to 80% (Maurizio, 1975a).

The small droplets of honeydew secreted by sap-sucking insects fall onto the surface of leaves and twigs where they may dry out, solidifying into manna. The honeydew or manna is then collected from these surfaces by bees for production of honey. While exposed on the surfaces as a sticky liquid, the honeydew picks up airborne particles such as dust, soot, pollen from wind-pollinated plants and green algae from the surface of tree. There is also a sooty mould fungus that grows on honeydew that is a characteristic constituent of the honeydew collected by bees (Maurizio, 1975b).

Pollen is also trapped in nectar that is collected by bees. In this situation, however, nectar is not exposed like the honeydew. The pollen arises from the anthers of the flowers producing the nectar: it falls in or is shaken in by the bees visiting the flowers. The pollen introduced by a visiting bee may also be from a completely different type of flower. Pollen is picked up on the bee's body as it brushes against the anthers of each flower.

The most widely used method of authenticating the particular floral sources of honeys is based on the identification of the pollen in honey. However, there are some features that can mislead the analyst attempting to determine the source of a honey. Different species of plants produce different proportions of pollen relative to the amount of nectar; variations in flower anatomy can also mean that different amounts of pollen get into the nectar. Anther dehiscence depends on temperature, so the amount of pollen in nectar may vary from season to season and between different locations within a species. The relative timing of nectar and pollen production varies with different species of flowers. In monoecious plants, the nectar yield may be different in the male and female flowers. Also, while the nectar is held in the bee's honey sac (crop), pollen is filtered out from it by the regulatory action of the proventriculus, which can remove 50–70% of the pollen within 15 min (Maurizio, 1975a). Thus the amount of pollen in the nectar also depends on the length of time that a bee is out foraging for each load, and the length of the flight back to the hive.

A further complicating factor arises when the bees visit flowers without nectar for the collection of pollen. Pollen serves as the source of protein, minerals, vitamins and sterols for the honey-bee colony, and bees have to forage from flowers over a considerably greater area to obtain enough pollen than to obtain sufficient nectar (Shuel, 1992). Bees forage chiefly on the most abundant species within their flight range, and collect from relatively few species of plants. However, they seem to prefer pollen from a mixture of species to that from a single species, possibly to obtain a good balance of amino acids. Much of the pollen is collected from plants that cannot be sources of honey: for example, wind-pollinated species and insect-pollinated species with flowers that do not have nectar. This pollen also can get into honey, by being shaken into the nectar when other flowers are visited, by being shaken into collected nectar in open cells in the honeycomb that bees crawl over when putting their loads of pollen into storage cells and by the action of house bees processing the nectar and pollen (see section 8.1.2.2).

Another way that honey can be contaminated with pollen unrelated to the nectar source is when extracted frames of honeycomb are re-used for another season's production: the bees re-fill the cells of the old comb that have pollen residues from the previous season.

8.1.2.2 Processing by bees. The processing of nectar or honeydew into honey begins as soon as the liquids are picked up by the foraging bees. Saliva is added from the bees' salivary and hypopharyngeal glands, and the enzymes in the secretions begin the process of ripening the collected liquid into honey. When a foraging bee returns to the hive, the contents of the honey sac are regurgitated and passed to one or more house bees, which add more saliva to the collected liquid. The house bees pass the collected liquid between themselves, with yet more saliva being added. The number of bees in this chain depends on the rate at which nectar or honeydew is being brought into the hive, and on the amount of water that has to be removed from the nectar or honeydew (Maurizio, 1975a).

The house bees remove water from the liquid by repeatedly regurgitating droplets onto their mouthparts where evaporation occurs, the water content being reduced in this way from the original 60–75% down to 35–40%. They then spread the liquid over the walls of the cells in the honeycomb for further drying to occur by the flow of warm air induced in the hive by the wing movement of other bees (Maurizio, 1975a). This working by the house bees is another way in which pollen unrelated to the nectar source may be introduced into honey. These same bees use their mouthparts to pack down and smooth out the pollen pellets brought in by foraging bees and feed pollen to the brood in the hive (Gary, 1992). The rapid spread of radioactivity from radiophosphorus-labelled sugar to all of the bees in a hive when it was fed (Nixon and Ribbands, 1952) is an indication of how much interaction and thus potential spread of pollen there is between bees in a hive.

When the water content of the honey being produced has dropped below 20% (range 24–25% for ling heather honey) the bees seal the cells by capping them with a thin covering of wax. When conditions of low temperature and high humidity prevail, it may not be possible for enough water to be removed from the honey to get the level below 20%, and the cells are then capped with a higher water content.

Enzymes added from the bees' saliva retain their activity so that changes continue to occur in the composition of the honey through this ripening period and on subsequent storage. The principal change in composition when nectar or honeydew ripens into honey comes from the activity of the invertase (sucrase) from the bees' saliva. This is more active than the invertase present in the plant secretions collected; indeed the sucrose, which may have been the major sugar, initially drops to 2–4% of the total sugar content, often to 1% (Maurizio, 1975a). In some cases, either where there is a heavy nectar flow being collected and there is sufficient time for extensive working by the house bees, or the sugar concentration in the nectar or honeydew collected is high and there is no need for extensive working, the invertase level in honey is considerably lower and the level of sucrose is consequently higher than usual (White, 1992a). Honey types in which this occurs are from acacia (*Robinia pseudoacacia*), lime (*Tilia* spp.), ling heather (*Culluna vulgaris*), orange (*Citrus aurantium*) and some of the Labiatae (Maurizio, 1962). However, the level of sucrose in these honeys can decrease, if the honey is stored at warm temperatures, through the continued action of the invertase that is present (White, 1992a). Even if bees are fed sucrose as the source for honey production, invertase added through the bees' saliva gives a final product with similar sugar composition to honey produced from nectar (Maurizio, 1962).

As a result of the transglucosylase action of the salivary invertase, significant quantities of oligosaccharides can be formed during the ripening of honey (Maurizio, 1962). The sugars that are formed are maltose, isomaltose, erlose and isomaltotriose (Siddiqui, 1970), and some higher oligosaccharides formed by repeated transfer of glucose onto the terminal glucose residue of erlose (Maurizio, 1962). For example, it has been found that after 2 years of storage at $26 \pm 3°C$, 9% of the monosaccharide content of honey was converted to additional oligosaccharides by continuation of this enzyme activity (White *et al.*, 1961).

The other enzymes added to honey in the bees' saliva also retain their activity through storage but are not responsible for any changes to the composition of the honey after the ripening process. Glucose oxidase, although active in producing gluconic acid and hydrogen peroxide while the honey is ripening, is not active in ripened honey; it does, however, become active again if the honey is diluted, possibly because of raising the pH from an inhibitory level by dilution (Molan, 1992b). Diastase in honey from the bees' saliva does not have any effect on the composition of honey at any stage as there is no starch in nectar or honeydew for it to act on. The level of its activity persisting in honey being marketed is, however, used as an indicator of any heating of the honey during processing after extraction.

8.1.2.3 Processing by apiarists. Although some honey is sold in the form of honeycomb, most is extracted from the comb for marketing. One way that this is achieved is to crush or press the combs, then strain the honey to remove pieces of wax and other debris. The contents of the pollen storage cells in the comb are crushed into the honey in this process and are not removed by straining. Thus the content of pollen in this form of honey can be very high; and much of it will not be associated with the nectar sources of the honey.

The other way of removing honey from the comb is to cut off the caps from the cells with a knife or some form of mechanical device, then either drain or centrifuge the honey from them. Although most of the pollen remains in its storage cells in the comb in this type of extraction process, some gets spread from pollen storage cells to honey storage cells by the cutting across the comb, again introducing to the honey pollen not associated with the nectar sources. A much larger amount of pollen is introduced into the honey salvaged by draining or centrifuging the 'cappings', which contain the crushed tops of both pollen and honey storage cells.

Thixotropic honeys need to be 'loosened' before they can be centrifugally extracted. This involves physical agitation to break up the gel, and this is commonly performed by plunging a sheet of needles repeatedly into the comb; an action which also breaks up pollen storage cells. This is probably why ling heather and manuka honeys, both thixotropic, have a high total pollen content (Maurizio, 1975b; Moar, 1985).

The filtering of honey that is often carried out as part of the processing for packaging for sale removes some or all of the pollen. Pollen grains are 5–200 μm in diameter and some can be removed by the fabric filter socks that are used for straining honey. Swiss food regulations prohibit strainers below 200 μm, which could remove pollen grains; this is to avoid misleading statements on the origin of honeys (Verordnung über den Verkehr mit Lebensmittein, Art. 217/3, 1971). The clarification of honey by filtration after the addition of diatomaceous earth filter aids completely removes the pollen, and thus prevents any identification of source by analysis of pollen. It also gives a product that technically cannot be described as pure honey, as diatoms from the filter aid added can be detected by microscopic examination (Maurizio, 1975b). To make honey more desirable for some markets, the colour, flavour and colloids may be removed by addition of activated carbon or bentonite, in conjunction with filtration with diatomaceous earth (Townsend, 1975). Additional particles will thus be added to the honey and a further adulteration in these processes involving filtration with diatomaceous earth is the addition of water to honey (1:4 w/w) to make filtration possible (Townsend, 1975).

In order to lower the viscosity to facilitate extraction, straining or filtration, the honey is usually warmed to a temperature of 32–40°C. This temperature is not much higher than that at which the honey is stored in the beehive, and for the short period involved in processing does not have much effect on the honey. However, there are adverse effects on honey when it is heated to higher temperatures for liquefaction or pasteurisation. Such excessive heating may occur if honey is pasteurised in bulk: the times required if rapid heating and cooling are involved are 7.5 min at 63°C or

1 min at 69°C (Townsend, 1975). Similarly, adverse effects may occur if honey is heated in bulk to 60–80°C to redissolve the crystals of glucose that may form on storage. Honey so treated may remain at fairly high temperatures for a further long period as it cools afterwards.

Ensuring that all of the crystals have dissolved is important if the honey is to be marketed in a liquid form, as any remaining will serve to nucleate further crystal growth. Another procedure used to obtain honey in a stable liquid form (i.e. honey that will not granulate on the shelf) is by the addition of water (Walsh, 1960). This dilution of the sugars, however, increases the risk of osmophilic yeasts being able to survive in the honey, so pasteurization is needed. Thus, unless a honey is from a plant source that gives a high ratio of fructose to glucose, then there is a high probability that it has been diluted and/or pasteurised.

Honey sold in the creamed form has crystal growth seeded with the very fine crystals of glucose obtained by grinding some crystallised honey. Subsequent growth of the crystals is allowed to occur at 14°C to get the desired smooth texture. Although the water content of the honey in which the crystals are dispersed increases as glucose monohydrate (water content of 9.1%) precipitates from a sugar solution (water content of about 18%), this does not generally give rise to problems of yeast fermentation and pasteurization is unnecessary (Dyce, 1975).

8.2 Authenticity of honey

8.2.1 Adulteration

Adulteration of honey is seen as a major problem, threatening the viability of the beekeeping industry (Rubenstein, 1977; Doner *et al.*, 1979; White, 1981). In addition, it is cheating the consumer who is expecting to get 100% natural product with its full complement of the complex of minor components that give honey its organoleptic qualities and biological activity. Extending honey with cheaper sugars and syrups increases the amount of honey on the market, which lowers the price obtained by apiarists for the pure, natural product. The extent to which this happens is great enough to cause serious concern to apiarists, leading to them taking action to attempt to stop the practice (White, 1981; Croft, 1983).

This is not only a recent problem: in the 1880s the USA domestic market was flooded with an adulterated product that contained only 15% honey, with disastrous effects on the sale of pure honey (Graham, 1979). In his review of the historical aspects of honey adulteration, Graham (1979) suggests that adulteration has been going on for as long as honey has been an item of commerce; he refers to the addition of concentrated grape juice to honey in ancient Roman times, and to the act passed in England in 1581, which referred to the selling of 'deceitful mixtures' as honey, and ordered the confiscation of honey found to be adulterated with any foreign substance. Further forms of deception referred to are the practices in the USA in the last half of the 19th century of selling pieces of honeycomb in containers filled with glucose syrup, and adding a few dead bees to the top of containers of adulterated honey to give the product more credibility.

The degree of deception can vary from the hard-to-detect practice of adding small amounts of water and sugar to the sale of a totally artificial product.

8.2.1.1 Extension with water. The food regulations of most countries do not allow the addition of anything to honey; water so added to honey is thus an adulterant and, in Germany, its addition is specifically prohibited. It is nevertheless common practice to add water to achieve a stable liquid honey. To limit the fraudulent extension of honey by dilution with water, food regulations often set standards for the maximum water content permitted. However, the natural water content of honey has a large variation, which has to be allowed for in the standards, and there are frequently exceptions for certain floral types of honey that naturally have higher levels. For example, the maximum content of water in honey permitted in the *Codex Alimentarius* is 21% for all except clover (*Trifolium* spp.) and heather (*Calluna* spp.) honeys for which it is 23%.

The water content of honey is usually determined by measuring the refractive index. Hand-held refractometers are commonly used by apiarists, these having a standard error of ±0.4%; however, the bench-type instrument gives greater accuracy (White, 1992a). The reliability of the method is discussed by White (1975b), who presents tables showing the water content corresponding to different refractometer readings. One table is for the conversion of Brix values (% w/w sucrose) to the correct values for honey, which vary from those for sucrose. The other table, based on experimental data obtained by Chataway (1932), shows the percentage water in honey that corresponds with various refractive index measurements at 20°C. (Measurements at higher temperatures need to be corrected to 20°C by the addition of 0.00023 to the measured refractive index for each degree above 20°C.) The water content of honey can also be determined by measuring the moisture loss on vacuum-oven drying, and also by measuring its relative density (specific gravity). A table showing the percentage water in honey that corresponds to various relative densities is also presented by White (1975b), who points out the difficulties in measuring relative density using both pycnometry and hydrometry, and the relative imprecision of both the density and drying methods compared with measurement of refractive index.

The limit of 21% water in honey allows some extension of the quantity of product available for sale. Indeed, the water content of 490 samples of honey in the USA was found to range from 13.4–22.9% (White, 1975a). This would not normally be detected but could, in principle, be observed using isotope-ratio mass spectrometry using ^{18}O:^{16}O and ^{2}H:^{1}H ratios (Doner *et al.*, 1992). Measurement of the ^{2}H:^{1}H ratio has been used successfully to detect the addition of water to orange juice (Bricout, 1973), and of the ^{18}O:^{16}O ratio to detect the water in syrup added to orange juice (Doner *et al.*, 1992). With the additional evaporation involved in the drying of the nectar and honeydew to become honey it is likely that the ratios in pure honey would be higher than in orange juice. However, the use of this technique with honey would require a database of the natural variations of the ratios in honey to be established first. If the variation is low enough for this to be a sensitive technique for detecting the addition of water, then it could also serve to detect the adulteration of honey with syrups as has been done with detection of adulteration of orange juice.

8.2.1.2 Extension with sugar and syrups. Although the view may be held by some that the addition of water to bring more concentrated honeys to a standard moisture content to achieve the desired physical characteristics is not a form of adulteration, the extension of honey by the addition of sugars or syrups is unarguably adulteration. Although in times past glucose was commonly used as a honey adulterant (Graham, 1979), it is not favoured because of its relatively low solubility, which would give excessive granulation in liquid honey or a hard creamed honey. Glucose is, however, available cheaply as corn syrup (a hydrolysate of corn starch) and, in this form, is a traditional adulterant of honey (Doner *et al.*, 1979). It is also possible that finely ground glucose crystals could be used instead of ground honey for seeding crystal growth in creamed honeys.

Sucrose is more soluble than glucose and a proportion may be added to honey without adverse effects on crystallisation but it does reduce the solubility of glucose. However, it is a minor natural component in most types of honey, so its presence at significantly high levels is an easily detected indicator of adulteration. The adulteration of honey with invert sugar cannot be detected by direct measurement because its constituents are the major natural components of honey and the adulterated product would also have similar physical properties to natural honey. The addition of invert sugar to honey is detected indirectly by measuring the high level of hydroxymethylfurfural that is in invert sugar as a result of reaction of fructose with the acid used for the hydrolysis of sucrose.

The most favoured extender of honey in more recent times has been high fructose corn syrup (HFCS), prepared by enzymically isomerising glucose to fructose in corn syrup. This product is readily available and has a fructose and glucose composition that is very similar to honey. As it is purified as part of the biotechnological process in which it is produced, addition of HFCS to honey cannot be detected easily. Up to 50% of an adulterated honey could be HFCS without the sugar composition being outside the legislative standards for honey, which encompass the variability in composition of honeys (White, 1987). It is available at less than the price of honey (Trumpeter, 1981) and is thus a tempting extender. Syrups with a higher proportion of fructose, and pure fructose itself, are more expensive and are unlikely to be used as adulterants.

8.2.1.3 Feeding sugar and syrups to bees. When bees are fed syrups or solid sugar (to maintain the colony in winter when there is insufficient nectar being collected) they process and store any surplus in honeycombs in the same way that they handle nectar or honeydew (Simpson *et al.*, 1974). Thus the adulterants already mentioned may be found in honey on sale when they have not been added deliberately to honey as an act of fraud but are present as a result of careless beekeeping practices. This is particularly so with HFCS, which is much more readily taken up and hoarded as honey by bees than sucrose is (Rinderer and Baxter, 1980). There are recommendations that bees should not be fed HFCS or other sugars when there is any chance that it will be stored in honeycomb that will be extracted, and the world's largest honey packer has recommended that HFCS not be fed to bees at all (Rinderer and Baxter, 1980; Editorial, 1981).

There is also the possibility of honey being sold that has been produced by deliberate heavy feeding of dry sugar or various syrups to bees. Honey produced in this way is pale-coloured and has a weak nondescript flavour (Rodgers, 1975). These are seen as desirable attributes by many consumers who dislike dark-coloured and strong-flavoured honeys and it would certainly be a useful product to blend with some of the less desirable stronger honeys. Although the honey from bees fed on sugar or syrups generally has a higher sucrose content than normal, it otherwise has a sugar composition similar to that of true honey (Maurizio, 1962). It also has the principal enzymes of honey present as these are derived from the saliva that bees add to the stored syrup in the same way as to nectar to honeydew (White, 1975a). Its main difference from true honey is in the lack of minor components that are normally derived from the plant sources of honey, some of which many consumers value for their perceived biological and nutritional properties.

8.2.1.4 Artificial honey. Products on sale and labelled as honey may be totally fraudulent if they are manufactured syrups made to simulate honey. The various sugar products used as adulterants of honey could be passed off as such if they had added to them an acidulent and honey flavouring. For example, a patent has been published for the manufacture of artificial honey by heating sucrose with lactic acid to hydrolyse it to a mixture of glucose and fructose, with an aqueous extract of corn pollen added to give it the flavour and aroma of bee honey (Sidorciuc and Popescu, 1972). A patent has also been published for the manufacture of synthetic honey flavouring by heating a solution of a monosaccharide with phenylalanine and some derivatives (Morton and Sharples, 1959). Nearly all phenylacetic esters are known to have a honey flavour (Jacobs, 1955).

8.2.2 Claims which may not be authentic

8.2.2.1 Natural honey. As discussed in section 8.1.1, if a product is described as honey then, according to most food regulations, it must be a completely natural product free from any additives. Thus the authenticity of natural honey may be challenged by testing for the addition of water and of the sugar adulterants metioned in the preceding section. More difficult is challenging the purity of honey in respect of possible additives to enhance its colour, flavour or keeping qualities. As a result of the large variability in the minor organic components of honey, it could be difficult to prove that a component (other than a synthetic dye) giving colour to the honey, or a flavour component, had not come naturally from a plant source. Even if benzoate is detected in honey, it does not necessarily mean that preservative has been added as it may have come from a plant source. Addition of sodium benzoate to honey has been proposed as a means of stopping yeast fermentation in honeys where there is a risk of spoilage because of high water content but it is said that it has never been widely used (Dyce, 1975). However, if it were suspected that there were food additives in a honey it may be possible to determine whether or not the chemical in question is a natural component of honey by use of site-specific nuclear magnetic resonance (NMR) spectroscopy to examine the site distribution of natural-abundance deuterium in the molecules, as has been used to distinguish natural and synthetic benzaldehyde flavourings (Hagedorn, 1992).

This technique should also be useful for determining whether any phenol and benzaldehyde (and benzoate arising from oxidation of benzaldehyde) present in honey are natural components from plant sources or the result of contamination of the honey by the apiarist using these substances to repel bees from the honeycombs when removing the combs from the hive. Sporns (1981) has published a method for determining the amount of phenol in honey using HPLC but attributes all phenol detected to the action of apiarists contaminating the honey. However, phenol at concentrations of 0.5–0.7 μg/g has been found in some New Zealand ling heather honeys (Tan *et al.*, 1989) and at concentrations of 1.2–2.8 μg/g in some New Zealand kamahi (*Weinmannia racemosa*) honeys (Tan, 1989), yet phenol is not used as a bee repellent in New Zealand. Takeba *et al.* (1990) also have reported the presence of low levels of phenol in honeys that had not been exposed to bee repellent. Benzaldehyde also could be in some honeys naturally rather than as a contaminant as it is known to be produced by plants as a secondary metabolite.

It is a philosophical point whether or not the occasional presence in honey of other substances such as soot, petrol hydrocarbons and insecticides that may be thought of as contaminants allows the description of the honey as 'natural' to be challenged. Although undesirable in a foodstuff, they are present in the honey by way of the natural production process of the bees in a polluted environment. Antibiotics may also arise in honey as a result of apiarists feeding them to bees to control diseases in the hive.

8.2.2.2 Raw honey. Honey is frequently on sale labelled as 'raw' or not having been heated in processing. This is to meet the consumer demand for unrefined and unprocessed foods that has increased in recent times. In some countries the heat-labile biological activity of honey is recognised as being desirable, and food regulations restrict the heating of honey during processing.

8.2.2.3 Botanical source. Many honeys are claimed to be from a particular floral source, or to have come from a particular type of flora, for example the 'forest honey' from the honeydew produced in the conifer forests, which is a favoured type in central Europe. In most cases, the source is stated because consumers prefer the characteristic flavour of the honey from a particular plant source. The distinct flavour of ling heather honey is preferred enough for this type of honey to fetch a higher price than other types (Sawyer, 1975). There may be other desired features associated with particular plant sources. For example, tupelo honey has such a high proportion of fructose that it does not granulate and is thus in demand for use in pharmaceutical preparations (Crane, 1975), and manuka honey is in demand for its unique antibacterial properties (Matheson, 1993).

In most cases, the identification of the source of a particular honey will have been made by the apiarist on the basis of the flavour, colour, aroma and texture of the honey, the location of the hive and the season in which the honey was produced (i.e. which species of plants would have been available to the bees as sources of the honey). There is also the consideration that no honey produced by bees flying free is likely to be unifloral. Therefore the claimed source of honeys is often challenged, particularly when it is of a type that is in demand.

8.2.2.4 Geographical source. As a result of tariffs and other restrictions on international trade, many countries require the country of origin of honey on sale to be stated on its label. In many instances, however, it may be more profitable for a trader to misrepresent the country of origin. This may be seen as a serious threat to the continuation of a national beekeeping industry (Croft, 1983) and consequently honeys are examined to determine if their country of origin is stated correctly.

8.2.2.5 Antibacterial activity. In recent times, there has been a marked increase in interest in the use of honey as an antibacterial substance as a result of publication of various findings on its effectiveness, and this has led to honey being sold with claims of antibacterial properties on the label. The antibacterial activity of honey has been recognised for a long time (for a review of the subject see Molan, 1992a; 1992b). There have been numerous reports in the medical literature of the effectiveness of honey as an antiseptic dressing for wounds, burns and ulcers (cited in Molan, 1992a), and the sensitivity of the major wound-infecting species of bacteria to honey has been demonstrated (Willix *et al.*, 1992). It has also been found that the non-peroxide antibacterial activity of manuka honey is effective in inhibiting the growth of *Helicobacter pylori*, the bacterium that is responsible for causing gastritis and duodenal ulcers, whereas the hydrogen peroxide activity of other honey does not (Al-Somai *et al.*, 1994). However, there is a very large variation in the antibacterial activity of different honeys, with many having insignificant levels of activity (Allen *et al.*, 1991) and it is therefore necessary to check that particular batches of honey have antibacterial activity.

8.3 Methods for determining authenticity

8.3.1 Detection of heating of honey

Excessive heating of honey is easily detected as it affects the flavour and colour through caramelisation of the sugar component. Lesser degrees of heating may be detected by measuring the amount of hydroxymethylfurfural formed, or by assessing the amount of denaturation of the honey enzymes.

8.3.1.1 Hydroxymethylfurfural. All honey contains some hydroxymethylfurfural (5–[hydroxymethyl]–2–furaldehyde), which is formed by the action, even at ambient temperature, of the acidity of honey on the monosaccharides present, especially fructose. The normal level in honey is low (mostly below 10 μg/g in 1554 honeys imported into Germany and Switzerland; Duisberg and Hadorn, 1966). White and Siciliano (1980) found that in 41 samples of comb honey the average level was very low (2.7 μg/g) but in 481 samples of commercially processed honey it was 6.2 μg/g, reflecting the various amounts of heating during the processing.

Normal large-scale commercial processing involving melting in drums, heating for filtering and pasteurisation, and storage for 1 year was found to increase the level by an average of 8.5 μg/g (White and Siciliano, 1980).

The formation of hydroxymethylfurfural in honey depends on the time and the temperature, there being a linear relationship between the temperature and the logarithm of the time taken to form a given amount (White, 1992a). Elevated levels thus indicate heating during processing but also may be the result of prolonged storage at warm temperatures. White (1980a) has reviewed the literature on the effects of heating on the content of hydroxymethylfurfural in honey. In a later discussion of the subject (White, 1994) he presents argument for the content of hydroxymethylfurfural being used as the definitive measure of the cumulative exposure of honey to heat in processing and storage. Although it is formed at different rates in different honeys, depending on their water content and acidity, the differences in the rates are small.

The initial European regional standards of the *Codex Alimentarius* specified a maximum permitted level of 40 μg/g to protect consumers from the detrimental effects of excessive heating on the quality of honey. This is a high limit for honey claimed to be 'raw' or unheated, representing about 2.4 h exposure to 80°C. However, it could also result from about 3 months' storage at 35°C and has been a problem for honey producers in subtropical climates (White, 1994). The new world-wide standard of 80 μg/g set in 1987 in the *Codex Alimentarius* recognises this problem.

The commonly used colorimetric method for the determination of hydroxymethylfurfural in honey, using barbituric acid and *o*-toluidine (Winkler, 1955) gives an unstable colour with temperature-sensitive development. An accurate ultraviolet absorption method has been developed (White, 1979a), which measures the difference spectrum after removal of hydroxymethylfurfural by reaction with bisulphite. An accurate HPLC method has also been developed (Wootton and Ryall, 1985).

8.3.1.2 Enzyme activity. Another indication that a honey has been heated is the loss of activity of the characteristic enzymes. Although there have been some attempts to use glucose oxidase activity as a measure of heating, this is not satisfactory because there are factors in honey that can degrade the hydrogen peroxide produced in the reaction, and the enzyme is inactivated by light as well as heat (see references in Molan, 1992b). The invertase and diastase activities have been used, however, although diastase activity is preferred as it is more stable (Rodgers, 1975). The most widely used method for determining diastase activity in honey is modified from that of Schade *et al.* (1958) as described in the *Codex Alimentarius*, and measures colorimetrically the rate of disappearance of a starch-iodine solution.

White (1975a) has given empirically derived equations for the half-life

($t_{1/2}$) of each enzyme over the temperature range 10–80°C, so that for diastase:

$$\log t_{1/2} = \frac{(^1/_T - 0.003000)}{0.000130} \qquad (8.1)$$

and for invertase:

$$\log t_{1/2} = \frac{(^1/_T - 0.003083)}{0.000113} \qquad (8.2)$$

where T is the temperature in K. The half-lives at the temperatures at which honey is likely to be stored fall on the same line as those at processing temperatures. Thus, for example, the loss of half of the activity of diastase that would indicate that a honey had been heated to 80°C for 1.2 h or 70°C for 5.3 h could instead be the result of storage at 30°C for 200 days. In this respect, the loss of enzyme activity is like the formation of hydroxymethylfurfural, a cumulative measure of the exposure of a honey to heat in processing and storage. However, it is a much less exact measure than formation of hydroxymethylfurfural because it cannot be known what the starting level of activity was before exposure to heat. The content of hydroxymethylfurfural in honey is near zero when the honey is first made by the bees but the activity of enzymes is very variable because different amounts of saliva are added by the bees under different conditions (see section 8.1.2). Thus a honey found to have a low invertase or diastase activity may have been heated but could, like citrus honey or clover honey, have had a low level of activity when made (White, 1994).

8.3.2 Determination of antibacterial activity

The most commonly used method of determining the antibacterial activity of honey has been the measurement of the 'inhibine number', which is the lowest concentration of a honey in a nutrient agar plate that will prevent the growth of a culture of *Staphylococcus aureus* on the plate. This method (described by Schade *et al.*, 1958) takes a large number of plates, is imprecise because it involves steps of concentration, and gives results that may vary if changes occur in the culture used. The use of an agar-well diffusion assay with a reference antibacterial substance such as phenol (as described by Allen *et al.*, 1991) overcomes these problems. Different measures of antibacterial activity will be obtained with different species of bacteria (Willix *et al.*, 1992). *Staphylococcus aureus* is the preferred species for testing because it is the most osmotolerant, so there is no antibacterial activity seen from the high sugar content of honey.

The consumer demand for manuka honey as an antibacterial substance (Matheson, 1993) is because of the unusual high level of non-peroxide antibacterial activity that is found in some batches of this type of honey. To ascertain authenticity where this type of activity is being claimed, it is necessary to test the antibacterial activity with hydrogen peroxide removed by the addition of catalase (Allen *et al.*, 1991).

8.3.3 Detection of adulteration

8.3.3.1 Detecting changes in composition. Taste testing revealed that natural honey, honey produced by feeding bees sucrose syrup, and the sucrose syrup itself could be distinguished easily from each other by their taste but that honey with 25% sucrose syrup added could not be easily detected (Simpson *et al.*, 1974). Honeys with a large proportion of syrups added have also been detected by their appearance (Rubenstein, 1977). However, it is difficult to detect when smaller proportions of a syrup are added, and this can be so even if chemical analysis is used as a means of detection.

The adulteration of honey with large amounts of a syrup would cause a detectable change from the normal honey concentration of the corresponding sugar or sugars but smaller amounts could still be within the normal range for honey. Since the natural level of sucrose in honey is low (about 1.5% on average), the adulteration of honey with quite small amounts of sucrose will take the level of sucrose above the normal range. The *Codex Alimentarius* sets a limit of 5% sucrose for most floral types of honey as a simple way of limiting adulteration with sucrose. However, the level of glucose in honey is much higher (about 30% on average) and is very variable, so quite a lot of glucose can be added before the glucose:fructose ratio becomes abnormally high. If invert sugar or high fructose corn syrup (HFCS) are used to adulterate honey then even very large proportions will not take the glucose and fructose levels outside the normal range and indirect means of detection are required.

The adulteration of honey with large amounts of a syrup would cause a detectable change from the normal honey concentration of any of the minor components of honey not present in the adulterant. A lower-than-normal level of honey components that are contributed by the bees would indicate adulterated or artificial honey; whereas a lower than normal level of only the honey components that are derived from nectar would indicate that the honey had been produced by feeding sugar to bees. Components that could be expected to be at abnormally low levels in artificial honey or heavily adulterated honey would be gluconic acid, acetylcholine, minerals, vitamins, amino acids and the enzymes glucose oxidase, diastase and invertase. Of these, the minerals, vitamins and amino acids (other than proline, which comes mostly from the bees) would be low in honey produced by bees fed on sugar or syrups, in which the content of amino acids is very low compared with that in honey (Davies, 1975). The acid phosphatase activity would also be low in honey produced by bees fed on sugar or syrups (White, 1975a). An abnormally low content of pollen grains in honey would also indicate that the honey has been adulterated or produced by feeding sugar or syrups to the bees. (Pollen would be completely absent if the honey had been filtered, or would have a preponderance of the smaller grains if it had been strained through a fine mesh.)

Characteristic honey oligosaccharides would also be decreased by adulteration. Maltose is the major oligosaccharide (Doner, 1977) but may be contributed to by syrups derived from starch hydrolysates. Erlose, present in small but significant quantities (Doner, 1977) is produced in honey by the action of the invertase added by the bees, and in honeydew by the insects' enzymes. It appears to be unique to these sources (White, 1987) so would not be present in artificial honeys. Some of the other oligosaccharides produced by the transglucosylase action of the insects' invertase (see section 8.1.2.2) could also be expected to be present in natural honey but not artificial honey. All of these would be present in honey produced by bees fed sucrose, however. Chromatographic methods for determining the sugar composition of honey (monosaccharides and oligosaccharides) have been reported, Bogdanov and Baumann (1988) using HPLC and Perez-Arquillué *et al.* (1994) using gas chromatography with flame ionisation detection (GC–FID).

The lack of protein in a honey was, in the early part of the century, used as a means of detecting artificial honey (White, 1975a; White and Rudyj, 1978b). The various standard methods of the time for measuring protein were used, and one author reported detection of adulteration by an immunological method using an antiserum raised against the protein from buckwheat honey. White and Rudyj (1978b) have also examined the possibility of measuring protein in honey to detect adulteration, using the Lowry method to determine protein after dialysing the honey samples to remove interfering free amino acids. The results obtained from the many honey samples of certified origin showed that it would be suitable for showing only gross adulteration: the coefficient of variance was 42.1%. A similar conclusion came from examining the possibility of measuring proline in honey to detect adulteration (White and Rudyj, 1978a), the coefficient of variance being found to be 40%.

8.3.3.2 Detection is limited by variability in composition. Doner *et al.* (1979) outlined the quest for sensitive methods for the detection of adulteration of honey. There are two approaches that have been taken to detect adulteration of honey with mixtures of fructose and glucose that are similar in composition to honey. One approach, as already discussed, is the detection of dilution of characteristic honey constituents by the adulterant. The other approach is the detection of components of the adulterants that are not normally in honey. The latter is inherently more sensitive than attempting to measure a small decrease in the level of a component. Also it is the only way of reliably detecting moderate degrees of adulteration when the very large variability in the composition of honey is taken into account. For dilution of a component to be conclusive evidence of adulteration, the component has to be at a level that is lower than would occur naturally. The variability seen in the composition of honey is shown in Table 8.1.

8.3.3.3 Detecting adulteration with sucrose. As already mentioned the addition of sucrose to honey can be detected simply by measuring the level of sucrose in the honey: a small proportion of sucrose added will raise noticeably the normally low natural level. The *Codex Alimentarius* sets a limit of 5% for the sucrose content of honey but recognises that in some floral types higher levels occur naturally: levels of 10% and 15% are permitted for these specified types. Thus it is necessary to ascertain the floral source of a honey before presuming that a high level of sucrose is caused by adulteration. A validated method for determining sucrose simultaneously with fructose and glucose by HPLC is described by White (1979b).

Measuring the level of sucrose in honey will not always enable detection of honey produced by feeding sucrose to bees. Although the level of sucrose in honey is generally higher than normal, this probably reflects the feeding of a concentrated sucrose syrup rather than the feeding of sucrose *per se*. As discussed in section 8.1.2.2, smaller amounts of salivary invertase are added by the bees if less 'working' is needed to get the sugar solution concentrated to <20% water. However, the sucrose will continue to be converted to fructose and glucose by invertase activity if the honey is

Table 8.1 The variation seen in the composition of honey[a]

Component	Mean concentration (%)	Range	No. of samples	CV[b]
Water	17.2	12.2–22.9	490	8.7
Fructose	38.4	30.9–44.3	439	4.6
Glucose	30.3	22.9–40.8	439	10.0
Sucrose	1.3	0.3–7.6	439	66.4
Maltose[c]	8.6	3.3–18.2	439	24.1
Higher sugars	1.4	0.1–3.9	439	81.6
Total acid[d]	0.57	0.17–1.17	490	35.1
Ash	0.17	0.02–1.03	490	88.8
Protein	0.17	0.058–0.79	740	42.1
Proline	0.048	0.015–0.14	482	38.5

[a]Source: White (1975a; 1975b; 1908b; 1992a), from certified samples representative of major commercial types of honey in the USA.
[b]CV = coefficient of variation (i.e. the standard deviation as a percentage of the mean).
[c]Including other reducing disaccharides.
[d]Calculated as gluconic acid and gluconolactone.

stored under warm conditions, so may become too low to detect. If bees are fed with dry sucrose, the amount of invertase added is higher than normal because the bees secrete saliva to dissolve the sugar (see references in White, 1975a), so again will be too low to detect. Thus some of the methods for detecting invert sugar in honey are better for detecting honey produced by feeding sucrose to bees.

8.3.3.4 Detecting adulteration with invert sugar. The traditional method of detecting the presence of invert sugar in honey is the detection of the hydroxymethylfurfural that is produced when sucrose is heated with acid to hydrolyse it to its constituent monosaccharides fructose and glucose. Although hydroxymethylfurfural is formed in natural honey, the levels in acid-inverted sucrose are much higher: values of 1700–6500 μg/g are cited by White (1980a). Adulteration of honey with 5% of these invert sugars would give levels of hydroxymethylfurfural in excess of the *Codex Alimentarius* limit of 80 μg/g. However, high levels of hydroxymethylfurfural in honey may be the result of excessive exposure to heat rather than to adulteration. It has been proposed that a level of 200 μg/g be taken as an indication of adulteration, and thus the need for further testing, as this represents a degree of heating that is not encountered in normal processing and would give an obvious deterioration in organoleptic quality (White and Siciliano, 1980). The view was stated that only excessively high levels (>500 μg/g) should be taken as being conclusive evidence of adulteration.

Nevertheless, there are invert syrups available commercially that are produced by enzymic hydrolysis of sucrose – a process that does not form any hyroxymethylfurfural. Detection of adulteration of honey with these poses more problems. One suggestion is that the fragments of disintegrated bacterial cells from the enzyme preparation used be detected by microscopy (Marek *et al.*, 1980). Another proposal is to use polarography to detect the presence of invert sugar in honey (Marek *et al.*, 1980); the ratio between the height of the oxygen peak and the height of the reduction wave of D-fructose is higher for artificial honeys than for natural honeys. A third possibility is to examine chromatographically the disaccharides and trisaccharides in the honey; if the pattern is unusual and like that produced by the transfructosylase action of yeast invertase, this indicates the presence of enzymically produced invert sugar (White, 1980b). The use of ^{13}C isotope analysis has also been reported for the qualitative and quantitative analysis of structurally similar disaccharides in honey (Low *et al.*, 1988).

If adulteration is carried out with invert sugar produced from cane sugar, it is best detected using isotope ratio mass spectrometry.

Beet sugar can be detected by way of its content of raffinose and some other galactose-containing oligosaccharides, which are present only at low levels in most honeys other than some produced from honeydew. White *et al.* (1986) have described a technique using galactose oxidase and a chromogen to measure colorimetrically the terminal galactose in these oligosaccharides after concentrating them from honey on a charcoal column. The mean galactose content of 23 commercial beet sugar products was 301 μg/g (s = 173) but only 31 μg/g (s = 44) for 81 honeys representative of the commercially significant floral sources in the USA. Only 9% of the honeys fell in the beet sugar range, although some were found that had a galactose content far in excess of that of beet sugar. It was proposed that any with a level above 80 μg/g be considered to be possibly adulterated, and be tested by other means, such as measurement of hydroxymethylfurfural.

A technique to detect the 'fingerprint' oligosaccharides, which are formed in the production of invert sugars, has more recently been described (Swallow and Low, 1994). The methodology, using anion-exchange liquid chomatography with pulsed amperometric detection, was able to detect the addition of 10% of various syrups to honey. Each type of syrup (invert sugar produced by acid or enzyme hydrolysis of beet sugar or cane sugar, and HFCS) contained characteristic oligosaccharides that were not detectable or present in much lower levels in the wide range of honeys they examined. The authors recognised the need for more pure honeys to be examined to establish baseline data but are of the opinion that this technique looks promising for determining the authenticity of honey.

8.3.3.5 Detecting adulteration with corn syrups by chromatography. Both conventional corn syrup and HFCS contain less hydroxymethylfurfural than honey, so cannot be detected in honey by testing for this component (White and Siciliano, 1980). However, these syrups have other constituents that can be detected in honey to reveal adulteration. Even after the extensive refining during its manufacture, HFCS contains about 4.4% oligosaccharides resulting from incomplete hydrolysis of the corn starch (Doner *et al.*, 1979a). These oligosaccharides are not found in honey naturally and thus if detected in honey serve as evidence of adulteration with a starch hydrolysate. They also allow the detection of conventional corn syrup, and syrups prepared from starches other than corn starch, as well as from C_3 plants that cannot be detected by carbon isotope ratios. A technique for concentrating selectively the oligosaccharides from honey on a charcoal-celite column then separating them by thin layer chromatography (TLC) has been reported by Kushnir (1979). Pure honeys give only one or two spots with R_f >0.35, whereas corn syrup gives a series of spots or streaks. The difference is sufficient for less than 10% of corn syrup in honey to be detectable (Doner *et al.*, 1979a). A method of determining these oligosaccharides in honey using reverse-phase HPLC has also been described (Lipp *et al.*, 1988).

The disaccharide isomaltose remaining from hydrolysis of the corn starch can also be used as an indicator of the presence of corn syrup in honey. Although both the 1→6 linkage (giving isomaltose) and the 1→4 linkage (giving maltose) are formed in honey by the transglucosylase action of the honey invertase, the proportion of isomaltose is small, whereas the proportion of isomaltose remaining in starch hydrolysates is larger. GC analysis of 80 certified samples representative of the types of honey produced commercially in the USA, and 35 samples of imported honey, showed the mean maltose:isomaltose ratios to be 0.34 (SD 0.23) and 0.39 (SD 0.17) respectively, whereas the mean ratio for 21 samples of HFCS was 2.09 (SD 1.12) (Doner *et al.*, 1979b). The authors tested this technique on pure honeys and honeys found by isotope ratio analysis to be adulterated, and found that the best discrimination came from taking a ratio of greater than 0.51 to indicate adulteration. This gave correct conclusions for 121 of the 124 pure honeys and 29 of the 37 adulterated honeys tested.

8.3.3.6 Detecting corn syrup and cane sugar by isotope ratio mass spectroscopy (IRMS). The sugar in the phloem sap of plants is derived from enzyme-catalysed reactions fixing CO_2 in photosynthesis and reflects the kinetic difference between CO_2 molecules containing ^{12}C and the heavier ^{13}C. The organic material of all plants contains a lower ratio of $^{13}C/^{12}C$ than occurs in atmospheric CO_2 but this is more so in the C_3 plants (those fixing CO_2 only through ribulose bisphosphate carboxylase) than in the C_4 plants (those fixing CO_2 through phosphoenolpyruvate carboxylase). The depletion of ^{13}C ($\delta^{13}C$) in organic matter compared with atmospheric CO_2 is used as an indicator of the type of plant source. It is defined as:

$$\delta^{13}C‰ = \left(\frac{^{13}C/^{12}C \text{ of sample}}{^{13}C/^{12}C \text{ of a standard}} - 1\right) \times 1000. \qquad (8.3)$$

Based on the Pee Dee Belemnite (PDB) standard, for C_3 plants it is usually −22 to −33‰, and for C_4 plants it is usually −10 to −20‰ (Doner *et al.*, 1979a). Sugar cane and corn are C_4 plants, however, almost all of the plants that are sources of honey are C_3 plants. Thus by assessing the change in $\delta^{13}C$ from the normal value for honey it is possible to determine the amount of cane sugar or corn syrup mixed in with the honey sugars.

White and Doner (1978a) measured the $\delta^{13}C$ values for 84 certified samples representative of the types of honey produced commercially in the USA, and 35 samples of imported honey. The mean value obtained for $\delta^{13}C$ was −25.4‰. Similar results have been obtained in Germany (Ziegler *et al.*, 1978) and Britain (Burroughs and Otlet, 1986). The coefficient of variation of 3.86% found by White and Doner (1978a) for the $\delta^{13}C$ values is the lowest found for any of the characteristics of honey (*cf.*

Table 8.1), giving this measure the potential of being the most sensitive technique for determining adulteration.

The mean $\delta^{13}C$ value found for HFCS from four commercial suppliers was −0.7‰ (SD 0.14) (White and Doner, 1978b). The $\delta^{13}C$ values of mixtures of honey and corn syrup have been demonstrated to fall linearly between the values for the pure honey and the pure syrup (White, 1980c). However, as White pointed out, in practice the proportion of syrup in a honey cannot be determined because the true $\delta^{13}C$ value of the honey before adulteration is unknown, only the mean value for honeys is known. An approximate estimate can be made from this mean value, which will be more exact the closer the $\delta^{13}C$ value of the adulterated honey is to −9.7‰, which is the value for pure corn syrup. Values of −12.2‰ and −11.47‰ ± 0.5 have been reported for the $\delta^{13}C$ of cane sugar (Doner and White, 1977).

White (1980c) proposed that $\delta^{13}C$ values less negative than −21.5‰ be taken as evidence of adulteration, this being the 4 SD limit from the mean, where the chance of a pure honey having this value naturally would be 1 in 25 000. Any with $\delta^{13}C$ values between −21.5‰ and −23.4% he proposed should be treated as suspicious and be subjected to testing by chromatography. By calculation, a $\delta^{13}C$ value of −21.5‰ would result from the addition of 25% corn syrup to an average honey, and a $\delta^{13}C$ value of −23.4‰ would result from the addition of 13% corn syrup (White, 1987). Thus the variation in the $\delta^{13}C$ in honeys prevents detection of low levels of adulteration. A further complication was the finding that some honeys had naturally occurring $\delta^{13}C$ values much less negative than the average. The mean $\delta^{13}C$ value of 15 citrus honeys was found to be −23.8‰, the citrus nectar being similar (White, 1987), and some honeys (from *Prosopis* spp. and *Acacia* spp.) from the desert regions of Mexico and the USA were found to have $\delta^{13}C$ values from −21‰ to −23‰ (White *et al.*, 1991).

To overcome this problem, White and Winters (1989) modified the procedure to use the protein in honey as an internal standard. Protein is precipitated from the honey with tungstic acid and the $\delta^{13}C$ value of the protein determined as well as that of the honey. The mean difference between the $\delta^{13}C$ of the protein from a honey and the $\delta^{13}C$ of the honey was found to be ±0.1‰ (range +1.9‰ to −0.7‰) for 50 certified USA honeys, including some citrus and *Acacia* honeys known to have less negative $\delta^{13}C$ values than usual. The $\delta^{13}C$ value for the protein thus shows the true value for the individual honey before adulteration, so the amount of adulteration can be more exactly determined, using the formula:

$$\text{Adulteration (\%)} = 100 \times \frac{(\delta^{13}C_{\text{protein}} - \delta^{13}C_{\text{honey}})}{(\delta^{13}C_{\text{protein}} - [-9.7])} \quad (8.4)$$

for adulteration with corn syrup (White and Winters, 1989). The range of +1.9‰ to −0.7‰ in the difference found between the protein and honey for unadulterated honeys by White and Winters thus corresponds to a maximum of 5% corn syrup, meaning that this technique will show conclusively any greater degree of adulteration. A smaller range was found by Roßmann *et al.* (1992) for 28 authentic honeys imported into Germany from various countries. They also found the mean difference in $\delta^{13}C$ between the protein and honey to be +0.1‰ but with a range of +1.1‰ to −0.9‰. White (1992b) has described an improved procedure for preparing the protein from honey, while Brookes *et al.* (1991) have reported successfully using an automated combustion technique, which gave results agreeing within 0.2‰, and a five-fold increase in throughput of samples.

Schmidt *et al.* (1993) have proposed that with the use of an internal standard it would be possible to detect adulteration with beet sugar (in larger amounts) in floral types of honey where the $\delta^{13}C$ is different from that of the beet sugar. This is feasible because there is quite a difference found in $\delta^{13}C$ values between different C_3 plants, possibly because of differences in isotope fractionation when CO_2 diffuses in through stomata that are open to a different extent.

8.3.4 Analysis of pollen content to determine the source of honey

8.3.4.1 Principles of pollen analysis. Pollen grains from different types of plants have a distinctive morphology that allows the family, genus or often the individual species of plant that produced the pollen to be identified by microscopic examination of the pollen. If scanning electron microscopy is used, even more can be identified at the level of the individual species. The pollen grains in honey, having entered in the various ways discussed in section 8.1.2, serve to show the types of plants that were around when the bees produced the honey. Thus analysis of the pollen content of honey (often referred to as mellissopalynology) is used extensively to determine the country of origin of honeys (based on the characteristic flora of different countries), and to check the authenticity of honeys claimed to be of a particular botanical origin. Much experience is required to be able to recognise the many different types of pollen that may be found but more rudimentary analysis can be carried out by those without experience by reference to guides.

Most publications on pollen analysis are intended for palaeobotanists working in fields such as oil exploration and the history of swamps. Nevertheless, books such as that of Moore and Webb (1978) are still useful in providing photomicrographs of different pollen forms to aid recognition. However, if such books are used for reference, it is necessary to carry out acetolysis on the pollen from honey to artificially fossilise the pollen grains (i.e. dissolve away the exine), a procedure that can destroy some thin-walled pollen grains and the algal cells and fungal hyphae that help identify honeydew sources (Louveaux *et al.*, 1978). Sawyer (1988) produced a book specifically on identifying pollen in honey but the examples in this are mostly British pollens. The International Commission for Bee Botany of the International Union of Biological Sciences has published the recommendations of experts in the field on the procedures to be used for analysis of pollen in honey (Louveaux *et al.*, 1978) but this publication directs the user to the literature on pollen analysis for identification of the pollen grains found. It is generally recommended that, where possible, reference slides are prepared from authentic pollen for comparison.

8.3.4.2 Limitations of pollen analysis. When honey is extracted by crushing the combs, or when a 'loosening' device is used before extraction, the contamination with pollen that was not picked up with the nectar (i.e. from pollen storage cells in the comb) makes it impossible to determine the botanical origin of the honey, although the types of pollen collected by the bees will still show the geographical origin. An unusually high amount of pollen in the honey will reveal this situation. When honey is filtered with diatomaceous earth there is no pollen left in the honey.

Pollen analysis is also of limited usefulness in revealing the source of honeys produced from secretions other than floral nectar. Extrafloral nectaries are a major source of honey from cotton plants (*Gossypium hirsutum*), castor-oil plants (*Ricinus communis*) (Crane, 1975) and rubber trees (*Hevea brasiliensis*) (Crane *et al.*, 1984). Although these secretions and the honeydew secreted by sap-sucking insects may contain pollen that falls in them, they do not necessarily contain pollen from the plant that was the source of the secretion. Also there are flowers such as those of the rewarewa (*Knightia excelsa*) that are pollinated by birds but have bees collect nectar from them without dislodging pollen from the anthers.

Conclusions from pollen analysis can be in error as a result of the variability of amounts of pollen from floral sources. The authenticity of thyme (*Thymus vulgaris*) honey is sometimes in dispute when the results of pollen analysis are in conflict with the opinion of the apiarist but this could well be because thyme has only female flowers on some plants and hermaphrodite flowers on others, and the proportion of each type of plant (hence the proportion supplying pollen as well as nectar) depends on the environment (Dommee *et al.*, 1978). A typical pollen spectrum of a unifloral honey may vary from country to country and season to season because it depends on what else is in flower at the time. There may also be variability resulting from bees foraging selectively from male or female flowers of dioecious species.

Erroneous conclusions can also result from the contamination of honey from one source with pollen from other sources as can arise in the various ways outlined in section 8.1.2. In a survey of the pollen content of New Zealand honeys, Moar (1985) found in the centrifugally extracted honeys 13 types of pollen from wind-pollinated species and five types from insect-pollinated plants that do not produce nectar. One citrus honey had 18% of its pollen content from *Actinidia* spp., which does not produce nectar. Although these types of pollen are excluded from consideration in the recommended procedure, they show clearly that pollen which is unrelated to the source of the nectar from which the honey is produced, is incorporated. The finding in honey of pollen substitute that is fed to bees (Fossel, 1968) also shows this.

The unreliability of pollen analysis in determining the botanical source of a honey is shown very clearly in the experiment performed by Adams and Smith (1981). They analysed the pollen content of freshly collected nectar taken from beehives twice a week over the honey production season. *Tilia* spp. was the predominant species over the first 2 weeks. Weighing of the hives on the days they were visited to take nectar samples revealed that most of the honey was deposited during this period. *Lotus* and *Lythrun* spp. became the predominant pollen types later. The average pollen count for the nectar over the season was 20.9% *Tilia*, with *Lotus* and *Lythrum* together making up 37.7%; however, the pollen count for the honey extracted centrifugally at the end of the season was only 6.7% *Tilia*, with *Lotus* and *Lythrum* together making up to 52.7%. It was suggested that in the processing of nectar in the hive the bees removed the *Tilia* pollen and that the honey became contaminated with pollen collected from *Lotus* and *Lythrum* by foragers. Pollen is not collected from *Tilia*, and in fact bees brush it off themselves if it gets on them (Howells, 1969). The honey, which observations had shown was produced mainly from *Tilia*, and which had the distinctive colour and aroma of lime (*Tilia*) honey, would not be accepted as such on the basis of pollen analysis: more than 20% of the pollen would have to be from *Tilia* for it to be acceptable (Louveaux *et al.*, 1978).

Another limitation of using pollen analysis to determine the botanical source of a honey is that pollen could feasibly be added fraudulently, as it is often harvested by means of traps at the hive entrance, which dislodge it from foraging bees that have been collecting it.

All of these potential sources of error in determining the sources of honey limit the usefulness of pollen analysis in determining conclusively the botanical origin of a honey; however, they do not affect the conclusions that may be reached on its geographical origin.

8.3.4.3 Qualitative analysis to determine the country of origin. The spectrum of types of pollen present in a honey tells which type of vegetation was in the area where the honey was produced. Vegetation zones have climatic and topographical boundaries rather than political boundaries, usually, however, an imported honey is from a country with vegetation sufficiently different for it to be detected as foreign. Thus pollen analysis is used frequently to determine the authenticity of honeys claimed to be locally produced, and successful prosecutions for misrepresentation have been based on such evidence.

It is important to consider not just the most prevalent types of pollen present but to look at the whole spectrum of types. Howells (1969) gives a good example of this in referring to a honey claimed to be British but containing 15% maize pollen. North American honeys frequently contain maize pollen but in the British honey this was accompanied in small proportions by typical British pollens. In North American honeys, the maize pollen is accompanied by quite different pollens. Enquiries revealed that there was a patch of maize growing in the locality where the honey was produced. This would have given misleading evidence if the high content of maize pollen had been all that was considered.

The minor contributors to the pollen spectrum are also useful in determining more precisely the geographical origin of honeys of floral types that are common to many countries. Sawyer (1975) gives the example of clover honey, which is produced in many countries. In addition to the *Trifolium repens* pollen common to all of them, is found pollen from: *T. pratense*, *Brassica napus*, *Caryophyllaceae*, *Cirsium* and other *Compositae* in clover honey from North America; *Myrtaceae*, *Lotus*, *Fagopyrum esculentum* and *Phacelia tanacetifolia* in clover honey from California; *Fagopyrum esculentum* and *Phacelia tanacetifolia* but not *Myrtaceae* and *Lotus* in clover honey from Poland and Eastern Europe; *Myrtaceae*, *Echium plantagineum*, *Acacia*, *Banksia* and various monocotyledons in clover honey from Australia and New Zealand. Maurizio (1975b) gives a detailed example of using the composition of the minor pollen constituents for the localization of sweet chestnut honey to the many different regions of Europe in which it is produced. Ling heather honey is another type that is produced in many countries that can have its origin localized by considering the minor pollen constituents: the types of heath or bell heather (*Erica* spp.) growing with the ling heather are characteristic of the different countries (Genier, 1966; Maurizio, 1973).

Determining the region in which a honey has been produced requires a good knowledge of the types of pollen that would be present. Reports on the types of pollen occurring in honeys in the various countries of the world may be found most easily by searching *Apicultural Abstracts*. (A computer facility for searching *Apicultural Abstracts* is provided by the International Bee Research Association, Cardiff, UK.) Not many countries export honey (Wilson, 1975), so familiarisation with the common pollen types likely to be present in imported honeys is feasible. Lists of the species of plants that are important sources of honey in the world have been published (Crane, 1975; Crane *et al.*, 1984). Consulting a world vegetation map such as that of Riley and Young (1966) would also be helpful.

8.3.4.4 Quantitative analysis to determine the floral source. Honeys produced naturally never come from a single botanical source. The term 'unifloral honey' is used to describe honey produced mostly from a single plant species: experimentally produced honey where the bees are restricted to a single type of flowering plant is termed 'single species honey' (Louveaux *et al.*, 1978). For a honey to be called 'unifloral' the pollen of the nominal species should generally be at least 45% of the total pollen count in the honey (Maurizio, 1975b). However, this percentage does not apply when a floral source gives nectar with a higher or lower number of pollen grains than the average. Centrifugally extracted honeys produced from most species of plants contain 20 000–80 000 pollen grains/10 g of honey but some can have up to 5 000 000, while in others it can be as low as 500. These are said to be over-represented and under-represented by their pollen in honey, and the International Commission for Bee Botany recommends different limits for the percentage of pollen of these species to be used to claim that a honey is unifloral if from one of these sources (Louveaux *et al.*, 1978). For example, unifloral citrus honey need have only 10% of its pollen from citrus but unifloral sweet chestnut honey needs at least 90% of the pollen content to be from *Castanea*.

The use of conversion coefficients for over-represented and under-represented species is discussed by Maurizio (1975b). These have been determined for many species of flowering plants with experimentally produced single-source honeys, and allow the pollen count for each species to be normalised so that it more truly represents the proportion of nectar.

Although the usual practice is to base the acceptance of a honey as being unifloral on the proportion of pollen present, the absolute amount of pollen of the nominal species is of more use than the proportion of that type of pollen. The other types of pollen present may have been incorporated as contaminants rather than be indicators of the admixture of other types of nectar. The pollen content of the nectar collected from the nominal species can be expected to be more consistent. Another consideration is that the proportion of pollen from the nominal species, but not the absolute amount, could be reduced markedly by incorporation of small proportions of nectars from species that are over-represented. Sawyer (1988) describes the use of pollen coefficients (number of grains/10 g of honey) in determining the sources of honeys.

Absolute numbers rather than proportions require the counting of the number of pollen grains in a measured amount of honey. This is usually performed using the direct counting methods described by Louveaux *et al.* (1978) but it is also possible to determine the absolute number of each type of pollen grain present by proportion to an 'internal standard'. In this method, the preparation from a measured amount of honey is 'spiked' with tablets containing a known number (about 10 000) of *Lycopodium* spores (Stockmarr, 1971).

8.3.5 Analysis of molecular components of honey to determine the floral source

The limitations of pollen analysis, with pollen analysis being time-consuming and very dependent on expert ability and judgement (Howells, 1969), have led to other methods being sought for the identification of the sources of honey. There is a shortage of people with the necessary experience to be able to identify the myriad of pollen types found in honey; Croft (1983) referred to there being only one person, retired, in Britain competent in the technique, and a similar situation exists in New Zealand. Therefore there is an interest in identifying minor chemical components of honey that would serve to determine floral sources. In the past, a few components have been recognised that could possibly be used as markers for particular floral sources; with the advent of modern instrumental analysis there is broader consideration of patterns of constituents that could serve as 'fingerprints' for honeys from the various different botanical sourcs. This is currently at the investigative stage but shows the potential of being a very useful technique once more data are collected to show the consistency of the patterns within types of unifloral honey, and to establish the uniqueness of the 'fingerprint' of each type.

8.3.5.1 Amino acids. The occurrence of amino acids in nectar is virtually if not actually universal (Baker and Baker, 1983). Intraspecific constancy of the amino acid composition of nectar has been demonstrated despite environmentally induced variation in the volume and concentration of the secretion (Baker and Baker, 1977). The amino acid composition of nectar is an inherited feature that appears in plant crosses (Baker and Baker, 1983) and it therefore could be expected that the amino acid composition of a honey will reflect its floral source.

Nevertheless, having reviewed the numerous reports on the amino acid content of honeys produced by sugar-fed bees, Davies (1975) concluded that the amino acid patterns of honeys are all generally similar because of the influence of the bees, but that variations arise from the types of nectar or honeydew from which the honeys are made. A further complication is that the contribution from the bees depends on the amount of 'working' of the nectar involved, which depends in turn on the water content, which will vary between different floral sources, and on the rate of drying, which will vary in different climates. Thus it may not be differences in the amino acid contents of the nectars that are the explanation of the finding of smaller amounts in Finnish honey than in imported honey (Komamine, 1960), or larger amounts in buckwheat honey than in lime (*Tilia* spp.) honey (Solov'eva and Bazarova, 1963).

With the advent of automatic amino acid analysers, true quantitative data became available and differences in the amounts of each amino acid present could be seen. Curti and Riganti (1966) found sufficient quantitative differences between the eight samples of honey that they analysed for them to suggest that it could be used to identify the botanical source of a honey. This suggestion was followed up by Davies (1975) who analysed 98 samples of honey from various sources and found that the ratio of amides:phenylalanine plotted against the ratio of aspartic acid:proline showed distinct groupings for honeys from Canada (clover), USA (clover), Australia (eucalypt–clover mix) and Yucatán, Mexico (indigenous flora). Davies concluded from this that some ratios between the contents of amino acids could be used to determine the source of a honey. This study was extended (Davies, 1976) by applying statistical procedures to the analytical data. The ratios of all of the amino acids to each other were computed and the 60 most consistent ones within the honeys from the same source were selected. The ratios from 16 samples of honey, several from each of the four sources, were then matched with these ratios to test for the probability of each of these honeys coming from one of the four sources. Data from 42 samples from other sources were also tested in the same way. At a probability level of 5% or less, 15 out of the 16 were assigned to the correct source, and only one of the 42 from other sources was incorrectly assigned to the test set. It was calculated that if honey from any of the four countries were mixed with 1–40% of honey from one of the other countries it would be detectable as it would no longer match its own group.

Speer and Montag (1986) also reported differences between various floral types of honey in their amino acid composition. Honeys from *Robinia* spp. and *Eucryphia* spp. were low in their total content of amino acids and in proline. Sweet chestnut honey had a high content of proline and also a higher content of aspartic acid than most other honeys. Heather honeys varied considerably in the amount of each amino acid present as well as the total quantity, while French and German heather honeys could be distinguished by the different ratios of proline:phenylalanine and proline:tyrosine. Bosi and Battaglini (1978) confirmed that the amino acid spectra of honeys from different sources vary but that the 26 unifloral honeys and four honeydew honeys that they studied had only two pairs from the same floral source; thus no conclusions could be drawn on whether the different patterns found were typical of particular sources or whether they were just the result of random variation. Gilbert *et al.* (1981) also found that the absolute amounts and relative proportions of the different amino acids are different in different honeys, and like other investigators, found that the variability within honeys from the same source was so great that it is not readily possible to determine the origin of a honey from the data. However, when they applied canonical variate analysis (a statistical technique that is widely used for discriminating between groups within which there is high sample variability) they were able to discriminate between honeys from different countries. With outliers removed, 10 samples from Australia were easily discriminated from 11 samples from Argentina, eight from Canada and 16 samples from the UK. By carrying out the analysis in two stages, with the Australian samples omitted from the second stage, it was possible to discriminate at the 95% confidence level between the honeys from the other three countries. It was thus concluded that the method could be used to distinguish honeys from different countries with their different flora but that the method needs testing on a much wider range of honeys of known botanical origin.

8.3.5.2 Trace organics. It is evident when examining a variety of honeys that a very wide range of aromas, flavours and colours exists. Consequently, it can be expected that the chemical analysis of the components responsible for these features of a unifloral honey could give a fingerprint indicating the floral source. Lothrop (1932) found methyl anthranilate to be responsible for the distinctive aroma of orange-blossom honey and found that it was not in any other type of honey examined. He proposed this as a specific test for orange-blossom honey. White (1966) supported this proposal after examining numerous samples of citrus honeys and non-citrus honeys but noted that confirmatory studies were required. Bonaga and Giumanini (1986) found 3-aminoacetophenone, not previously found occurring naturally, was the major volatile component of unifloral sweet chestnut honey of high purity. Its aroma was dominant in this type of honey. Blank *et al.* (1986) found that the terpenes linden ether (3,9–epoxy–1,4–[8]–*p*–menthadiende) and *cis*-rose oxide are major odourants of linden (*Tilia cordata*) honey. They were present in all of the samples of linden honey (and in linden blossom) but not in any of the honeys from *Robinia* spp., *Erica* spp., orange blossom, rape and wild tobacco. The odourless terpene *trans*–8–*p*–menthen–1,2–diol was also found in all of the samples of linden honey but none of the others. They suggested that these compounds could be used as indicators of honey from a linden source after further research to establish the lowest concentrations to be expected in a honey described as linden. Häusler and Montag (1989) suggested that the odourless compound (*S*)–(+)–dehydrovomifoliol may be useful as an indicator of a heather honey. It was found at a level of 186–264 μg/g in the four French heather honeys examined, and at 56 μg/g in the one Spanish sample. The eight other floral honeys from various countries contained 0.03–6.02 μg/g. Dichloromethane extracts of six different types of Australian honey were examined by Graddon *et al.* (1979) using capillary GC–MS, and it was noted that some compounds were present at high concentrations in some honeys and absent in others. Only the *Eucalyptus* honeys contained hexenyl butyrate, various methoxy aromatics were found only in the honey from *Bursaria spinosa*, while acetoin was found only in the honeys from the *Eucalyptus* and *Banksia* species.

The shortcoming of using single compounds as markers of floral sources is that they may well be found in honeys from other species of plants. Bicchi *et al.* (1983) found that the characteristic floral odour of Piedmontese multifloral honeys was caused principally by phenylacetaldehyde but that this compound was also found in unifloral honeys from *Robinia* and clover. However, the pattern of volatile components (52 were identified in the study) seen on capillary GC–MS showed no meaningful difference over three successive seasons in the multifloral honey produced from the same locality. The patterns in the other two types of honey also did not differ significantly over two successive seasons but were distinctly different for each of the three types of honey and the different components contributed significantly to the distinctive aromas of the different types of honey. Thus the whole pattern was a distinctive 'fingerprint' of each type of honey, even though single compounds occurred in more than one type.

The use of whole chromatograms as 'fingerprints' was first suggested by Dörrscheidt and Friedrich (1962) who carried out some of the early packed-column GC–FID analyses of honey. Although they found 31 peaks on the chromatograms, only four of these were common to the six different types of honey examined. Cremer and Riedmann (1964) used capillary GC to extend the work of Dörrscheidt and Friedrich, and found 50 compounds in 10 different honeys, with only three of these compounds common to all. Bonaga and Giumanini (1986) found 50 compounds in unifloral sweet chestnut honey and identified 40 of them; 19 had not previously been found in honey. The authors cautioned that these did not necessarily provide a unique 'fingerprint' for sweet chestnut honey but with such a large number of compounds it appears likely that they may. The very large number of trace organic components found in honeys gives a lot of scope for each floral type of honey to have a unique 'fingerprint': 54 different hydrocarbons, 47 alcohols, 20 aldehydes, 27 ketones, 65 acids, 42 esters, 10 furans, diethyl ether and 1,1–diphenylhydrazine have been identified in honey by various researchers (references in Tan, 1989). There are many additional peaks that have been found on GC that have been reported as unidentified.

An attempt has been made by Wilkins, Tan and co-workers to determine 'fingerprints' for some New Zealand unifloral honeys. Analysis of diethyl ether liquid–liquid extracts of more than 200 samples of honey collected over four seasons showed the distribution and relative concentration of the array of components present to be sufficiently diagnostic to identify the floral sources of the honeys (Tan *et al.*, 1989b). The diacids, aromatic compounds and degraded carotenoids were found to be more related to the floral source than the aliphatic acids and hydrocarbons which are associated with the beeswax. It was found that the floral source of samples of honey could be routinely identified prior to the results of pollen analysis being obtained, the GC traces being distinctive enough for even an untrained person to recognise the different patterns.

The distinctive features of the GC analyses have been published for honeys from: clover (*Trifolium repens*), kanuka (*Leptospermum ericoides*) and Manuka (*Leptospermum scoparium*) (Tan *et al.*, 1988); ling heather (*Calluna vulgaris*) (Tan *et al.*, 1989a); thyme (*Thymus vulgaris*) and willow (*Salix* spp.) (Tan *et al.*, 1990); nodding thistle (*Carduus nutans*) (Wilkins *et al.*, 1993b). A follow-up study of more samples of manuka honey from a later season confirmed the earlier conclusions (Wilkins *et al.*, 1993a).

In order to confirm the reliability of this technique in the identification of floral sources, it will be necessary to quantitatively analyse a large number of samples of each floral type so that the variation in concentration of the organic substances characteristic of a particular floral type can be determined, especially from season to season. In addition, it is important to analyse a very wide range of floral types to establish that these compounds are indeed characteristic of a particular floral source and not occurring widely. Such information requires long-term studies and samples of honeys of high purity. It will also be necessary to establish the effects of storage of honey on the 'fingerprints', as increases in some compounds (Cremer and Riedmann, 1965) and some loss of volatile components (Wootton *et al.*, 1978) have been reported for honeys that have been stored.

8.3.6 Distinguishing honeydew sources

Honey produced from honeydew sources may be distinguished quite easily from honey from floral sources on several criteria but the botanical source of the honeydew is not so easy to determine.

8.3.6.1 Microscopy. Honeydew characteristically contains spores of sooty moulds (which give honey produced from honeydew its dark colour), pieces of fungal hyphae, and algal cells, especially *Pleurococcus* spp. In general, honey produced from honeydew has at least three times more particles of fungal or algal origin present (excluding spores of plant pathogens) than it has pollen grains (Louveaux *et al.*, 1978). Honeys produced from honeydew can also be distinguished by their content of unusually high numbers of pollen grains from wind-pollinated species.

The pollen that gets picked up by honeydew serves as an indicator of the region in which the honeydew was produced. Knowledge of which sources of honeydew there are in different regions can enable the pollen content to indicate the botanical source as well if this is in question.

8.3.6.2 Presence of characteristic sugars. Honeydew may be produced on many types of plants: ash, beech, cedar, elm, hickory, linden, maple, oak and poplar trees and various fruit trees (White, 1992a); larch and spruce trees (Maurizio, 1975a); pineapple plants (Gray, 1952). However, the major changes in the sugar composition of the phloem sap that occur as it passes through the gut of the sap-sucking insects, and the fairly uniform composition of the sugars in the phloem sap of different plants, means that it is the type of insect rather than the type of plant that is indicated by the different sugar compositions of various honeydew honeys. Nevertheless, the association of certain types of insect with certain species of plants means that compositional difference still may be useful indicators of the botanical source.

Melezitose and *l*-kestose (*O*–α–D–glucopyranosyl–[1→2]–*β*–D–fructofuranosyl–[1→2]–*β*–D–fructofuranoside) occur in honey from honeydew in considerable proportions but not to an appreciable extent in honey from nectar sources (Siddiqui, 1970). Melezitose granulates readily, and may granulate in the comb if enough is present (White, 1992a). It is present in larch and spruce honeydew and in lime (*Tilia* spp.) honeydew as 4–11% of the total sugar (Maurizio, 1962; 1975a). In other types of honeydew honey there is instead another trisaccharide that is characteristic of honeydew honey, namely erlose (Siddiqui, 1970; White, 1992a).

As discussed in section 8.2.1.1, raffinose may also be present at high levels in some honeydews. Although raffinose and melezitose may occur in some nectars, its occurrence is fairly uncommon and the quantities low (except for the unsually high level of raffinose in sunflower nectar). Likewise, erlose may be formed by the action of enzymes added to honey by the bees but it is formed in small quantities. Thus it should be possible to distinguish honeydew honeys from nectar honeys, and to distinguish different types of honeydew honeys, by the presence of relatively high levels of one or more of these trisaccharides. There is, however, the need for quantitative analytical data to be collected first on authentic samples of various types of honeydew honey.

8.3.6.3 Other characteristics. Melezitose and erlose are strongly dextrorotatory, and the quantities in honeydew honey make it characteristically dextrorotatory while floral honey is laevorotatory (White, 1980b). Thus polarimetry can be used as a simple test to distinguish honeys produced from honeydew. White (1980b) examined this long-standing method and proposed that the dividing line between the two types of honey be at −2°S rather than 0°.

Another simple test that has been used to distinguish honeys produced from honeydew is the measurement of conductivity. Vorwohl (1964) reported the conductivity of floral honeys to be from 0.6 to 1.46 ohm/cm (heather honey 7.7), while for honeydew honeys it was from 6.3 to

16.4 ohm/cm. The conductivity of honey depends on the concentration of mineral salts, organic acids and peptide material present. The high mineral content of honeydew honey is recognised in the higher limits set in the *Codex Alimentarius* for ash content of honeydew honey. A comparison of the average composition of honeydew and floral honeys (White, 1992a) shows honeydew honey to be lower in its content of fructose and glucose, and higher in oligosaccharides, pH, titratable acidity, ash (minerals) and nitrogen. A more sophisticated test to distinguish honeydew honey has been proposed by Kirkwood *et al.* (1960; 1961) on the basis of their examination of the differences between the two types of honey. They derived an equation for discriminating between the two types of honey:

$$X = -8.3x_1 - 12.3x_2 + 1.4x_3 \quad (8.5)$$

where x_1 is pH, x_2 is the content of ash (minerals) in the honey (%), and x_3 is the content of reducing sugars in the honey (%). They stated that floral honeys give values of X greater than 73.1, and honeydew honeys give values less than 73.1

8.4 Uncertainty resulting from variability in the natural product

The very large amount of variation that occurs naturally in the composition and physical features of honey (discussed in section 8.3.3 and outlined in Table 8.1) makes it very difficult to establish with certainty whether or not a honey in question is authentic. For instance, high water content may be a natural consequence of humidity in the hive rather than dilution by a honey processor; a high sucrose content may be the result of low invertase activity rather than adulteration with sucrose. Adulteration with small proportions of syrups may not change a honey beyond the normal range, even when several different features are tested, yet a completely natural honey may itself be an 'outlier' beyond the normal range and be wrongly condemned. There is less uncertainty when the known deviations of certain types of honey from the typical values are taken into account, and the *Codex Alimentarius* permits higher levels of sucrose in some specified floral types. However, there is also a problem of uncertainty in identifying the floral type because of the variability in pollen content. For example, ling heather is not one of the floral sources generally recognised as being under-represented by its pollen in honey but there are frequently problems in getting New Zealand ling heather honey accepted in other countries because of its low content of ling heather pollen (Moar, 1985). Variability in pollen content appears to be a feature of this floral type of honey, as it has also been reported by Maurizio and Louveaux (1964) that in parts of Europe an almost pure ling heather honey is produced in which ling heather pollen is secondary in proportion. It is important that the idiosyncrasies of each type of honey be recognised whenever possible, and be taken into account whenever decisions are made on authenticity. It is desirable that as much uncertainty as possible be removed by the generation of much more reference data for the commonly traded types of honey.

References

Adams, R.J. and Smith, M.V. (1981) Seasonal pollen analysis of nectar from the hive and of extracted honey. *J. Apicultural Res.* **20(4)**, 234–48.

Ali, A.T.M.M., Al-Humayyd, M.S. and Madan, B.R. (1990) Natural honey prevents indomethacin- and ethanol-induced gastric lesions in rats. *Saudi Med. J.* **11**, 275–79.

Al-Somai, N., Coley, K.E., Molan, P.C. *et al.* (1994) Susceptibility of *Helicobacter pylori* to the antibacterial activity of manuka honey. *J. Royal Soc. Med.* **87(1)**, 9–12.

Allen, K.L., Molan, P.C. and Reid, G.M. (1991) A survey of the antibacterial activity of some New Zealand honeys. *J. Pharm. Pharmacol.* **43(12)**, 817–22.

Baker, H.G. and Baker, I. (1977) Intraspecific constancy of floral nectar amino acid complements. *Botanical Gazette* **138(2)**, 183–191.

Baker, H.G. and Baker, I. (1983) A brief historical review of the chemistry of floral nectar. In: *The Biology of Nectaries* (Eds B. Bentley and T. Elias). Columbia University Press, New York, pp. 126–52.

Baker, H.G., Opler, P.A. and Baker, I. (1978) A comparison of the amino acid complements of floral and extrafloral nectars. *Botanical Gazette* **139**, 322–32.

Bicchi, C., Belliardo, F. and Frattini, C. (1983) Identification of the volatile components of some Piedmontese honeys. *J. Apicultural Res.* **22(2)**, 130–36.

Blank, I., Fischer, K.-H. and Grosch, W. (1989) Intensive neutral odourants of linden honey. *Zeitschrift für Lebensmittel-Untersuchung und-Forschung* **189**, 426–33.

Bogdanov, S. and Baumann, E. (1988) Bestimmung von Honigzuckern mit HPLC. *Mitteilungen aus dem Gebiete der Lebensmitteluntersuchung und Hygiene* **79**, 198–206.

Bonaga, G. and Giumanini, A.G. (1986) The volatile fraction of chestnut honey. *J. Apicultural Res.* **25(2)**, 113–20.

Bosi, G. and Battaglini, M. (1978) Gas chromatographic analysis of free and protein amino acids in some unifloral honeys. *J. Apicultural Res.* **17(3)**, 152–66.

Bricout, J. (1973) Control of authenticity of fruit juices by isotopic analysis. *J. Assoc. Official Analytical Chem.* **56(3)**, 739–42.

Brookes, S.T., Barrie, A. and Davies, J.E. (1991) A rapid $^{13}C/^{12}C$ test for determination of corn syrups in honey. *J. Assoc. Official Analytical Chem.* **74(4)**, 627–29.

Burroughs, L.F. and Otlet, R.L. (1986) The proline contents and stable carbon isotope ratio of genuine United Kingdom honey. *J. Assoc. Public Analysis* **24**, 91–93.

Chataway, H.D. (1932) Determination of moisture content in honey. *Canad. J. Res.* **6**, 532–47.

Church, J. (1954) Honey as a source of the anti-stiffness factor. *Fed. Proc. Am. Physiol. Soc.* **13(1)**, 26.

Crane, E. (1975) The flowers honey comes from. In: *Honey: A Comprehensive Survey* (Ed. E. Crane). Heinemann, London, pp. 3–76.

Crane, E., Walker, W. and Day, R. (1984) *Directory of Important World Honey Sources*, International Bee Research Association, London.

Cremer, E. and Riedmann, M. (1964) Identifizierung von gaschromatographisch getrennten Aromastoffen in Honigen. *Zietschrift für Naturforschung B* **19(1)**, 76–77.

Cremer, E. and Riedmann, M. (1965) Gaschromatographisch Untersuchungen zur Frage des Honigaromas. *Monatshefte für Chemie* **96(2)**, 364–68.

Croft, L. (1983) UK beekeeping under threat. In: *The Sunday Times*, London, 20 February, p. 22.

Curti, R. and Riganti, V. (1966) Richerche sigli aminoacidi del miele. *Rassegna Chimica* **18(6)**, 278–82.

Davies, A.M.C. (1975) Amino acid analysis of honeys from eleven countries. *J. Apicultural Res.* **14(1)**, 29–39.
Davies, A.M.C. (1976) The application of amino acid analysis to the determination of geographical origin of honey. *J. Food Technol.* **11**, 515–23.
Dommee, B., Assouad, M.W. and Vadeyron, G. (1978) Natural selection and gynodioecy in *Thymus vulgaris. Botanical J. Linnean Soc.* **77**, 17–28.
Doner, L.W. (1977) The sugars of honey – a review. *J. Sci. Food Agric.* **28**, 443–56.
Doner, L.W. and White, J.W. (1977) Carbon-13/carbon-12 ratio is relatively uniform among honeys. *Science* **197** 891–92.
Doner, L.W., Kushnir, I. and White, J.W. (1979a) Assuring the quality of honey. Is it honey or is is syrup? *Analyt. Chem.* **51(2)**, 224A–232A.
Doner, L.W., White, J.W. and Phillips, J.G. (1979b) Gas-liquid chromatographic test for honey adulteration by high fructose corn sirup. *J. Assoc. Official Analytical Chem.* **62(1)**, 186–89.
Doner, L.W., Brause, A.R. and Petrus, D.R. (1992) $\delta^{18}O$ measurements in water for detection of sugar beet-derived syrups in frozen concentrated orange juice: collaborative study. *J. AOAC Internat.* **75(6)**, 1107–11.
Dörrscheidt, W. and Friedrich, K. (1962) Trennung von Aromastoffen des Honigs mit hilfe der Gas-Chromatographie. *J. Chromatogr.* **7(1)**, 13–18.
Duisberg, H. and Hadorn, H. (1966) Weiche Anforderungen sind an Handelshonige zu stellen? Vorschläge auf Grund der statistichen Auswertung von ca. 1600 Honig-Analysen. *Mitteilungen aus dem Gebiete der Lebensmitteluntersuchung und Hygiene* **57(5)**, 386–407.
Durkee, L.T. (1983) The ultrastructure of floral and extrafloral nectaries. In: *The Biology of Nectaries* (Eds B. Bentley and T. Elias). Columbia University Press, New York, pp. 1–29.
Dustmann, J.H. (1971) Über die Katalaseaktivität in Bienenhonig aus der Tracht deer Heidekrautgewächse (Ericaceae). *Zeitschrift für Lebensmittel-Untersuchung und -Forschung* **145**, 294–95.
Dyce, E.J. (1975) Producing finely granulated or creamed honey. In: *Honey: A Comprehensive Survey* (Ed. E. Crane). Heinemann, London, pp. 293–306.
Editorial (1981) *American Bee Journal* **121(12)**, 860.
Fasler, A. (1975) Honey standards legislation. In: *Honey: A Comprehensive Survey* (Ed. E. Crane). Heinemann, London, pp. 329–54.
Fossel, A. (1968) Pollenersatzmittel im mikroskopischen Befund von Frühtrachthonigen. *Zeitschrift für Bienenforschung* **9(5)**, 206–11.
Gary, N.E. (1992) Activities and behaviour of honey bees. In: *The Hive and the Honey Bee*, revised edn. (Ed. J.M. Graham). Dadant & Sons, Hamilton, Illinois, pp. 269–372.
Genier, G. (1966) Le pollen des Ericaceae dans le miels français. *Annales de l'Abeille (Paris)* **9(4)**, 271–321.
Gilbert, J., Shepherd, M.J., Wallwork, M.A. *et al.* (1981) Determination of the geographical origin of honeys by multivariate analysis of gas chromatographic data on their free amino acid content. *J. Apicultural Res.* **20(2)**, 125–35.
Graddon, A.D., Morison, J.D. and Smith, J.F. (1979) Volatile constituents of some unifloral Australian honeys. *J. Agricultural Food Chem.* **27(4)**, 832–37.
Graham, J. (1979) Honey adulteration is not a new problem. *Am. Bee J.* **119(3)**, 228–29.
Gray, R. (1952) Composition of honeydew excreted by pinapple mealybugs. *Science* **115**, 129–33.
Hagedorn, J.L. (1992) Differentiation of natural and synthetic benzaldehydes by ^2H nuclear magnetic resonance. *J. Agricultural Food Chem.* **40(4)**, 634–37.
Häusler, M. and Montag, A. (1989) Isolation, identification and quantitative determination of the isoprenoid (S)-(+)-dehydrovomifoliol in honey. *Zeitschrift für Lebensmittel-Untersuchung und-Forschung* **189**, 113–15.
Haydak, M.H., Crane, E., Duisberg, H. *et al.* (1975) Biological properties of honey. In: *Honey: A Comprehensive Survey* (Ed. E. Crane). Heinemann, London, pp. 258–66.
Howells, V.W. (1969) Some reflections on the pollen analysis of honey. *J. Assoc. Public Analysts* **7**, 88–93.
Jacobs, M.B. (1955) Flavouring with honey. *Am. Perfumer Essential Oil Rev.* **66(1)**, 46–47.
Kandil, A., El-Banby, M., Abdel-Wahed, K. *et al.* (1987) Curative properties of true floral and false non-floral honeys on induced gastric ulcer. *J. Drug Res. (Cairo)* **17**, 103–106.
Kirkwood, K.C., Mitchell, T.J. and Smith, D. (1960) An examination of the occurrence of honeydew in honey. *Analyst, London* **85(1011)**, 412–16.

Kirkwood, K.C., Mitchell, T.J. and Ross, I.C. (1961) An examination of the occurrence of honeydew in honey. Part II. *Analyst, London* **86(1020)**, 164–65.
Komamine, A. (1960) Amino acids in honey. *Acta Chemica Fennica* **B33**, 185–87.
Kushnir, I. (1979) Sensitive thin-layer chromatographic detection of high-fructose corn syrup and other adulterants in honey. *J. Assoc. Official Analytical Chem.* **62(4)**, 917–20.
Lipp, J., Ziegler, H. and Conrady, E. (1988) Detection of high fructose and other syrups in honey using high-pressure liquid chromatography. *Zeitschrift für Lebensmittel-Untersuchung und-Forschung* **187(4)**, 334–38.
Lothrop, R.E. (1932) Specific test for orange honey. *Indust. Engineering Chem. – Analytical Edn.* **4** 395–96.
Louveaux, J., Maurizio, A. and Vorwohl, G. (1978) Methods of melissopalynology. *Bee World* **59(4)**, 139–57.
Low, N.H., Brisbane, T., Bigam, G. *et al.* (1988) Carbon-13 nuclear magnetic resonance for the qualitative and quantitative analysis of structurally similar disaccharides. *J. Agric. Food Chem.* **36(5)**, 953–57.
Marek, M., Bacilek, J. and Jary, J. (1980) Polarographic determination of the adulteration of honey with invert sugar. *J. Apicultural Res.* **19(4)**, 255–60.
Matheson, A. (1993) Honey research brings commercial success. *Bee World* **74(3)**, 145–46.
Maurizio, A. (1962) From the raw material to the finished product: honey. *Bee World* **43**, 66–81.
Maurizio, A. (1973) The heather honeys of Europe. *Bee World* **54(3)**, 111–16.
Maurizio, A. (1975a) How bees make honey. In: *Honey: A Comprehensive Survey* (Ed. E. Crane). Heinemann, London, pp. 77–105.
Maurizio, A. (1975b) Microscopy of honey. In: *Honey: A Comprehensive Survey* (Ed. E. Crane). Heinemann, London, pp. 240–57.
Maurizio, A. and Louveaux, J. (1964) Pollens des plantes melliferes d'Europe. *Pollen et Spores* **3**, 219–346.
Moar, N.T. (1985) Pollen analysis of New Zealand honey. *NZ J. Agric. Res.* **28**, 39–70.
Molan, P.C. (1992) The antibacterial activity of honey. 1. The nature of the antibacterial activity. *Bee World* 73(1), 5–28.
Molan, P.C. (1992b) The antibacterial activity of honey. 2. Variation in the potency of the antibacterial activity. *Bee World* **73(2)**, 59–76.
Moore, P.D. and Webb, J.A. (1978) *An Illustrated Guide to Pollen Analysis*, Hodder and Stoughton, London.
Morton, I.D. and Sharples, E. (1959) Process of preparing a honey flavour. *U.S. Patent No.* 2 916 382.
Nixon, H.L. and Ribbands, C.R. (1952) Food transmission within the honeybee community. *Proc. Royal Soc. London, Series B* **140**, 43–50.
Percival, M.S. (1961) Types of nectar in angiosperms. *The New Phytologist* **60**, 235–81.
Perez-Arquillué, C., Conchello, P., Ariño, A. *et al.* (1994) Quality evaluation of Spanish rosemary (*Rosmarinus officinalis*) honey. *Food Chem.* **51(2)**, 207–10.
Riley, D. and Young, A. (1966) *World Vegetation*, Cambridge University Press, London.
Rinderer, T.E. and Baxter, J.R. (1980) Honeybee *Apis mellifera* hoarding of high-fructose corn syrup and cane sugar syrup. *Am. Bee J.* **120(12)**, 817–18.
Rodgers, P.E.W. (1975) Honey quality control. In: *Honey: A Comprehensive Survey* (Ed. E. Crane). Heinemann, London, pp. 314–25.
Roßmann, A., Lüllmann, C. and Schmidt, H.-L. (1992) Massenspektometrische Kohlenstoff- und Wasserstoff-Isotopen-Verhältnismessung zur Authentizitätsprüfung bei Honigen. *Zeitschrift für Lebensmittel-Untersuchung und-Forschung* **195(4)**, 307–11.
Rubenstein, R.M. (1977) Honey adulteration testing. *Am. Bee J.* **117(12)**, 736–37.
Sawyer, R.W. (1975) Melissopalynology in the determination of the geographical and floral origin of honey. *J. Assoc. Public Analysts* **13**, 64–71.
Sawyer, R.W. (1988) *Honey Identification*, Cardiff Academic Press, Cardiff, UK.
Schade, J.E., Marsh, G.L. and Eckert, J.E. (1958) Diastase activity and hydroxymethylfurfural in honey and their usefulness in detecting heat alteration. *Food Res.* **23**, 446–63.
Schmidt, H.-L., Butzenlechner, M., Rossmann, A. *et al.* (1993) Inter- and intramolecular isotope correlations in organic compounds as a criterion for authenticity identification and origin assignment. *Zeitschrift für Lebensmittel-Untersuchung und-Foreschung* **196(2)**, 105–10.

Shuel, R.W. (1992) The production of nectar and pollen. In: *The Hive and the Honey Bee*, revised edn. (Ed. J.M. Graham). Dadant & Sons, Hamilton, Illinois, pp. 401–36.
Siddiqui, I.R. (1970) The sugars of honey. *Adv. Carbohydrate Chem. Biochem.* **25**, 285–309.
Sidorciuc, D. and Popescu, N. (1972) [Man-made honey]. *Rumanian Patent No. 53959.*
Simpson, J., Moxley, E. and Greenwood, S.P. (1974) Can honey from sugar-fed bees be distinguished from natural honey by its flavour? *Bee World* **55(1)**, 10–14.
Solov'eva and Bazarova, V.I. (1963) The content of free amino acids in some unifloral honeys [in Russian]. *Voprosy Pitaniya* **22**, 69–70.
Speer, K. and Montag, A. (1986) Verteilung freier Aminosären in Honigen unter besonderer Berücksichtigung der deutschen un der französischen Heidehonige. *Deutsche Lebensmittel-Rundschau* **82(8)**, 248–53.
Sporns, P. (1981) High-pressure liquid chromatographic determination of phenol in honey. *J. Assoc. Official Analytical Chem.* **64(2)**, 337–39.
Stockmarr, J. (1971) Tablets with spores used in absolute analysis. *Pollen et Spores* **13**, 615–21.
Swallow, K.W. and Low, N.H. (1994) Determination of honey authenticity by anion-exchange liquid-chromatography. *J. AOAC Internat.* **77(3)** 695–702.
Takeba, K., Matsumoto, M., Shida, Y. *et al.* (1990) Determination of phenol in honey by liquid chromatography with amperometric detection. *J. Assoc. Official Analytical Chem.* **73(4)**, 602–04.
Tan, S.T. (1989) A chemical investigation of some New Zealand honeys. DPhil. Thesis, University of Waikato, Hamilton, New Zealand.
Tan, S.T., Holland, P.T., Wilkins, A.L. *et al.* (1988) Extractives from New Zealand honeys. 1. White clover, manuka and kanuka unifloral honeys. *J. Agric. Food Chem.* **36(3)**, 453–60.
Tan, S.T., Wilkins, A.L., Holland, P.T. and McGhie, T.K. (1989a) Extractives from New Zealand unifloral honeys. 2. Degraded carotenoids and other substances from heather honey. *J. Agric. Food Chem.* **37(5)**, 1217–21.
Tan, S.T., Wilkins, A.L., Holland, P.T. *et al.* (1990) Extractives from New Zealand honeys. 3. Unifloral thyme and willow honey. *J. Agric. Food Chem.* **38(9)**, 1833–38.
Tan, S.T., Wilkins, A.L., Molan, P.C. *et al.* (1989b) A chemical approach to the determination of the floral sources of New Zealand honeys. *J. Apicultural Res.* **28(4)**, 212–22.
Townsend, G.F. (1975) Processing and storing liquid honey. In: *Honey: A Comprehensive Survey* (Ed. E. Crane). Heinemann, London, pp. 269–92.
Trumpeter, J.N. (1981) Clarifying some of the myths and confusion about HFCS, sugar and honey. *Am. Bee J.* **121(12)**, 857–63.
Verordmung über den Verkehn mit Lebensmittein (1971) Art. 217/3.
Vorwohl, G. (1964) Die Beziehungen zwischen der elektrischen Leitfähigkeit der Honige und ihrer trachtmässigen Herkunft. *Annales de l'Abeille* **7(4)**, 301–9.
Walsh, R.S. (1960) Anti-granulation method for honey packers. *NZ J. Agric.* **10(3)**, 229–31.
White, J.W. (1966) Methyl anthranilate content of citrus honey. *J. Food Sci.* **31(1)**, 102–04.
White, J.W. (1975a) Composition of honey. In: *Honey: A Comprehensive Survey* (Ed. E. Crane). Heinemann, London, pp. 157–206.
White, J.W. (1975b) Physical characteristics of honey. In: *Honey: A Comprehensive Survey* (Ed. E. Crane). Heinemann, London, pp. 207–39.
White, J.W. (1979a) Spectrophotometric method for hydroxymethylfurfural in honey. *J. Assoc. Official Analytical Chem.* **62(3)**, 509–14.
White, J.W. (1979b) Methods for determining carbohydrates, hydroxymethylfurfural, and proline in honey: collaborative study. *J. Assoc. Official Analytical Chem.* **62(3)**, 515–26.
White, J.W. (1980a) Hydroxmethylfurfural content of honey as an indicator of its adulteration with invert sugars. *Bee World* **61(1)**, 29–37.
White, J.W. (1980b) Detection of honey adulteration by carbohydrate analysis. *J. Assoc. Official Analytical Chem.* **63(1)**, 11–18.
White, J.W. (1980c) High-fructose corn syrup adulteration of honey: confirmatory testing required with certain isotope ratio values. *J. Assoc. Official Analytical Chem.* **63(5)**, 1168.
White, J.W. (1981) The honey industry council fights adulteration. *Am. Bee J.* **121(6)**, 452–53.

White, J.W. (1987) Wiley led the way: a century of federal honey research. *J. Assoc. Official Analytical Chem.* **70(2)** 181–89.
White, J.W. (1992a) Honey. In: *The Hive and the Honey Bee*, revised edn. (Ed. J.M. Graham), Dadant & Sons, Hamilton, Illinois, pp. 869–925.
White, J.W. (1992b) Internal standard stable carbon isotope ratio method for determination of C-4 plant sugars in honey: collaborative study, and evaluation of improved protein preparation procedure. *J. AOAC Internat.* **75(3)**, 543–48.
White, J.W. (1994) The role of HMF and diastase assays in honey quality evaluation. *Bee World* **75(3)**, 104–17.
White, J.W. and Doner, L.W. (1978a) The $^{13}C/^{12}C$ ratio in honey. *J. Apicultural Res.* **17(2)**, 94–99.
White, J.W. and Doner, L.W. (1978b) Mass spectrometric detection of high-fructose corn sirup in honey by use of $^{13}C/^{12}C$ ratio: collaborative study. *J. Assoc. Official Analytical Chem.* **61(3)**, 746–50.
White, J.W. and Rudyk, O.N. (1978a) The proline content of United States honeys. *J. Apicultural Res.* **17(2)**, 89–93.
White, J.W. and Rudyj, O.N. (1978b) The protein content of honey. *J. Apicultural Res.* **17(4)**, 234–38.
White, J.W. and Siciliano, J. (1980) Hydroxmethylfurfural and honey adulteration. *J. Assoc. Official Analytical Chem.* **63(1)**, 7–10.
White, J.W. and Winters, K. (1989) Honey protein as internal standard for stable carbon isotope ratio detection of adulteration of honey. *J. Assoc. Official Analytical Chem.* **72(6)**, 907–11.
White, J.W., Meloy, R.W., Probst, J.L. *et al.* (1986) Detection of beet sugar adulteration of honey. *J. Assoc. Official Analytical Chem.* **69(4)**, 652–54.
White, J.W., Riethof, M.L. and Kushnir, I. (1961) Composition of honey, VI. The effect of storage of carbohydrates, and diastase content. *J. Food Sci.* **26(1)** 63–71.
White, J.W., Riethof, M.L., Subers, M.H. *et al.* (1962) Composition of American honeys. In: *US Department of Agriculture Technical Bulletin 1261*.
White, J.W., Subers, M.H. and Schepartz, A.I. (1963) The identification of inhibine, the antibacterial factor in honey, as hydrogen peroxide and its origin in a honey glucose-oxidase system. *Biochimica et Biophysica Acta* **73**, 57–70.
White, J.W., Bryant, V.M. and Jones, J.G. (1991) Adulteration testing of Southwestern Desert honeys. *Am. Bee J.* **131(2)**, 123–26, 135.
Wilkins, A.L., Lu, Y. and Molan, P.C. (1993a) Extractable organic substances from New Zealand unifloral manuka (*Leptrospermum scoparium*) honeys. *J. Apicultural Res.* **32(1)**, 3–9.
Wilkins, A.L., Lu, Y. and Tan, S.T. (1993b) Extractives from New Zealand honeys. 4. Linalool derivatives and other components from nodding thistle (*Carduus nutans*) honey. *J. Agric. Food Chem.* **41(6)**, 873–78.
Willix, D.J., Molan, P.C. and Harfoot, C.J. (1992) A comparison of the sensitivity of wound-infecting species of bacteria to the antibacterial activity of manuka honey and other honey. *J. Appl. Bacteriol.* **73**, 388–94.
Willson, R.B. (1975) World trading in honey. In: *Honey: A Comprehensive Survey* (Ed. E. Crane). Heinemann, London, pp. 355–77.
Winkler, O. (1955) Beitrag zum Nachweis und zur Bestimmung von Oxymethylfufural in Honig und Kunsthonig. *Zeitschrift für Lebensmittel-Untersuchung und-Forschung* **102(3)**, 161–67.
Wootton, M., and Ryall, L. (1985) A comparison of *Codex Alimentarius* Commission and HPLC methods for 5–hydroxymethyl–2–furaldehyde determination in honey. *J. Apicultural Res.* **24(2)**, 120–24.
Wootton, M., Edwards, R.A. and Faraji-Haremi, R. (1978) Effect of accelerated storage conditions on the chemical composition and properties of Australian honeys 3. Changes in volatile components. *J. Apicultural Res.* **17(3)**, 167–72.

Wykes, G.R. (1952) An investigation of the sugars present in the nectar of flowers of various species. *The New Phytologist* **51**, 210–15.
Ziegler, H., Stichler, W. Maurizio, A. *et al.* (1978) Die Verwendung stabiler Isotope zur Charakterisierung von Honigen, ihrer Herkunft und ihrer Verfälschung. *Apidologie* **8(4)**, 337–47.

9 Authenticity of coffee

J. PRODOLLIET

9.1 Introduction

The coffee tree is an evergreen and roughly pyramidal shrub. It can grow to a height of 3–6 m or more. The leaves, arranged in pairs along the branches, are slender and lance-shaped, with a bright, waxy sheen. The blossoms are white, and smell intensely of jasmine. Coffee itself appears as small, cherrylike berries, deep red when ripe (Kolpas, 1993).

A coffee cherry consists of six different parts (Figure 9.1): (i) the 'skin' or exocarp, red (or yellow); (ii) the 'pulp' or mesocarp, yellow-white; (iii) the 'mucilage', yellow-white, (iv) the 'parchment' or endocarp, pale yellow; (v) the 'silverskin' or spermoderm, silvery; and (vi) the 'bean' or endosperm, brown-yellow to green-bluish when dry. Skin, pulp and parchment constitute the pericarp. When dry, the pericarp and the silverskin are called 'husks' or 'hulls' and represent about half of the total dry matter of the cherry. Furthermore, a coffee cherry may contain one, two or, rarely, more beans.

Coffee trees thrive between latitude 30 degrees north and latitude 30 degrees south, at altitudes ranging from sea level to more than 2000 m, in a global-encircling geographic region known as the 'coffee belt'. The main coffee-producing countries are Brazil, Colombia, Indonesia, Mexico and the Ivory Coast.

More than 60 species of the genus *Coffea* L. (Rubiaceae) are known. The most important are: (a) *Coffea arabica* Linnaeus or Arabica coffee; (ii) *Coffea canephora* Pierre ex Forehner or Robusta coffee; (iii) *Coffea liberica* Hiern or Liberica coffee; and (iv) *Coffea dewevrey* De Wild and Durand or Excelsa coffee (ISO, 1989). Arabica and Robusta coffees account, respectively, for about 75% and 24% of the world production, whereas Liberica and Excelsa coffees contribute less than 1%.

A great variety of coffee products are now available on the market. International coffee trade is conducted almost exclusively with green coffee. However, consumers are nowadays offered roasted coffee beans, roasted and ground coffee, as well as liquid and dried coffee extracts (soluble coffee). Furthermore, coffee can be mixed with coffee substitutes, and also sold as roasted and ground blends or as dried extracts. Whole-bean roasted coffee may also be soaked with liquid-flavouring agents to

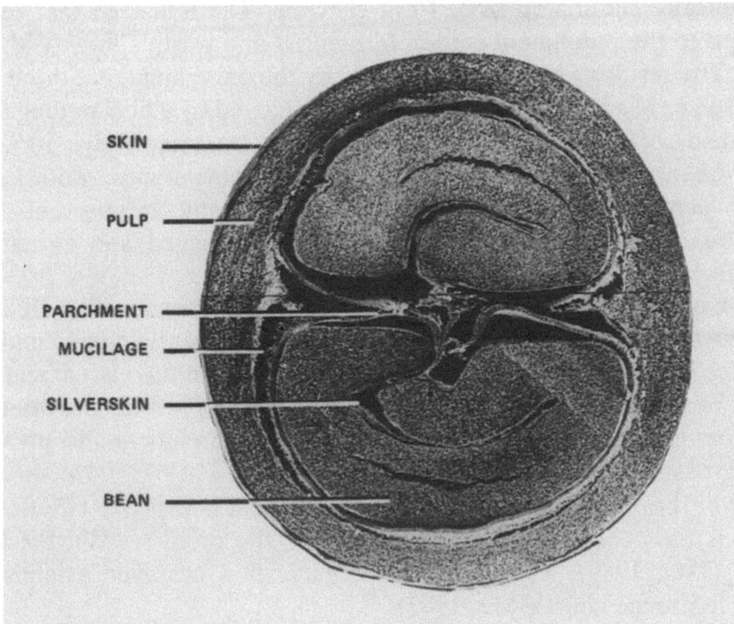

Figure 9.1 Transverse section of a coffee cherry containing two beans. (Courtesy Miss E. Dentan.)

produce flavoured coffees. Finally, dried coffee extracts mixed with milk solids (café au lait, cappuccino) exist. Decaffeinated forms of each of these coffee products are also available.

Whatever its final use, coffee remains an expensive raw material and, over the years, many fraudsters have been tempted to replace it with cheaper substitutes and to falsify the product declaration. Different types of adulteration, specific for each coffee product category, have been reported. The following sections describe these adulteration practices, as well as the analytical methods developed to detect them. Each section also gives the most relevant information on processing, definition and regulation of the original products.

9.2 Authenticity of green coffee

Once harvested, coffee cherries can be processed in one of two ways. In the 'dry process', the fresh ripe cherries are dried in the sun for up to 3 weeks. The husks are then removed mechanically to liberate the coffee beans. In the 'wet process', the outer skin and part of the mucilage are mechanically eliminated from fresh cherries. The mucilage that remains attached to the parchment is then fermented and washed away with clean water. The resulting beans, still covered by the parchment, are dried either in the sun or in a dryer. The parchment is removed by a final milling to give the naked coffee beans. Neither post-harvest treatments give 10% clean coffee beans, and impurities or 'defects', such as stones, wood, sticks, husks, parchments, or whole cherries, may still be present. After dehulling, the beans are cleaned, calibrated, bagged and stored until roasting.

What is green coffee? The International Coffee Organisation (ICO), an inter-governmental body that used to regulate the world supply and price of coffee beans, defines green coffee as 'all coffee in the naked bean form before roasting' (ICO, 1982a). In the USA, it is defined as 'the dried seed free from external layers (skin, pulp, mucilage, parchment and silverskin) before roasting' (Federal Specification, 1981).

Within the International Oranisation for Standardisation (ISO), green coffee is the "commercial term designating the dried seed of the coffee plant" (ISO, 1989). Moreover, this organisation has even established a defect reference chart (ISO, 1993).

No official international standards for the quality and authenticity of green coffee have been issued. However, in commercial practice, the most frequently accepted quality factors are the moisture content, the number of defects, amount of foreign matter, and the bean size (Clarke, 1979; 1988). Analytical methods for the determination of these parameters are described in several ISO standards (ISO, 1978a; 1978b; 1980; 1985a; 1991). Obviously, a simple visual examination of the product prevents any fraudulent addition of cheaper coffee substitutes. Nevertheless, it is important to know the country of origin and the species of coffee for the purpose of fair international trade (McClumpha, 1988). Moreover, Article 43 of the former International Coffee Agreement of 1983 stipulates that "every export of coffee by a member [of ICO] shall be covered by a valid Certificate of Origin" (ICO, 1982b). Suspicions were aroused that green coffee was not from the origin specified on the contract (Dyszel, 1985; Krivan *et al.*, 1993). Indeed, a country that had a coffee crop exceeding its export limit might have attempted, by transhipment through a non-ICO country and false labelling, to circumvent this limit (Dyszel, 1985). Furthermore, Arabica coffee is more expensive than Robusta. For these reasons, different techniques to authenticate its origin have been reported.

9.2.1 Geographic origin

In-cup taste and aroma are markedly influenced by the country of origin of the coffee. However, because of the large number of producing countries and intracountry variability, no taster can reliably identify them. Moreover, opinions of the tasters are subjective and not always accepted unanimously (Krivan *et al.*, 1993). The same discussion is true when considering the appearance (e.g. size, shape and colour) of the green beans; more objective methods are thus required for determination of the origin of green coffee.

Surprisingly, very few methods have been reported. Thermogravimetric analysis coupled with atmospheric pressure chemical ionisation mass spectrometry was applied to 13 green coffees from nine countries. Comparison of the ion curves obtained for each sample showed that intracountry, intercountry and intervariety variations are too important for an unambiguous determination (Dyszel, 1985). Krivan *et al.* studied the mineral and trace element profile of 10 green coffees from eight countries (Krivan *et al.*, 1993). Manganese was found to be the best indicator, followed by the elements C, Co, Cs, Na and Rb. Many more samples should be analysed to validate the method, and a multivariate analysis would certainly improve the results. Finally, it has been shown that multivariate stable isotope analysis can distinguish between African and American coffees (Dunbar and Wilson, 1982; Danho *et al.*, 1992). The technique consists of extracting and purifying the caffeine from finely ground green beans, and determining its stable isotope ratios ($^{2}H/^{1}H$), ($^{13}C/^{12}C$), ($^{15}N/^{14}N$), and ($^{18}O/^{16}O$) using mass spectrometry. Alternatively, the ($^{2}H/^{1}H$) ratios of each of the hydrogen atoms present in the caffeine molecule can be measured by deuterium nuclear magnetic resonance (Danho *et al.*, 1992). Unfortunately, no further differentiation could be made among the two groups of African (16 samples) and American (7 samples) coffees in the latter study (Danho *et al.*, 1992).

9.2.2 *Botanic origin*

Arabica beans are usually larger, flatter, heavier and more deeply grooved than Robusta beans. There are also important differences in flavour between the two types of coffee. Arabica is more aromatic with more perceptible acidity but less body. The Robusta species produces coffee with a bitter, full-bodied taste but little acidity. Therefore, a serious organoleptic examination of a green coffee bag should prevent any adulteration of Arabica beans with cheaper Robusta beans. Nevertheless, the appearance of the coffee beans, within the same species, also varies according to the country of origin. Moreover, some washed Robusta coffees approach the taste quality of Arabica. Consequently there is still room for non-declared substitution and, as for the determination of the geographic origin (see section 9.2.1), more objective methods are required.

Arabica and Robusta coffee undoubtedly differ in chemical composition (Clarke and Macrae 1985; Viani, 1986; 1993). However, the variations observed are small, and no major constituent can be used as a reliable marker for the presence of a small amount of Robusta beans in Arabica coffee. Most reported searches of discriminatory elements concentrated on the unsaponifiable matter of coffee oil.

In particular, Δ^5–avenasterol was found in higher concentration in Robusta beans and proposed for quantification of the relative proportion of the two coffee species in a blend. It can be seen clearly from Table 9.1 that interspecies variability is relatively important, especially in the study of Arabica beans by Tiscornia *et al.* (1979). The technique, when applied to blends of unknown composition, will therefore only detect large amounts of Robusta beans. In addition to Δ^5–avenasterol, 24–methylenecholesterol was also found (Mariani and Fedeli, 1991) in higher levels in Robusta (1.5–2.4% total sterols, average = 1.9%, $n = 15$) than in Arabica (0–0.4% total sterols, average = 0.2%, $n = 16$). Even a multidimensional approach, when considering these two sterols together with sitostanol (higher in Arabica), does not allow the detection of adulteration of 20% Robusta in Arabica (Mariani and Fedeli, 1991).

Certain diterpenes have proved to be more reliable tracers for the botanical origin of coffee. In Robusta, kahweol is either absent (Nackunstz and Maier, 1987), or present in only trace amounts (Pettitt, 1987; Speer and Montag, 1989). More interestingly, 16–*O*–methylcafestol was isolated from Robusta bean oil (Pettit, 1987; Speer, 1989b, Speer and Mischnick, 1989; Speer and Mischnick – Lübbecke, 1989). Its concentration in the green Robusta bean varies over a relatively narrow range (Table 9.2), and it is virtually absent in Arabica. From all reported data, 16–*O*–methylcafestol was detected in trace amounts (0.01 g/kg) in only one sample of Arabica (Speer and Montag, 1989).

Therefore, this diterpene appears to be a very good marker for Robusta

Table 9.1 Δ^5–Avenasterol in green coffee (% total sterols)[a]

Reference		Arabica	Robusta
Duplatre *et al.* (1984)	Range	1.8–2.9	10.1–12.5
	Average	2.2 ($n = 8$)	11.8 ($n = 6$)
Mariani and Fedeli (1991)	Range	1.6–4.1	5.1–12.6
	Average	2.7 ($n = 16$)	9.1 ($n = 15$)
Picard *et al.* (1984)	Range	–	7.1–14.7
	Average	–	10.7 ($n = 12$)
Tiscornia *et al.* (1979)	Range	2.0–14.6	–
	Average	6.3 ($n = 20$)	–
All four studies	Range	1.6–14.6	5.1–14.7
	Average	4.2 ($n = 44$)	10.2 ($n = 33$)

[a]n = Number of samples analysed.

Table 9.2 16–*O*–Methylcafestol in green Robusta coffee (g/kg db)[a]

Reference		Robusta
Speer (1989a)	Range	0.5–1.4
	Average	not reported (n = 16)
Speer and Montag (1989)	Range	0.6–1.3
	Average	1.0 (n = 15)
Speer *et al.* (1991a)	Range	0.6–1.8
	Average	1.1 (n = 20)
Speer *et al.* (1991b)	Range[b]	0.67–1.54
	Average[b]	1.12 (n = 10)

[a]n = number of samples analysed.
[b]Results expressed in g/kg.

and should allow the detection of low levels (e.g. 5%) of this species when blended with Arabica coffee.

Key aroma compounds in freshly roasted and ground (R&G) Arabica and Robusta beans (or their brews) have been identified using gas chromatography (GC). It is noteworthy that the studies reported were not all aimed at authenticating the botanic origin of green coffee. Multivariate analyses of 32 peaks from the headspace profile of coffee brews (Wada *et al.*, 1987) and of 13 peaks from the headspace profile of R&G beans (Holscher and Steinhart, 1992) differentiated between the two species. However, the techniques only seem suitable for the detection of large amounts of Robusta. The same conclusions can be drawn from the work of Tressl *et al.*, who found 20 times more 4-ethylguajacol (Tressl *et al.*, 1978a), and four times more ethylfuraneol and maltol (Tressl *et al.*, 1978b) in R&G Robusta aroma. Moreover, the variability of the concentration of these potential markers in Arabica and Robusta beans from different origins has not been reported. An earthly smelling compound, 2–methylisoborneol, has been identified using gas chromatography–mass spectrometry (GC–MS) as a key component for the typical Robusta flavour (Vitzthum *et al.*, 1990). Its concentration varies from 200–600 ng/kg in R&G Robusta, whereas it was beneath the detection limit of the method (50 ng/kg) in R&G Arabica. Again, many more samples should be analysed, to determine more precisely the concentration range of 2–methylisoborneol in both species. Higher levels (~20 times) of unidentified sulphur-containing compounds were also observed in the aroma complex of R&G and brewed Robusta (Nurok *et al.*, 1978). These differences were claimed to allow the detection of as little as 1% Robusta in a blend. In the author's opinion, the performance of the technique is highly exaggerated and does not take into account the (reported) variabilities of the proposed markers. It should be stressed here that all these techniques can only be applied to the problem of green coffee species identification, provided that all the samples (used for the database or unknown) are roasted, ground, stored, brewed and analysed under exactly the same conditions. It is well known that all these parameters influence greatly the aroma profile.

9.3 Authenticity of roasted coffee

During the roasting process, green coffee is gradually heated to about 200°C for several minutes. The stone-hard blue-to yellow-green beans then develop a dark brown colour, a brittle structure, and characteristic flavour and aroma (Viani, 1986).

Roasted coffee is defined by the ICO as "green coffee roasted to any degree and includes ground coffee" (ICO, 1982a), and by ISO as "coffee obtained by roasting green coffee" (ISO, 1989). As for green coffee, no official international standards for the quality and authenticity of roasted coffee have been issued.

Whole cherry roasted coffee is now available in almost any shop, and specific claims on either the geographic or botanic origin are usually made on the product label. The same arguments used for the correlation of organoleptic attributes to the origin of green coffee also apply to roasted coffee (see sections 9.2.1 and 9.2.2). Taste testing is even less indicative, as the degree of roasting influences markedly the taste of coffee, introducing an extra variability between coffee of the same origin. More objective methods are therefore required. Conversely, addition of cheaper non-coffee substitutes can be detected easily by macroscopic examination (Dick, 1995a).

9.3.1 Geographic origin

According to Bicchi *et al.* (1993) roasted coffee from Costa Rica, Honduras, Guatemala and Nicaragua can be distinguished clearly using principal component analysis on 26 peaks observed in their headspace GC aroma profile. The differentiation is not at all evident, and it is unlikely that this technique can be applied to a sample of unknown origin (see section 9.2.2). One can reasonably speculate that multivariate mineral and stable isotope analysis, already described for the authentication of green beans (see section 9.2.1), would also apply to roasted coffee with the same chance of success.

9.3.2 Botanic origin

Sterols are stable upon roasting (Duplatre *et al.*, 1984; Mariani and Fedeli, 1991). Therefore, taking into account the results given in section 9.2.2, we should not expect them to be better indicators for Robusta in roasted beans than in green coffee. This is confirmed by the relatively disappointing results of a discriminant analysis of campesterol, stigmasterol, β–sitosterol and Δ^5–avensterol values performed on 36 Arabica and 23 Robusta samples (Consiglieri *et al.*, 1991).

The diterpene, 16–*O*–methylcafestol, which appears to be the best marker for Robusta green coffee so far identified (see section 9.2.2), is also stable upon roasting (Speer, 1989a; Speer and Montag, 1989; Speer *et al.*, 1991a). Consequently, it also applies to roasted coffee.

Speer and Montag succeeded in detecting 2% Robusta in model mixtures of Arabica and Robusta roasted beans (Speer and Montag, 1989). The same authors analysed 21 commercial roasted coffees and found 16–*O*–methylcafestol in nine samples. It is noteworthy that none of the claimed 'pure Arabica' coffee contained any trace of the diterpene. It was also reported more recently that there are marked differences in the diterpene alcohol profile of Arabica and Robusta roasted coffee. When using characteristic ratios between non-identified diterpene alcohols, appearing in the GC trace of the unsaponifiable matter, 5–10% Robusta could be detected (Frega *et al.*, 1994).

Finally, a good discrimination of model mixtures of Arabica and Robusta roasted beans was obtained by applying principal component analysis to 42 peaks observed either in their headspace GC aroma profile or in the HPLC chromatograms of the extracts (Bicchi *et al.*, 1993). However, the technique is certainly inappropriate for the analysis of unknown samples as the degree of roasting and the storage conditions, which greatly influence the results, are not under the analyst's control.

9.4 Adulteration in roasted and ground coffee

Grinding of the roasted coffee beans is performed in various types of mills, from low-capacity household equipment to large, industrial roll mills. The process creates a myriad of tiny surfaces from which hot water can more easily extract the roasted coffee's flavour and body elements.

Therefore, the particle size of ground coffee is another key factor of the in-cup quality of coffee and must be related to the brewing method applied (Koplas, 1993; Viani, 1986).

The ICO definition of roasted coffee given in section 9.3 already includes R&G coffee (ICO, 1982a). ISO defines R&G coffee simply as a "product obtained by grinding roasted coffee" (ISO, 1989). No official international standards for the quality and authenticity of R&G coffee have been issued.

As with roasted coffee beans, R&G coffee is also commercially available to the consumers and, likewise, information on the geographic or botanic origin are sometimes given on the product label. Again, the in-cup taste and aroma of coffee depend upon origin (see sections 9.2.1 and 9.2.2), degree of roasting (see section 9.3) and, for a specific brewing method, on the grinding process. Furthermore, the shape and size of the whole beans (see section 9.2.2) cannot be used as indicators of botanic origin, as the product has been ground.

However, the bad news for people concerned with coffee authenticity is coffee substitutes. Basically almost anything from the vegetable kingdom can be roasted, ground and used in its own right as a coffee substitute (Maier, 1987), or sold as coffee/coffee substitute blend (see section 9.6) given correct labelling. The problem is that coffee substitutes can be added fraudulently to sound R&G coffee, and the resulting product sold as 'pure' R&G coffee.

Adulteration with chicory, barley, rye, corn, coffee husks/parchments, cocoa hulls, soya beans, cashew nuts, chick peas, acorns, figs, sugar beet, sugar or caramelised sugar etc. have been reported (Bose *et al.*, 1976; Cesar *et al.*, 1984; Copestake *et al.*, 1986; Ferreira *et al.*, 1971; Lee, 1980; Lopez, 1983; Macrae *et al.*, 1987; Miya and Shirose, 1977). In this connection, the results of Lopez's study, involving the analysis of 38 280 samples of Brazilian R&G coffee from 1976 to 1980, are quite impressive (Table 9.3). Microscopy is certainly the most simple and most diagnostic technique for detecting and identifying coffee substitutes in R&G coffee. However, confirmatory or semiquantitative methods have also been developed.

9.4.1 *Geographic and botanic origin*

Methods proposed for the authentication of whole roasted beans should also apply to R&G coffee, as their results are essentially unaffected by grinding.

Table 9.3 Type and number of adulterated samples detected in Brazilian R&G coffee from 1976 to 1980[a,b]

Year	1976 $n = 8762$	1977 $n = 9127$	1978 $n = 9232$	1979 $n = 6360$	1980 $n = 4799$
Caramel	31	43	66	31	59
Caramel and barley	4	5	5	3	11
Caramel and cocoa hulls	8	0	2	2	0
Barley	23	31	33	7	14
Corn	47	39	47	24	14
Soya	4	1	1	2	2
Sugar	15	9	8	0	2
Cocoa hulls	6	3	0	0	0

[a]Source: Lopez (1983).
[b]n = Total number of samples analysed during the year.

9.4.2 *Detection of chicory*

The presence of R&G chicory is detected easily using microscopy. Details of sample preparation, microscopic features and examples of photomicrographs can be found elsewhere (Dick, 1995c; Macrae *et al.*, 1987; *Swiss Food Manual*, 1973).

The amount of soluble solids that can be extracted from R&G chicory is much higher than from R&G coffee. This provides the basis of a very simple test in which the unknown sample is sprinkled onto the surface of cold water held in a tall vessel. Coffee floats, if not over-roasted, but chicory particles begin to sink within a few seconds and often leave behind a distinctive brown colour (Dick, 1995a; Macrae *et al.*, 1987).

The mineral composition of chicory has been found to be significantly different from that of coffee. In particular, the ratios K:Na and Cu:Rb are markedly lower in chicory (Table 9.4; Ferreira *et al.*, 1971). According to the authors, however, only amounts of chicory in excess of 40% can be detected.

More recently, as little as 1% of R&G chicory has been detected in a model mixture using photoacoustic spectroscopy (Cesar *et al.*, 1984). Nevertheless, a variability study of photoacoustic effect of R&G coffee and chicory of different origin, roasting degree and particle size should be performed to determine the 'true' detection limit for unknown samples.

The large differences in the monosaccharide profiles of R&G coffee and chicory form the basis of the most commonly applied and most sensitive detection procedure. In particular, very low levels of both free and total fructose and glucose are found in sound R&G coffee (see Table 9.5). Conversely, the main constituent polysaccharide of chicory is a polyfructan called inulin. Under controlled acid hydrolysis, inulin yields large amounts

Table 9.4 Selected mineral ratios in coffee, barley, chicory, and chick pea[a,b]

Ratio		Coffee $n = 48$	Barley $n = 24$	Chicory $n = 18$	Chick pea $n = 18$
Zn/Rb	Range	0.5–1.1	10–30	0.1–0.5	0.1–0.5
	Average	0.81	25	0.36	0.30
Sr/Cu	Range	0.1–0.5	2–5	0.1–0.5	0.1–0.5
	Average	0.33	4	0.32	0.28
K/Na	Range	160–250	120–160	5–7	100–120
	Average	185	138	6	115
Cu/Rb	Range	1.2–3.8	0.1–0.3	0.1–0.3	1.2–2.0
	Average	2.8	0.18	0.16	1.6
Cu/Mn	Range	0.5–1.1	0.3–0.5	0.5–1.0	10–20
	Average	0.85	0.38	0.70	16

[a]Source: Ferreira *et al.* (1971).
[b]n = Number of samples analysed.

Table 9.5 Fructose and glucose in pure roasted coffee (%db)[a]

Reference		Fructose	Glucose	Method	Hydrolysis[b]
Bheema Rao *et al.* (1986)	Range[c]	0.18–0.24	–	Colorimetry	Yes
	Average[c]	0.22 (*n* = 7)	–		
Kazi (1979)	Range[c]	0.10–0.80	–	Colorimetry	Yes
	Average[c]	0.53 (*n* = 50)	–		
Kröplien (1971)	Range	0.02–0.03	0.01–0.02	TLC	No
	Average	0.03 (*n* = 5)	0.01 (*n* = 5)		
Kröplien (1974)	Range	<0.01–0.90	<0.01–0.90	TLC	No
	Average	0.14 (*n* = 42)	0.21 (*n* = 42)		
Lutman (1982)	Range	–	0.02–0.05	GC	Yes
	Average	–	0.03 (*n* = 11)		
Noyes and Chu (1993)	Range	–	–	GC	No
	Average	0.01 (*n* = 21)	0.001 (*n* = 21)		
Prodolliet *et al.* (1995c)[d]	Range	–	0.55–1.28	HPAE–PAD[e]	Yes
	Average	–	0.88 (*n* = 24)		

[a] *n* = Number of samples analysed.
[b] 'No' results expressed as free sugar, 'Yes' results expressed as total sugar.
[c] Results expressed as ketohexose.
[d] Unreported data from Prodolliet *et al.* (1995c).
[e] High-performance anion-exchange chromatography with pulsed amperometric detection.

of fructose and a little glucose. Reported values of fructose and glucose content in R&G chicory are given in Table 9.6. When comparing Tables 9.5 and 9.6, it can be seen clearly that total fructose is the marker of choice of the presence of chicory. Its detection limit in R&G coffee can be estimated to be around 3%. In principle, any analytical methods can be used to determine total fructose, provided that the hydrolysis conditions are strong enough to completely convert inulin or fructo-oligomers into fructose but not so excessive to cause fructose decomposition (Berger *et al.*, 1991).

9.4.3 Detection of cereals

R&G cereals such as barley, rye and corn also exhibit typical microscopic features (Lopez, 1983; Macrae *et al.*, 1987; *Swiss Food Manual*, 1973). Like chicory, R&G cereals can also be detected using the simple solubility test described in section 9.4.2. Cereal particles that sink to the bottom of the vessel can be isolated and examined microscopically for a more accurate identification (Dick, 1995a; Lee, 1980).

Starch is the main constituent of all cereal products. It is partially degraded to dextrins and glucose oligomers during roasting (Pazola, 1979; 1987b). However, the total amount of the glucose polymers is still very high, typically 50–68% for roasted barley and 60–74% for roasted rye (Pazola and Cieslak, 1979; Pazola, 1987b). Equivalent levels of total glucose are, therefore, formed upon acid hydrolysis of roasted cereals. This is much higher than the maximum quantities of total glucose reported in sound R&G coffee (see Table 9.5). It can be estimated that about 2% of R&G cereals can be detected in an unknown sample, using total glucose as tracer.

Alternatively, according to Ferreira fraudulent addition of 6% of barley can be identified using Zn, Rb, Sr and Cu as tracers (Ferreira *et al.*, 1971). Indeed, the ratios Zn:Rb and Sr:Cu are much higher in barley than in coffee (see Table 9.4). An even lower detection limit (1%) was claimed when photoacoustic spectroscopy was applied to model mixtures (Cesar *et al.*, 1984). As in the case of chicory (see section 9.4.2), the technique needs further validation.

9.4.4 Detection of coffee husks/parchments

Coffee husks or parchments are byproducts of the post-harvest treatment of green coffee. They are mainly used as fuel, composting material or animal feed. However, they can also be used to adulterate coffee. Indeed, adulteration of R&G coffee with R&G coffee husks or parchments has been reported in Brazil (Miya and Shirose, 1977). For this reason, Miya and Shirose submitted model mixtures of R&G coffee containing different

Table 9.6 Fructose and glucose in roasted chicory (%db)[a]

Reference		Fructose	Glucose	Method	Hydrolysis[b]
Bheema Rao *et al.* (1986)	Range[c]	33.4–34.7	–	Colorimetry	Yes
	Average[c]	34.0 (n = 4)	–		
Frei *et al.* (1991)	Range	2.40–5.81	0.95–2.44	Enzymatic	No
	Average	3.64 (n = 20)	1.50 (n = 20)		
Frei *et al.* (1991)	Range	27.9–57.2	2.84–6.78	Enzymatic	Yes
	Average	46.1 (n = 20)	5.34 (n = 20)		
Kazi (1979)	Range[c]	45.9–49.7	–	Colorimetry	Yes
	Average[c]	48.0 (n = 125)	–		

[a] n = Number of samples analysed.
[b] 'No' results expressed as free sugar, 'Yes' results expressed as total sugar.
[c] Results expressed as ketohexose.

amounts of R&G coffee husks and coffee parchments to a taste panel. It was concluded that the presence of 2.5% of both materials can be detected, when the brew is compared with that obtained with the same reference R&G coffee used for the preparation of the model mixtures (Miya and Shirose, 1977). It is unlikely that coffee husks or parchments can be detected so sensitively in unknown samples for which no reference coffee is available.

The same conclusion can be drawn from the study of Cesar *et al.* (1984), in which the presence of 1% R&G coffee parchment was discerned in a model mixture using photoacoustic spectroscopy.

R&G coffee beans and husks also present very specific microscopic feature. Roasted coffee displays small clusters (Figure 9.2a) of globular, thick-walled parenchyma cells (Figure 9.2b), whereas roasted husks appear as small plates (Figure 9.3a) of long, thin-walled cells (Figure 9.3b).

In principle, the methods reported for the detection of coffee husks/parchments in soluble coffee (see section 9.5.4) should also apply to R&G coffee.

9.4.5 Detection of chick pea

The Cu:Mn ratio in chick pea has been found to be significantly higher than in coffee (see Table 9.4; Ferreira *et al.*, 1971). According to the authors, addition of 6% chick pea in coffee can be detected.

9.4.6 Detection of cashew nut shells

Adulteration of R&G coffee with cashew nut shell powder was reported in India. The substitute can be detected easily by a microscopic examination of the unknown sample (Bose *et al.*, 1976).

Specific constituents of cashew nut shell, anacardic acid, cardol and anacardol, were also identified using TLC. Normally they are absent from coffee. As little as 5% shell can be detected using these three markers (Sengupta *et al.*, 1974).

9.4.7 Detection of palm nut

Faudulent addition of R&G palm nut in R&G coffee was suspected in Italy. It was found that the lauric acid content, expressed on total fatty acid basis, varies between 0.2 and 0.4% in roasted coffee, and 38–40% in roasted palm nut (Foschini and Usai, 1968). By coupling this technique with microscopic analysis, it should be possible to identify the presence of very low quantities (~2%) of R&G palm nut.

9.5 Adulteration in soluble (instant) coffee

In soluble coffee production, R&G coffee is semicontinuously extracted with water at 100–180°C. The amount of soluble solids that can be removed (extraction yield) is in the range of 35–55%. The cooled extract is then dried either at high temperature in a spray-drying tower or at low temperature by freeze–drying (lyophilisation) to produce soluble (or instant) coffee. Both processes yield a finished product with 1–5% moisture. Volatile aromatic substances, which are lost during the grinding

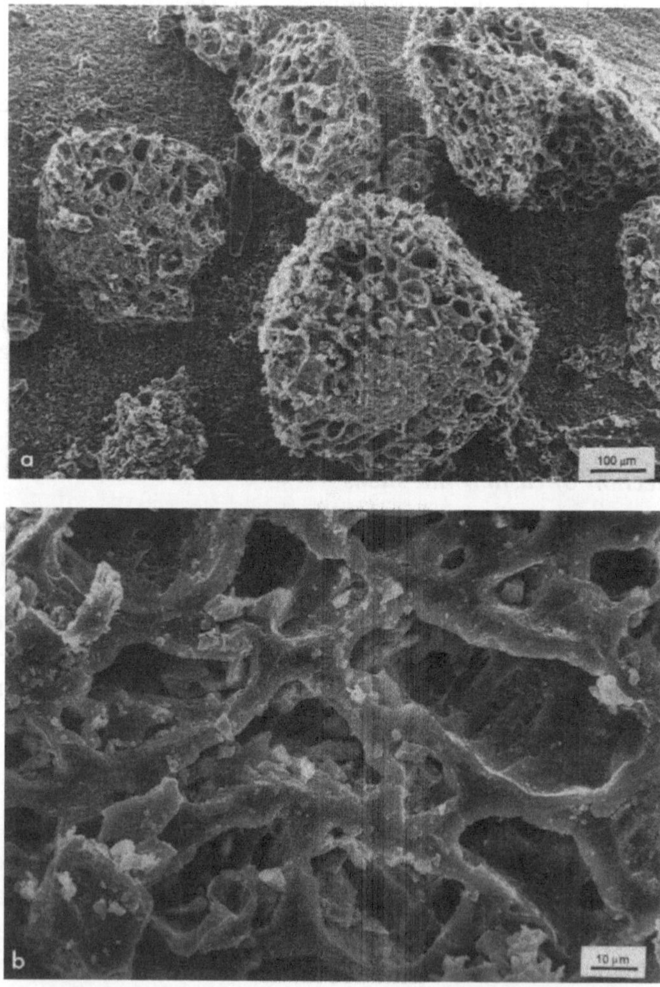

Figure 9.2 Scanning electronmicrographs of roasted and ground coffee: (a) clusters; (b) parenchyma cells. (Courtesy Miss E. Dentan.)

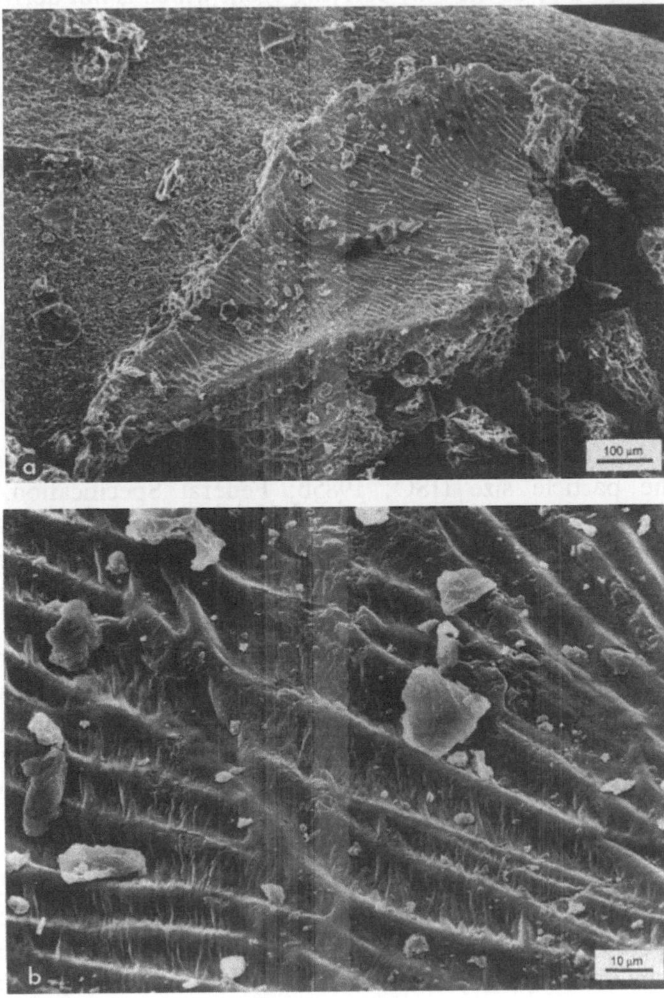

Figure 9.3 Scanning electronmicrographs of roasted and ground coffee husks: (a) plates; (b) cells. (Courtesy Miss E. Dentan.)

of roasted coffee and extraction, are recovered and emulsified with oil from pressed roasted coffee or spent grounds. They are then reconstituted by spraying the emulsion on the powder (aromatisation), at a level of 0.3–1%, during the filling operation (Viani, 1986).

ICO defines soluble coffee as "the dried water-soluble solids derived from roasted coffee" (ICO, 1982a). ISO uses the terms 'instant coffee', 'soluble coffee' or 'dried coffee' extract and defines them as "dried, water-soluble product, obtained exclusively from roasted coffee by physical methods using water as the only carrying agent which is not derived from coffee" (ISO, 1989). Within the European Union, geographically a more restricted organisation than ICO and ISO, the terms 'soluble coffee', 'instant coffee', 'dried coffee extract' or 'dried extract of coffee' all mean "coffee extract in powder, granular, flake, cube or other solid form, of which the coffee-based dry matter content is not less than 90% by weight".

It is also stipulated that "this product may not contain any substances other than those derived from its extraction" (EEC, 1985). However, in the USA, instant coffee may contain ascorbic acid (Federal Specification, 1981).

Moisture content is obviously an important requirement of any soluble coffee legislation. The convenient method to assess this parameter is to determine the percentage loss of mass of the material at 70°C under reduced pressure (EEC, 1979; Federal Specification, 1981; ISO, 1983a). Other national and international standards are also available for determination of the particle size (ISO; 1985b; Federal Specification, 1981), insoluble matter (ISO, 1985c; Federal Specification, 1981) and ascorbic acid (Federal Specification, 1981). However, no official international standards have been issued to check the authenticity of soluble coffee.

As in the case of R&G coffee, there are two main issues. Firstly, claims on either the geographic or the botanic origin of coffee are becoming more popular for high-quality commercial soluble coffee. For this particular problem, taste-testing is of no help, because of the enormous influence of processing on the individual taste and aroma characteristics of each coffee. Secondly, soluble solids of coffee substitutes can be added fraudulently to sound soluble coffee, and the resulting product sold as 'pure' soluble coffee. Indeed, adulteration with cereals, maltodextrins, coffee husks/parchments and caramelised sugar has been reported (Blanc *et al.*, 1989; Blanc and Parchet, 1989; Davis *et al.*, 1990; Prodolliet *et al.*, 1991; 1995a; 1995c). Obviously, detection of the presence of coffee substitutes is not feasible by microscopy or by other physical means traditionally used for the authentication of green, roasted and R&G coffee (see preceding sections). Therefore, different approaches are required.

9.5.1 Geographic and botanic origin

No methods have been reported so far for the determination of the origin of coffee in soluble coffee. It is unlikely that a multivariate mineral analysis, as proposed to assess the geographic origin of green and roasted coffee (see sections 9.2.1 and 9.3.1), can be applied to soluble coffee. Indeed, variations in the mineral composition of the processing water would influence that of the finished product (Macrae *et al.*, 1987). Danho's technique, based on the multivariate stable isotope analysis of caffeine could, however, be considered (Danho *et al.*, 1992).

Table 9.7 Carbohydrate profile of pure soluble coffee (%db)[a]

Reference		Mannitol	Free fructose	Free glucose	Sucrose	Total fructose	Total glucose	Total xylose	Method
Berger *et al.* (1991)	Range	—	—	—	—	<0.01–0.58	0.51–1.51	—	Enzymatic
	Average	—	—	—	—	0.24 (*n* = 21)	0.88 (*n* = 21)	—	
Blanc *et al.* (1989)	Range	—	0.05–0.42	0.00–0.43	0.06–0.30	—	0.57–1.51	0.11–0.28	HPLC
	Average	—	0.12 (*n* = 43)	0.12 (*n* = 43)	0.17 (*n* = 13)	—	0.97 (*n* = 43)	0.17 (*n* = 43)	
Davis *et al.* (1990)	Range	0.02–0.22	0.09–0.52	0.08–0.54	0.01–0.64	—	0.61–1.74	0.00–0.32	HPLC + GC
	Average	0.09 (*n* = 56)	0.29 (*n* = 56)	0.30 (*n* = 56)	0.20 (*n* = 56)	—	1.15 (*n* = 56)	0.20 (*n* = 56)	
Kröplien (1974)	Range	—	<0.01–0.48	<0.01–0.30	—	—	—	—	TLC
	Average	—	0.14 (*n* = 31)	0.08 (*n* = 31)	—	—	—	—	
Noyes and Chu (1993)	Range	—	—	—	—	—	—	—	GC
	Average	0.037 (*n* = 21)	0.360 (*n* = 21)	0.038 (*n* = 21)	0.16 (*n* = 21)	—	—	—	
Prodolliet et al. (1995)[b]	Range	0.01–0.28	0.00–1.00	0.00–0.56	0.00–0.74	—	0.46–2.04	0.01–0.39	HPAE–PAD[c]
	Average	0.07 (*n* = 589)	0.25 (*n* = 589)	0.16 (*n* = 589)	0.16 (*n* = 392)	—	1.26 (*n* = 589)	0.21 (*n* = 589)	
Promayon *et al.* (1976)	Range[d]	—	0.04–0.50	0.01–0.50	—	—	—	—	Enzymatic
	Average[d]	—	0.20 (*n* = 6)	0.18 (*n* = 6)	—	—	—	—	
Sabbagh *et al.* (1977)	Range[d]	—	0.02–0.13	0.15–0.31	0.02–0.30	—	—	—	GC
	Average[d]	—	0.08 (*n* = 10)	0.20 (*n* = 10)	0.12 (*n* = 10)	—	—	—	
Trugo and Macrae (1982)	Range	—	traces–1.04	traces–0.48	—	—	—	—	HPLC
	Average	—	0.23 (*n* = 7)	0.07 (*n* = 7)	—	—	—	—	
Trugo and Macrae (1985)	Range	—	0.23–1.29	n.d.–1.03	—	—	—	—	HPLC
	Average	—	0.48 (*n* = 13)	0.37 (*n* = 13)	—	—	—	—	

[a]*n* = Number of samples analysed; n.d. = not detected.
[b]Partially reported data.
[c]High-performance anion-exchange chromatography with pulsed amperometric detection.
[d]Results expressed on the products as is.

The author doubts strongly that any method based on the analysis of the unsaponifiable matter or the headspace aroma profile, as proposed for the detection of Robusta in green or roasted Arabica, would be applicable to soluble coffee. These parameters are indeed influenced markedly by the extraction conditions and the aromatisation process.

9.5.2 Detection of chicory

Pure soluble coffee is characterised by low free fructose, free glucose, mannitol, sucrose, and total (after acid hydrolysis) glucose and xylose levels (Table 9.7). High amounts of total galactose and total mannose are also typical for soluble coffee. Examples of chromatograms of free carbohydrate and total carbohydrate profiles, obtained using high-performance anion-exchange (HPAE) chromatography with pulsed amperometric detection (PAD), are given in Figures 9.4 and 9.5, respectively.

Reported values for total fructose (or total ketose) in dried chicory powder ranged from 56–72% (Kazi, 1979; Newman, 1981; Pazola, 1987a). Therefore, by considering total fructose as a marker, as little as 2–3% soluble chicory solids could be detected. This is very close to the detection limit of 1–5% determined by Berger *et al.* (1991), based on the enzymatic analysis of model mixtures. Moreover, deliberate addition of 3% chicory can be detected very clearly in the HPAE–PAD chromatogram of the mixture (Figure 9.6), which exhibits a large total fructose peak. In this

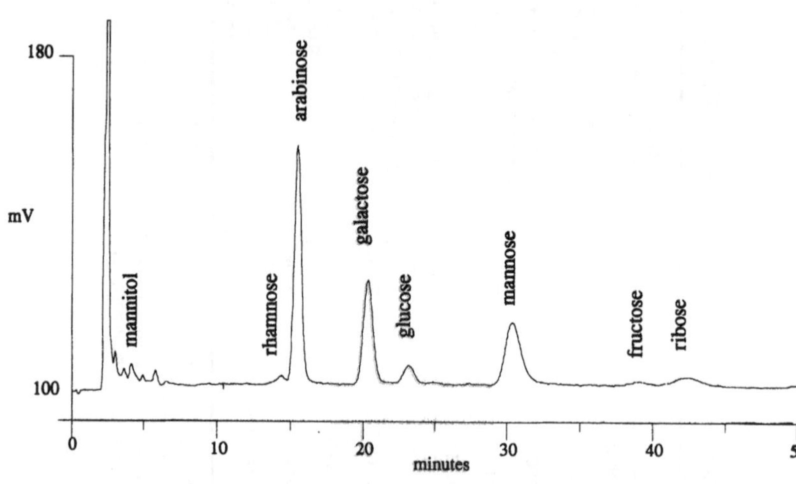

Figure 9.4 HPAE–PAD typical free carbohydrate profile of a pure soluble coffee. Reproduced with permission from Prodolliet, J. *et al.* (1995a) and the *Journal of AOAC International.*

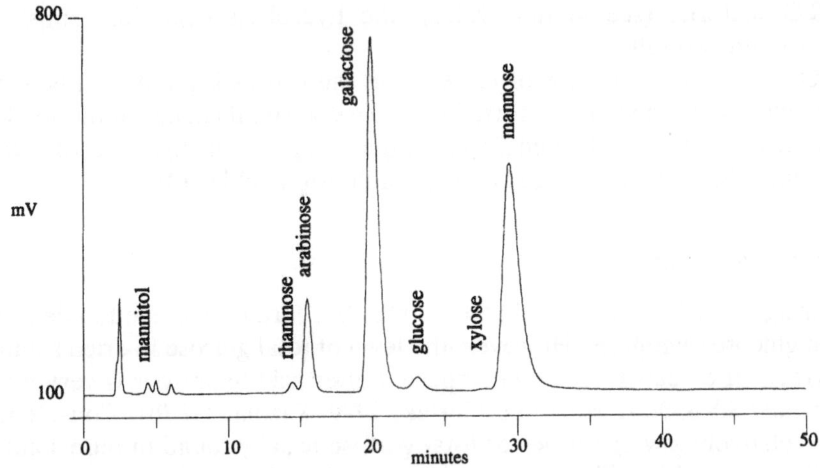

Figure 9.5 HPAE–PAD typical total carbohydrate profile of a pure soluble coffee. Reproduced with permission from Prodolliet, J. *et al.* (1995a) and the *Journal of AOAC International*.

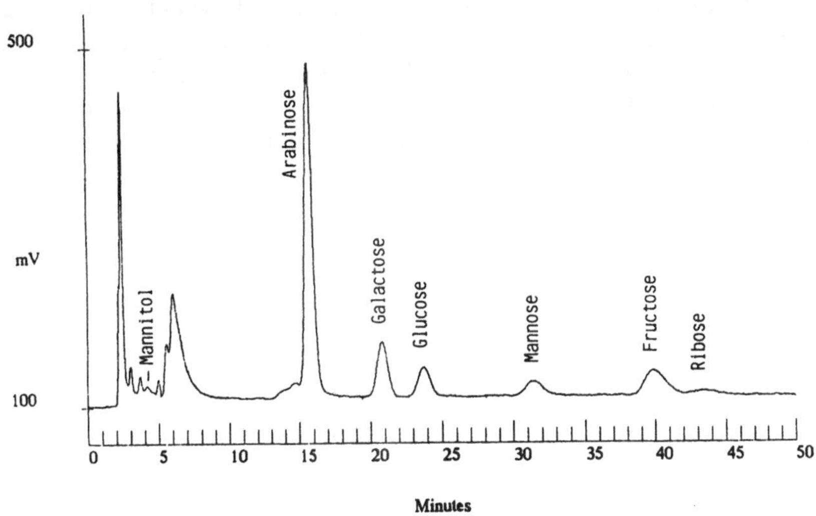

Figure 9.6 HPAE–PAD total carbohydrate profile after mild hydrolysis of a soluble coffee containing 3% chicory. Reproduced with permission from Prodolliet, J. *et al.* (1991) and the *ASIC*.

particular case, the total fructose content on the product was 2.19% (Prodolliet *et al.*, 1991), a result about two-times higher than the highest value ever measured in sound soluble coffee (see Table 9.7). Alternatively, total fructose can be determined colorimetrically (Bheema Rao *et al.*, 1986; Kazi, 1979; Newman, 1981). As for the analysis of total fructose in

R&G material (see section 9.4.2), the hydrolysis conditions must be controlled carefully.

Owing to the influence of minerals in the processing water, it is to be expected that the values reported in Table 9.4 would change in the soluble product. Therefore, Ferreira's technique proposed in the case of R&G products (see section 9.4.2) is not suitable for soluble coffee.

9.5.3 *Detection of cereals*

During extraction of roasted cereals, starch is further degraded to dextrins and glucose oligomers. However, the level of total glucose in dried soluble extracts of cereals that can be expected after acid hydrolysis is very high, typically 80–85% (Pazola and Geslak, 1979; Pazola, 1987b). Considering the relatively low quantities of total glucose usually found in pure soluble coffee (see Table 9.7), it can be estimated that about 3% of soluble solids of cereal origin can be detected in an unknown sample using this tracer. The analysis of model mixtures, containing 5% of a cereal blend (barley, rye), 5% of malted barley and 2.5% of the same cereal blend, allowed determination of a similar (2.5%) detection limit (Berger *et al.*, 1991). The presence of maltodextrins, eventually added directly to the coffee extract before drying, would obviously be detected more easily. Figure 9.7 gives the HPAE–PAD total carbohydrate profile of a model mixture containing 3% of cereal solids. In this example, the total glucose content of the

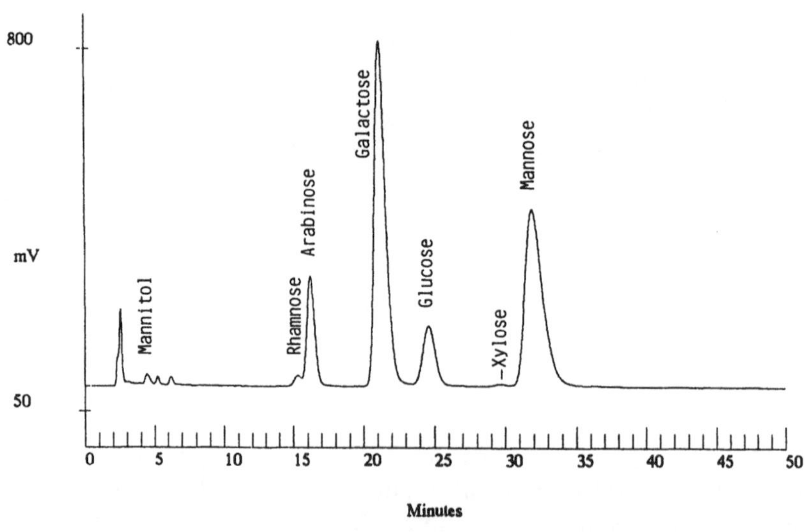

Figure 9.7 HPAE–PAD total carbohydrate profile of a soluble coffee containing 3% cereals. Reproduced with permission from Prodolliet, J. *et al.* (1991) and *ASIC*).

Table 9.8 Carbohydrate profile of commercial soluble coffee (% db)

Reference	Class		Mannitol	Sucrose	Total glucose	Total xylose	Method
Blanc *et al.* (1989)	A	Range	—	0.06–0.30[a]	0.57–1.51	0.11–0.28	HPLC
	(n = 43)	Average	—	0.17	0.97	0.17	
	B	Range	—	0.00–1.27[b]	1.20–5.77	0.45–1.95	
	(n = 65)	Average	—	0.45	3.11	0.98	
	C	Range	—	0.24–3.00[c]	10.7–50.8	0.15–1.20	
	(n = 14)	Average	—	1.22	24.6	0.66	
Davis *et al.* (1990)	A	Range	0.02–0.22	0.01–0.64	0.61–1.74	0.00–0.32	GC
	(n = 56)	Average	0.09	0.20	1.15	0.20	
	B	Range	0.10–1.85	0.00–1.43	0.90–5.71	0.25–3.15	
	(n = 68)	Average	0.52	0.36	2.45	0.94	
	C	Range	0.10–1.18	0.19–5.59	3.69–38.1	0.01–3.24	
	(n = 21)	Average	0.48	1.39	15.9	0.80	
Prodolliet et al. (1995c)	A	Range	0.01–0.28	0.00–0.74[d]	0.46–2.04	0.01–0.39	HPAE–PAD[e]
	(n = 589)	Average	0.07	0.16	1.26	0.21	
	B	Range	0.01–1.55	0.00–1.29f	0.75–20.42	0.40–3.15	
	(n = 81)	Average	0.39	0.27	3.08	0.84	
	C	Range	0.04–0.41	0.06–15.7[g]	1.01–42.5	0.08–0.40	
	(n = 30)	Average	0.21	2.85	9.09	0.25	

[a]n = 13; [b]n = 43; [c]n = 12; [d]n = 392; [e]High-performance anion-exchange chromatography with pulsed amperometric detection; [f]n = 47; [g]n = 11; where n = number of samples analysed.

product was 3.41% (Prodolliet *et al.*, 1991), a result much higher than the maximum value observed in pure soluble coffee (see Table 9.7).

The carbohydrate profile has been determined in commercial samples from a large number of producing and consuming countries, and all sold as pure soluble coffee. Some of the results are gathered in Table 9.8. The products could be separated into three different classes, each class having a characteristic profile. Clearly, high levels of total glucose were found in class C samples, compared with the pure soluble coffees (class A). It was, therefore, concluded that class C samples were adulterated with cereals, maltodextrins or caramelised sugar. Considering the highest values found for certain products (e.g. 42.5% or 50.8%) one could seriously doubt the presence of any coffee material. Class B products will be discussed in the next section.

The ratios Zn:Rb and Sr:Cu have been suggested as tracers for R&G barley in R&G coffee (see section 9.4.3). It is unlikely that the technique can be applied to soluble coffee because of the influence of the mineral composition of the processing water.

9.5.4 *Detection of coffee husks/parchments*

Coffee husks or parchments are certainly one of the most evident adultering materials that can be used by fraudsters. Their mannitol and total xylose contents differ, however, from that of coffee and provide the basis of all reported detection methods. Indeed, large quantities of the two carbohydrates, and to a certain extent of total glucose, were found in raw, dried, and autoclave-extracted coffee husks (Table 9.9); they are present in rather low concentrations, however, in pure soluble coffee (see Table 9.7). Interestingly, the level of mannitol in Robusta husks is about seven times lower than in dried Arabica husks and is similar to that found in sound soluble coffee. However, the total xylose concentration is equivalent in both species. Therefore, total xylose is the marker of choice for the presence of coffee husks/parchments.

In the three surveys of commercial soluble coffee described (see section 9.5.3), products characterised by high levels of mannitol, total glucose and total xylose were clearly identified (Tables 9.8, class B). It was concluded that these samples were adulterated with coffee husks and/or parchments. The number of class B adulterated samples relative to the total number of tested soluble coffee was impressive, *circa* 53% (Blanc *et al.*, 1989), 55% (Davis *et al.*, 1990) and 12% (Prodolliet *et al.*, 1995c) respectively. Analysis of the total xylose in a wide selection of green coffee beans and the assessment of its fate during processing, allowed the derivation of a maximum total xylose limit of 0.40%, above which a soluble coffee should be considered as adulterated (Prodolliet *et al.*, 1995c). It was also

Table 9.9 Mannitol and total xylose (% db) in coffee husks[a]

Reference	Type of husks		Mannitol	Total glucose	Total xylose	Method
Blanc *et al*. (1989)	Arabica, extract[b]		—	15.7	5.00	HPLC
	Arabica, extract[c]		—	4.32	5.74	
Davis *et al*. (1990)	Arabica, sun-dried	Range	1.61–2.03	—	—	GC
		Average	—	—	—	HPLC + GC
Noyes *et al*. (1991)	Arabica, raw	Range	0.20–1.50	—	2.58–8.38	HPLC (xylose)
		Average	0.91 (n = 8)	—	4.40 (n = 8)	GC (mannitol)
	Robusta, raw	Range	0.07–0.21	—	2.80–5.19	
		Average	0.14 (n = 3)	—	4.35 (n = 3)	
Prodolliet et al. (1995a)	Arabica, sun-dried	Range	1.17–1.43	6.64–11.5	4.67–5.93	HPAE–PAD[d]
		Average	1.30 (n = 5)	10.5 (n = 6)	5.20 (n = 6)	
	Robusta, sun-dried	Range	0.18–0.21	3.32–3.59	4.87–5.59	
		Average	0.19 (n = 5)	3.42 (n = 5)	5.15 (n = 5)	

[a] n = Number of samples analysed. [b] Extracted from sun-dried husks (one sample only). [c] Extracted from medium roasted husks (one sample only). [d] High-performance anion-exchange chromatography with pulsed amperometric detection.

observed, in the same study, that the samples (class A) considered as non-adulterated (see Table 9.8; Prodolliet *et al.*, 1995c) exhibited mannitol and total glucose levels below 0.30 and 2.10%, respectively. Moreover, Davis *et al.* (1990) reported that a mannitol content exceeding 0.30% constitutes further evidence that a soluble coffee contains undeclared material.

Different analytical methods were developed for determination of the husks/parchments tracers. Blanc *et al.* (1989) applied HPLC with post-column derivatisation with Tetrazolium Blue for the analysis of both free and total carbohydrate profiles. The technique is nevertheless unsuitable for mannitol. A similar method was applied by Noyes *et al.* (1991) for the total carbohydrate profile only. Mannitol and free sugars were analysed using GC. The same analytical approach was applied by Davis *et al.* (1990), although the HPLC analytical column and derivatisation agent were different from that used by Noyes *et al.* (1991). Enzymatic procedures have also been reported (Prodolliet *et al.*, 1991; 1995a). Finally, Prodolliet *et al.* (1991; 1995a; 1995b; 1995c) developed the so-called HPAE–PAD procedure in which all carbohydrates are separated in one single run on a special anion-exchange resin, and detected using pulsed amperometry. Figure 9.8 shows a HPAE–PAD total carbohydrate profile of a commercial soluble coffee adulterated with coffee husks/parchments. The HPAE–PAD technique was tested collaboratively on six soluble coffee samples by 11 laboratories (Prodolliet *et al.*, 1995b). The precision of the method was good for carbohydrate levels above 0.3%. The procedure has been officially approved for publication as an ISO international standard (ISO,

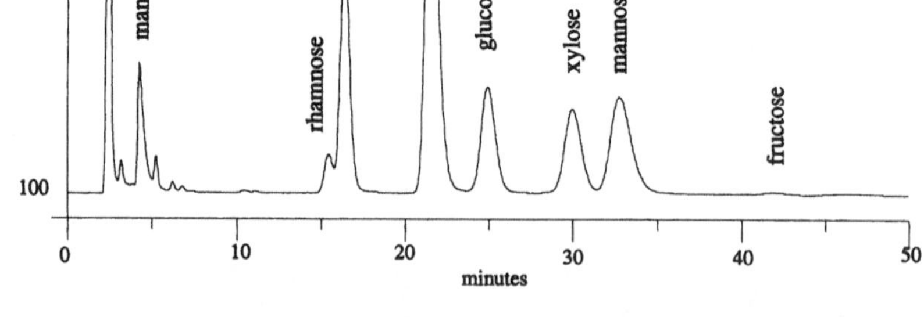

Figure 9.8 HPAE–PAD total carbohydrate profile of a soluble coffee containing coffee husks/parchments. Reproduced with permission from Prodolliet *et al.* (1995a) and the *Journal of AOAC International.*

1995) and has been adopted as first action (Method 995.13) by the Association of Official Analytical Chemists (AOAC).

Another carbohydrate, psicose, has been found as a tracer of the presence of coffee husks (Blanc and Parchet, 1989). According to Blanc and co-workers, the level of psicose in sound soluble coffee should not exceed 0.005%. The monosaccharide was analysed in 122 commercial samples. Some of the tested products contained a high proportion of psicose (0.094–0.307%) and were identified as soluble coffee adulterated with coffee husks (Blanc and Parchet, 1989).

9.5.5 Detection of caramelised sugar

Caramelised sugar is obtained by heating sugars, mainly sucrose, invert sugar and glucose, alone or with a catalyst, usually a small amount of an ammonium salt (Pazola, 1987c). Adulteration of R&G coffee with caramelised sugar has already been reported (Copestake *et al.*, 1986). Products obtained by co-roasting of coffee with sugar are popular in Spain under the name 'torrefacto'. 'Torrefacto'-soluble coffee also exists. Lack of clear labelling would be a fraud. It is very likely that the same type of fraud occurs in soluble coffee. Addition of caramelised sugar should markedly increase the levels of sucrose, total fructose and total glucose, depending on the extent of caramelisation.

Commercial samples, all sold as pure soluble coffee, exhibited abnormally high concentrations of sucrose and total glucose (see Table 9.8, class C). It was concluded that these products were adulterated with caramelised sugar, cereals or maltodextrins (Blanc *et al.*, 1989; Davis *et al.*, 1990; Prodolliet *et al.*, 1995a; 1995c). Indeed, these coffee substitutes can not be distinguished by the methods used.

According to Berger and co-workers, 3.5% of caramelised sugar could be detected using total glucose alone as a marker (Berger *et al.*, 1991). However, the variability of the total glucose content in different samples of dried caramel sugar extracts should be studied in order to give more credit to the limit.

9.6 Adulteration in coffee/coffee-substitute blends

Many coffee substitutes have already been tested and have found a commercial application. Nowadays, the most frequently used coffee substitutes are roasted chicory, cereals (e.g. barley, malted barley, rye), sugar beet and figs (Maier, 1987). They can be sold in their own right, or more frequently as blends with other coffee substitutes (R&G or soluble coffee). However, they can also be mixed with coffee and sold as coffee/ coffee-substitute blends. Both R&G and soluble blends are available on the market. In some countries, legislation is strict with respect to labelling of coffee/coffee-substitute blends (Macrae *et al.*, 1987). Manufacturers sometimes indicate the exact quantities of each material used. When such information is not given, most regulations (e.g. the European Union) require that ingredients are listed in order of decreasing weight. As coffee substitutes are much cheaper than coffee, fraudsters may be tempted to add more of the less expensive material than suggested or indicated on the product label. Therefore, there is a need for analytical methods that can quantify accurately the amount of each ingredient in the blend.

Chemical methods rely on the determination of specific constituents present either in the coffee material or in the coffee substitutes. The ideal marker should be present in only one of the commodities and not in the others; this is not that easy to find. Furthermore, variations in the concentration of the specific constituent should be as narrow as possible (ideally constant) in the ingredient to be quantified. Unfortunately, the analyst deals with natural products usually exhibiting large natural variations in chemical composition. Obviously, processing conditions will also influence the concentration of the marker. Finally, in the case of soluble blends, there is an other variability factor, namely the extraction yield. These products can be manufactured by co-extraction of roasted coffee and the roasted substitute(s), or by blending the separate soluble extracts before drying. In either case, differences in the extraction yields involved will influence considerably the composition of the final blend, and complicate interpretation of any quantitative data (Macrae *et al.*, 1987).

Nevertheless, several attempts to quantify coffee or coffee substitute materials by chemical methods have been reported (Table 9.10). Two strategies were used for quantification. Firstly, the constituent specific to one material is determined in selected samples of that material, preferably from various origins and prepared under different processing conditions. The average value is then taken for estimation of the amount of material in blends of unknown composition. Secondly, one (hopefully) representative sample of each ingredient of the blend is selected. The chemical marker is then analysed in model mixtures containing different proportions of each ingredient. A calibration curve, giving the concentration of the marker as a function of the amount of material containing the marker, is then used for subsequent quantification of that material in blends of unknown composition. Some authors applied their technique to commercial blends, for which the exact composition was declared on the label (Promayon et al., 1976; Resmini *et al.*, 1978; Lutman, 1982; Ferreira *et al.*, 1987). Not surprisingly, in most cases, agreement with the declared values was poor. This can be attributed to the large variability of the chosen marker in the material analysed. In conclusion, unless the levels of such markers are constant, the chemical methods are reliable only if both the coffee and the

Table 9.10 Chemical methods for the quantitative analysis of R&G and soluble coffee/coffee-substitute blends

Reference	Type of blend	Ingredient	Marker	Method
Bheema Rao *et al*. (1986)	R&G coffee/chicory	Chicory	Total ketohexoses	Colorimetry
Blanc (1978)	Soluble coffee/chicory	Chicory	Free and total fructose Total sugars	Enzymatic Colorimetry
Ferreira *et al*. (1987)	R&G cofee/barley/chicory	Coffee	Caffeine	HPLC
Kazi (1979)	R&G coffee/chicory R&G coffee/fig	Chicory Fig	Total ketoses Total ketoses	Colorimetry Colorimetry
Lutman (1982)	R&G coffee/barley	Barley	Total glucose	GC
Newman (1981)	Soluble coffee/chicory/glucose syrup	Chicory Glucose syrup Coffee	Total ketoses Total glucose Caffeine	Colorimetry Reductometry Spectrophotometry
Promayon (1976)	Soluble coffee/chicory	Chicory	Free fructose	Enzymatic
Resmini *et al*. (1978)	R&G coffee/barley	Barley	Total reducing sugars	Luff-Schoorl
Vilar and Ferreira (1971)	Soluble coffee/barley/chick pea/barley	Coffee	Chlorogenic acid	Spectrophotometry

coffee substitute(s) used in the preparation of the blend are available. This is obviously never the case in official control laboratories, which definitively need alternative procedures. However, the chemical methods are certainly very useful to coffee/coffee-substitute manufacturers, for process control at the factory level.

Alternative methods, such as Fourier transform infrared (FT–IR) and near-infrared (NIR) spectroscopy have been used for the determination of the coffee content in soluble coffee/coffee-substitute blends (Fábian *et al.*, 1994). This type of technique is only suitable for unknown samples provided that a very large database containing a spectrum of coffee and coffee substitutes from various origins and processing conditions is available.

9.7 Authenticity of decaffeinated products

Decaffeination is usually performed on the green coffee beans and less frequently on soluble coffee. The solvents used range from water, methylene chloride, ethyl acetate, edible fats and oils, to supercritical carbon dioxide (Viani, 1986). Decaffeinated coffee can be processed into all types of known coffee products by the same methods used for regular non-decaffeinated coffee.

ICO defines decaffeinated coffee as “green, roasted or soluble coffee from which caffeine has been extracted” (ICO, 1982a). The ISO definition is more simple “coffee from which caffeine has been extracted”, with a note to say that “a maximum residual caffeine content would usually be stated in a specification for decaffeinated coffee” (ISO, 1989). The level of residual caffeine is obviously the most important requirement for the quality of decaffeinated coffee. Often, national legislations set the maximum allowable caffeine content in coffee products (mainly soluble coffee). Analytical methods for the determination of this parameter are described in several national standards (Dick, 1995b; EEC, 1979; ISO, 1983b).

Another possible type of adulteration that might occur with decaffeinated coffee is a mislabelling of the decaffeination process used. Indeed, the water process is more appealing for the consumer than the organic solvent process. It has been strongly suspected that decaffeinated roasted coffee labelled as water processed was in fact decaffeinated with organic solvents (Lee, 1982). Organoleptically, it is impossible to differentiate them. However, detection of traces of residual solvent in the product would be indicative of the solvent process.

9.8 Conclusions

There are still many unanswered questions, and many to come. For example, was coffee really roasted at the declared date (coffee freshness)? Was it really roasted or grown at altitude, as sometimes claimed on the label? Was it extracted according to the current legislation, which may require the use of water and coffee only? Was it flavoured with natural or synthetic flavours? Or was it really grown without the use of chemicals (organic coffee)? Obviously new analytical methods are required to face the old problems and the new challenges. This is the good news for analytical chemists but, what is the message to the consumer?

Is there any coffee in coffee products? After such a comprehensive listing of adulteration practices, this seems difficult to believe. However, the answer is definitively the affirmative. Coffee is a good beverage, and

Figure 9.9 Café Martin, 1921 poster by Cappiello, 'Avoid counterfeits'.

we like it. We also like the image and the dreams coffee carries. Therefore, when we drink coffee, we want to be sure that it is an authentic product of reputed quality, and not a pale imitation. It is the responsibility of all partners of the coffee industry, from farmers to distributors, to keep or to win the consumers' confidence. The great majority are honest and trustful. Very few are not. Cappiello, in his famous 1921 poster, said "Avoid counterfeits" (Figure 9.9). We think that this cannot be avoided but we can now definitely make life much tougher for the bad guys.

References

Berger, A., Brülhart, M. and Prodolliet, J. (1991) Detection of adulteration in pure soluble coffee by enzymatic sugar determination. *Lebensmittel-Wissenschaft und- Technologie* **24(1)**, 59–62.

Bheema Rao, M., Vittal Rao, A.S., Abraham, K.O. *et al.* (1986) Estimation of chicory in coffee-chicory mixture. *Indian Coffee* **1**, 13–19.

Bicchi, C.P., Binello, A.E., Legovich, M.M. *et al.* (1993) Characterization of roasted coffee by S–HSGC and HPLC–UV and principal component analysis. *J. Agric. Food Chem.* **41(12)**, 2324–28.

Blanc, M. (1978) The chicory root – the evolution of fructose during technological treatments. *Lebensmittel-Wissenschaft und- Technologie* **11(1)**, 19–22.

Blanc, M. and Parchet, J.M. (1989) Identification and quantification of a new carbohydrate in coffee chemistry. *ASIC, 13th Colloque*, Paipa, pp. 191–95.

Blanc, M.B., Davis, G.E., Parchet, J.-M. *et al.* (1989) Chromatographic profile of carbohydrates in commercial soluble coffees. *J. Agric. Food Chem.* **37(4)**, 926–30.

Bose, P.K., Chatterjee, T.K., Das, S.P. *et al.* (1976) Detection of cashew nut shells in tea, coffee and chicory. *J. Ind. Acad. Forensic Sci.* **15(2)**, 42–44.

Cesar, C.L., Vargas, H., Lima, C.A.S. *et al.* (1984) On the use of photoacoustic spectroscopy for investigating adulterated or altered powdered coffee samples. *J. Agric. Food Chem.* **32(6)**, 1355–58.

Clarke, R.J. (1979) International coffee standardisation and legislation. *Food Chem.* **4**, 81–96.

Clarke, R.J. (1988) International Standardisation. In: *Coffee, Volume 6: Commercial and Technicolegal Aspects*, chap. 6 (Eds. R.J. Clarke and R. Macrae). Elsevier Applied Science, London and New York, pp. 105–43.

Clarke, R.J. and Macrae, R. (Eds. (1985) *Coffee, Volume 1: Chemistry*. Elsevier Applied Science, London and New York.

Consiglieri, G., Fonseca, G. and Amendola, F. (1991) Multivariate analysis for the classification of coffee blends. *Rassegna Chimica* **43(4)**, 171–77.

Copestake, J.G., Thiele, G. and Farrington, J. (1986) The economics of coffee adulteration in Bolivia. *Tropical Sci.* **26**, 129–34.

Danho, D., Naulet, N. and Martin, G.J. (1992) Deuterium, carbon and nitrogen isotopic analysis of natural and synthetic caffeines. Authentication of coffees and coffee extracts. *Analusis* **20**, 179–84.

Davis, G.E., Garwood, V.W., Barfuss, D.L. *et al.* (1990) Chromatographic profile of carbohydrates in commercial coffees. 2. Identification of mannitol. *J. Agric. Food Chem.* **38(6)** 1347–50.

Dick, R.H. (1995a) Coffee and Tea, AOAC Official Method 920.90, Roasted coffee – macroscopic examination. In: *Official Methods of Analysis of AOAC International*, 16 edn. (Ed. P. Cunniff). AOAC International, Arlington, p. 2.

Dick, R.H. (1995b) Coffee and Tea, AOAC Official Method 979.11, Caffeine in roasted coffee – chromatographic–spectrophotometric method. In: *Official Methods of Analysis of AOAC International*, 16 edn. (Ed. P. Cunniff). AOAC International, Arlington, p. 3.

Dick, R.H. (1995c) Coffee and Tea, AOAC Official Method 920.95, Chicory in roasted coffee – Infusion Method. In: *Official Methods of Analysis of AOAC International*, 16th edn. (Ed. P. Cunniff). AOAC International, Arlington, p. 3.

Dunbar, J. and Wilson, A.T. (1982) Determination of geographic origin or caffeine by stable isotope analysis. *Anal. Chem.* **54(3)**, 590–92.

Duplatre, A., Tisse, C. and Estienne, J. (1984) Contribution à l'identification des espèces Arabica et Robusta par l'étude de la fraction stérolique. *Annales des Falsifications, de l'Expertise Chimique et Toxicologique* **828**, 259–70.

Dyszel, S.M. (1985) Characterization of green coffee beans by combined thermogravimetric analysis/atmospheric pressure chemical ionization mass spectrometry. *Thermochimica Acta* **87**, 89–98.

EEC (1985) EEC Council Directive of 19 December 1985 amending Directive 77/436/EEC on the approximation of the laws of the Member States relating to coffee extracts and chicory extracts (1985), 85/573/EEC. *Official Journal of the European Communities*, **L372**, 22–24.

EEC (1979) EEC Council Directive of 13 November 1979 laying down Community methods of analysis for testing coffee extracts and chicory extracts (1979), 79/1066/EEC. *Official Journal of the European Communities*, **L327**, 17–28.

Fábián, Z., Izvekov, V., Salgó, A. *et al.* (1994) Near-infrared reflectance and Fourier transform infrared analysis of instant coffee mixtures. *Analytical Proc. Incl. Analytical Comm.* **31**, 261–63.

Federal Specification (1981) HHH–C–575D of 19 October 1981. Coffee, Instant.

Ferreira, L.A.B., Oliveira, E.F., Vilar, H. *et al.* (1971) Contribution to the identification of coffee purity. *ASIC, 5th Colloque*, Lisbonne, pp. 79–84.

Ferreira, M.A., Barros, P. and de Oliveira Fernandes, J. (1987) Doseamento de cafeina por cromatografia liquida de alta pressão. *Revista Portuguesa de Farmácia* **37(3)**, 17–30.

Foschini, A. and Usai, A. (1968) L'Individuazione della noce di palma dum torrefatta nei torrefatti polverizzati di caffè. *Rassegna Chimica* **4**, 147–51.

Frega, N., Bocci, F. and Lercker, G. (1994) High-resolution gas chromatographic method for determination of Robusta coffee in commercial blends. *J. High Resolution Chromatogr.* **17**, 303–307.

Frei, F., Duquaine, S. and Prodolliet, J. (1991) Determination of sugars in dried and roasted chicory. *Nestlé Internal Report.*

Holscher, W. and Steinhart, H. (1992) Investigation of roasted coffee freshness with an improved headspace technique. *Zeitschrift für Lebensmittel-Untersuchung und- Forschung* **195**, 33–38.

International Coffee Organization, ICO (1982a) International Coffee Agreement 1983 **Article 3**, 5.

International Coffee Organization, ICO (1982b) International Coffee Agreement 1983 **Article 43**, 43.

ISO International Standard (1995) Instant coffee – determination of free and total carbohydrate contents – method using high-performance anion-exchange chromatography. **ISO 11292: 1995**.

ISO International Standard (1978a) Green coffee – determination of moisture content (basic reference method). **ISO 1446 : 1978**.

ISO International Standard (1978b) Green coffee – determination of moisture content (routine method). **ISO 1447 : 1978**.

ISO International Standard (1980) Green coffee – olfactory and visual examination and determination of foreign matter and defects. **ISO 4149 : 1980**.

ISO International Standard (1983a) Instant coffee – determination of loss in mass at 70°C under reduced pressure. **ISO 3726 : 1983**.

ISO International Standard (1983b) Coffee – determination of caffeine content (reference method). **ISO 4052 : 1983**.

ISO International Standard (1985a) Green coffee – determination of proportion of insect-damaged beans. **ISO 6667 : 1985**.

ISO International Standard (1985b) Instant coffee – size analysis. **ISO 7532 : 1985**.

ISO International Standard (1985c) Instant coffee – determination of insoluble matter content. **ISO 7534 : 1985**.

ISO International Standard (1989) Coffee and its products – vocabulary. **ISO 3509 : 1989**.

ISO International Standard (1991) Green coffee – size analysis: manual sieving. **ISO 4150: 1991**.

ISO International Standard (1993) Green coffee – defect reference chart. **ISO 10470 : 1993**.

Kazi, T. (1979) Coffee mixtures – spectrophotometric method for the determination of coffee content. *Food Chem.* **4**, 73–80.

Kolpas, N. (1993) *A Cup of Coffee – From Plantation to Pot, A Coffee Lover's Guide to the Perfect Brew*, Michael Friedman, New York.

Krivan, V., Barth, P. and Feria Morales, A. (1993) Multi-element analysis of green coffee and its possible use for the determination of origin. *Mikrochimica Acta* **110**, 217–36.

Kröplien, U. (1971) Monosaccharides in coffees and coffee substitutes. *ASIC, 5th Colloque*, Lisbonne, 217–23.

Kröplien, U. (1974) Monosaccharides in roasted and instant coffees. *J. Agric. Food Chem.* **22(1)**, 110–16.

Lee, S. (1980) Chicory, vegetables and cereals – is there any coffee in there? *Tea & Coffee Trade J.* **152(5)**, 8, 48.

Lee, S. (1982) Fraud detected in current decaf coffee bean marketing. *Tea & Coffee Trade J.* **154(6)**, 6, 43.

Lopez, F.C. (1983) Quantitative determination of substances most frequently used in adulteration of ground roasted coffee. *Revista do Instituto Adolfo Lutz,* **43(1/2)**, 3–8.

Lutman, F. (1982) Riconoscimento dell'orzo eventualmente aggiunto al caffé torrefatto e macinato. *Bollettino dei Chimici dell'Italiana dei Laboratori Provincialli*, **33(S4–S5)**, 573–80.

Macrae, R., Beaumont, J. and Vaughan, J.G. (1987) Detection and analysis. In: *Coffee Volume 5: Related Beverages*, chap. 6 (Eds R.J. Clarke and R. Macrae). Elsevier Applied Science, London and New York, pp. 149–78.

Maier, H.G. (1987) Introduction. In: *Coffee Volume 5: Related Beverages*, chap. 1 (Eds R.J. Clarke and R. Macrae). Elsevier Applied Science, London and New York, pp. 1–18.

Mariani, C. and Fedeli, E. (1991) Sterols of coffee grain of Arabica and Robusta species. *La Rivista Italiana delle Sostanze Grasse* **68**, 111–15.

McClumpha, A.D. (1988) The trading of green coffee. In: *Coffee Volume 6: Commercial and Technicolegal Aspects*, chap. 1 (Eds R.J. Clarke and R. Macrae). Elsevier Applied Science, London and New York, pp. 1–27.

Miya, E.E. and Shirose, I. (1977) Effects of adulteration on the quality of coffee. *Boltim do Instituto de Technologia de Alimentos* **49**, 123–49.

Nackunstz, B. and Maier, H.G. (1987) Determination of cafestol and kahweol in coffee. *Zeitschrift für Lebensmittel-Untersuchung und- Forschung,* **184**, 494–99.

Newman, J.R. (1981) Estimation of coffee, chicory and glucose solids contents of instant coffee products. *J. Assoc. Public Analysts* **19**, 59–64.

Noyes, R.M. and Chu, C.M. (1993) Material balance on free sugars in the production of instant coffee. *ASIC, 15th Colloque*, Montpellier, pp. 577–82.

Noyes, R.M., McCarthy, J.P. and Oram, C.P. (1991) The variation of xylose, mannitol, and free sugar levels in instant coffee. *ASIC, 14th Colloque*, San Francisco, 202–210.

Nurok, D., Anderson, J.W. and Zlatkis, A. (1978) Profiles of sulfur-containing compounds obtained from Arabica and Robusta coffees by capillary column gas chromatography. *Chromatographia* **11(4)**, 188–92.

Pazola, Z. and Cieslak, J. (1979) Changes in carbohydrates during the production of coffee substitute extracts, especially in the roasting process. *Food Chem.* **4**, 41–52.

Pazola, Z. (1987a) The chemistry of chicory and chicory-product beverages. In: *Coffee Volume 5: Related Beverages*, chap. 2 (Eds R.J. Clarke and R. Macrae). Elsevier Applied Science, London and New York, pp. 19–57.

Pazola, Z. (1987b) The chemistry of cereal-based beverages. In: *Coffee Volume 5: Related Beverages*, chap. 3 (Eds R.J. Clarke and R. Macrae). Elsevier Applied Science, London and New York, pp. 59–104.

Pazola, Z. (1987c) The chemistry of non-cereal-based beverages. In: *Coffee Volume 5: Related Beverages*, chap. 4 (Eds R.J. Clarke and R. Macrae). Elsevier Applied Science, London, pp. 105–119.

Pettitt, Jr., B.C. (1987) Identification of the diterpene esters in Arabica and Canephora coffees. *J. Agric. Food Chem.* **35(4)**, 549–51.

Picard, H., Guyot, B. and Vincent, J.-C. (1984) Study on the sterol compounds of coffee. *Coffee canephora* oil. *Café Cacao Thé* **28(1)**, 47–62.
Prodolliet, J., Blanc, M.B., Brülhart, M. *et al.* (1991) Determination of carbohydrates in soluble coffee by high-performance anion-exchange chromatography with pulsed amperometric detection. *ASIC, 14th Colloque*, San Francisco, pp. 211–19.
Prodolliet, J., Bruelhart, M., Lador, F. *et al.* (1995a) Determination of free and total carbohydrate profile in soluble coffee. *J. AOAC Internat.* **78(3)**, 749–61.
Prodolliet, J., Bugner, E. and Feinberg, M. (1995b) Determination of carbohydrates in soluble (instant) coffee by high-performance anion-exchange chromatography with pulsed amperometric detection: collaborative study. *J. AOAC Internat.* **78(3)**, 768–82.
Prodolliet, J., Bruelhart, M., Blanc, M.B. *et al.* (1995c) Adulteration of soluble coffee with coffee husks/parchments. *J. AOAC Internat.* **78(3)**, 761–67.
Promayon, J., Barel, M., Fourny, G. *et al.* (1976) Chicory content determination tests in soluble mixtures of coffee and chicory. *Café Cacao Thé* **20(3)** 209–18.
Resmini, P., Volonterio, G. and Prati, F. (1978) Determinazione quantitativa del caffè in miscele con suoi surrogati. *La Rivista della Società Italiana de Scienza dell'Alimentazione* **7(2)**, 123–26.
Sabbagh, N.K., Faria, J.B. and Yokomizo, Y. (1977) Determination of monosaccharides in Brasilian instant coffees. *Coletanea do Instituto de Tecnologia de Alimentos* **8**, 55–73.
Sengupta, P., Sen, A.R., Ghoshdastidar, N. *et al.* (1974) Detection of chashew nut shells in coffee, tea, and chicory. *J. Assoc. Official Analytical Chemists* **57(3)**, 761–62.
Speer, K. (1989a) 16-*O*-Methylcafestol – a new diterpene in coffee. In: *Agriculture, Food Chemistry and the Consumer – Proceedings of the 5th European Conference on Food Chemistry, volume 1, the Consumer and Food, Analytical Food Chemistry and Agricultural Production, Biotechnology and Bioassays*, pp. 302–306.
Speer, K. (1989b) 16-*O*-Methylcafestol – a new diterpene in coffee. Methods for the determination of the 16-*O*-methylcafestol in crude and treated coffees. *Zeitschrift für Lebensmittel-Untersuchung und-Forschung* **189**, 326–30.
Speer, K. and Mischnick, P. (1989) 16-*O*-Methylcafestol – a new diterpene in coffee. Discovery and identification. *Zeitschrift für Lebensmittel-Untersuchung und- Forschung* **189**, 219–22.
Speer, K. and Mischnick-Lübbecke, P. (1989) 16-*O*-Methylcafetol – ein neues Diterpen im Kaffee. *Lebesmittelchemie und Gerichtliche Chemie* **43**, 43.
Speer, K. and Montag, A. (1989) 16-*O*-Methylcafestol – ein neues Diterpen im Kaffee. Erste Ergebnisse: Gehalte in Roh- und Röstkaffees. *Deutsche Lebensmittel-Rundschau* **85(12)**, 381–84.
Speer, K., Tewis, R. and Montag, A. (1991a) 16-*O*-Methylcafestol – a quality indicator for coffee. *ASIC, 14th Colloque*, San Francisco, pp. 237–44.
Speer, K., Tewis, R. and Montag, A. (1991b) 16-*O*-methylcafestol – a new diterpene in coffee. Free and bound 16-*O*-methylcafestol. *Zeitschrift für Lebensmittel-Untersuchung und- Forschung* **192**, 451–54.
Swiss Food Manual (1973) Röstkaffee – Schwimmprobe zum Nachweis von Kaffee-Surrogaten. **Method 35B/02**, 19.
Tiscornia, E., Centi-Grossi, M., Tassi-micco, C. *et al.* (1979) Sterols fraction of coffee seeds oil (Coffea arabica L.) *La Rivista Italiana delle Sostanze Grasse* **56**, 283–92.
Tressl, R., Grünewald, K.G., Köppler, H. and Silwar, R. (1978a) Phenols in roasted coffees of different varieties. I. *Zeitschrift für Lebensmittel-Untersuchung und- Forschung* **167**, 108–10.
Tressl, R., Bahri, D., Köppler, H. *et al.* (1978b) Diphenols and caramel compounds in roasted coffees of different varieties. II. *Zeitschrift für Lebensmittel- Untersuchung und-Forschung* **167**, 111–14.
Trugo, L.C. and Macrae, R. (1982) The determination of carbohydrates in coffee products using high-performance liquid chromatography. *ASIC, 10th Colloque*, Salvador, pp. 187–92.
Trugo, L.C. and Macrae, R. (1985) The use of the mass detector for sugar analysis of coffee products. *ASIC, 11th Colloque*, Lomé, pp. 245–51.
Viani, R. (1986) Coffee. In: *Ullmann's Encyclopedia of Industrial Chemistry, vol. A7*, 5th edn., VCH, Weinheim, pp. 315–39.

Viani, R. (1993) The composition of coffee. In: *Caffeine, Coffee, and Health* (Eds S. Garattini). Raven Press, New York, pp. 17–41.

Vilar, H. and Ferreira, L.A.B. (1971) How chlorogenic acid may contribute to the quantitative evaluation of coffee in soluble coffees mixed with substitutes. *ASIC, 5th Colloque*, Lisbonne, 120–27.

Vitzthum, O.G., Weisemann, C., Becker, R. *et al.* (1990) Identification of an aroma key compound in Robusta coffees. *Café Cacao Thé* **34(1)**, 27–33.

Wada, K., Sasaki, H., Shimoda, M. *et al.* (1987) Objective evaluation of various trade varieties of coffee by coupling of analytical data and multivariate analyses. *Agric. Biol. Chem.* **51**(7), 1753–60.

10 Authentication of eggs and egg products

P. BRERETON

Egg quality has traditionally been seen as either a food safety issue or a concern of industry. Until recently, there has been less emphasis on ensuring egg quality from a regulatory perspective. However, the egg industry produces annually in excess of 500 000 billion eggs throughout the world. In addition, consumers have become more discerning and seek a 'superior' and well-defined product. Therefore, it is easy to see that financial pressures and incentives may be placed on egg producers. The authenticity of egg products may thus be questioned as opportunities for adulteration may exist to provide the advantages of commercial gain to unscrupulous producers, processors and retailers.

10.1 Egg production

It is estimated that the value of eggs produced in England and Wales in 1993 was in the region of £400 million. These eggs were produced from some 27 million laying hens. Laying flocks range in size from a few hens to large intensive units.

10.1.1 Intensive egg production

In intensive units hens are kept in wire-floored cages, usually four to five hens per cage. The eggs roll down the sloping floors of the cages to moving belts, which carry them to an assembly area where they are washed, graded by candling and packaged for transport to retail markets.

10.1.2 Non-intensive egg production

There has been a considerable move by the consumer towards purchase of eggs produced by non-intensive means. This is in recognition of animal welfare concerns as well as consumer perception that intensively produced eggs are an inferior food product. There has, therefore, been a considerable increase in the number of free range and perchery flocks in the UK in recent years. The approximate production for 1994 is given in Figure 10.1.

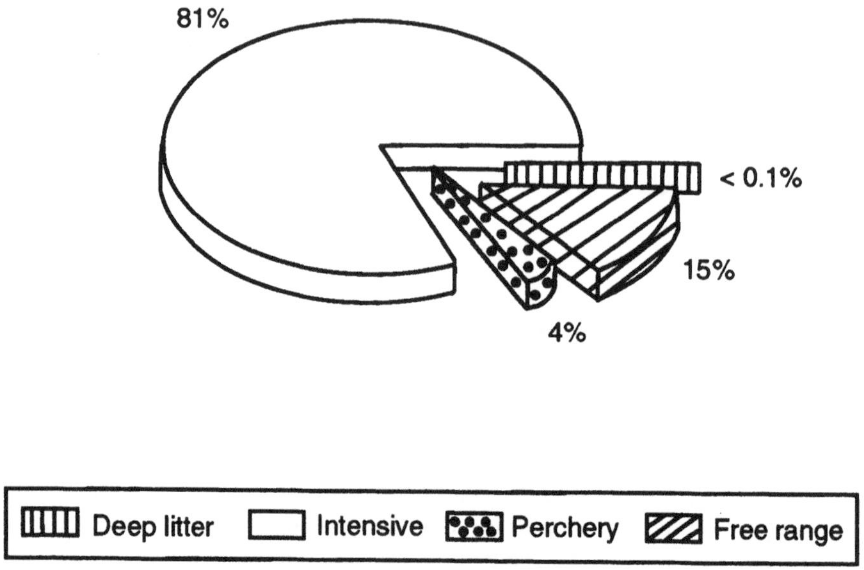

Figure 10.1 UK egg production 1994.

European Commission Regulation (EEC 1274/91) defines types of non-intensively produced eggs. In that Regulation there are four descriptions of eggs, which may be marked on the pack, to describe the type of production:

10.1.2.1 Free-range production. In this category:

- hens have continuous daytime access to open-air runs
- the ground to which the hens have access is mainly covered with vegetation
- the maximum stocking density is not greater than 1000 hens per hectare of ground available to the hens (one hen/10 m^2)
- the interior of the building must satisfy the conditions specified in deep-litter and perchery production.

10.1.2.2 Semi-intensive production. Here:

- hens have continuous access to open-air runs
- the ground to which the hens have access is mainly covered with vegetation
- the maximum stocking density is not greater than 4000 hens/hectare of ground available to the hens (one hen/2.5 m^2)
- the interior of the building must satisfy the conditions specified in deep-litter and perchery production.

10.1.2.3 Deep-litter production. In this category:

- the maximum stocking density is not greater than seven hens/m^2 of floor space available to the hens,
- at least one-third of this floor area is covered with a litter material such as straw, wood shavings, sand or turf
- a sufficiently large part of the floor area available to the hens is used for the collection of bird droppings.

10.1.2.4 Perchery (barn eggs) production. Here:

- the maximum stocking density is not greater than 25 hens/m^2 of floor space available to the hens
- the interior of the building is fitted with perches of a length sufficient to ensure at least 15 cm of perch space for each hen.

As the consumer is prepared to pay a higher price for non-intensively produced eggs for what is perceived as a superior product, there is considerable scope for misleading labelling claims. Indeed, this authenticity issue is one of the major concerns of the regulatory authorities. As the definitions of egg types only differ in terms of the physical accommodation of the hens, there is considerable scope for unscrupulous producers to sell intensively produced eggs as free-range ones. The enforcement agencies have few analytical methods available that can be used to differentiate the various egg types. In the UK, the relevant enforcement agency is the Ministry of Agriculture Fisheries and Food through its Egg Marketing Inspectorate (EMI). They regularly perform audits and inspections of premises to check labelling claims and account for egg production.

10.2 Grading

Eggs throughout the world are usually graded both in terms of size and quality. In the European Union (EU) eggs are graded for freshness as follows (EEC 1274/91):

- Class A or 'fresh eggs'
- Class B or 'second quality or preserved eggs'
- Class C or 'down-graded eggs intended for the food industry'

EEC 1274/91 makes a provision that 'loose sales' such as farm gate eggs need not be graded but require 'best before' date.

10.2.1 Grade A eggs

Grade A eggs are graded by weight (Table 10.1) and have the following characteristics:

- airspace height does not exceed 6 mm, 4 mm at the time of packing for eggs marked 'extra' (fresh)
- shell and cuticle are normal, clean and undamaged
- egg white is clear, limpid, of gelatinous consistency and free of extraneous matters of any kind
- yolk is visible on candling as a shadow only, without clearly discernible outline, not moving appreciably away from the centre of the egg on rotation and free of extraneous matter of any kind
- germ cell – there is imperceptible development
- odour – the egg is free of extraneous odour.

In addition, Grade A eggs are not to be washed, cleaned, preserved or refrigerated at less than 5°C. From 1 June 1994 Class A eggs must be labelled with a 'best before' date. Quality grade and size of eggs are independent considerations; both should be noted on the egg carton label. Eggs should be regraded if they eventually become outside the specification. Eggs belonging to Grades A and B are required to have a label on the box showing the best before data by which the consumer should use the eggs.

10.2.2 Grade B eggs

Grade B eggs include refrigerated and preserved eggs as well as eggs not fitting into the Grade A classification, and have the following characteristics:

- airspace height not exceeding 9 mm
- shell and cuticle are normal, clean and undamaged
- egg white is clear, limpid, of gelatinous consistency and free of extraneous matter of any kind
- egg yolk is visible on candling as a shadow only (this characteristic is not required of eggs preserved in lime), and free of extraneous matter of any kind

Table 10.1 Grading for size for Grade A eggs

Grade size	Weight(g)
Grade 0	75+
Grade 1	70–75
Grade 2	65–70
Grade 3	60–65
Grade 4	55–60
Grade 5	50–55
Grade 6	45–50
Grade 7	under 45

- germ cell – imperceptible development
- odour – the egg is free of extraneous odour.

10.2.3 Grade C eggs

Grade C eggs are eggs that do not meet the requirement for Grades A and B. They may only be passed to the food industry or non-food industry for processing and are marketed in packs containing a yellow band or label.

10.2.4 Industrial eggs

Industrial eggs (e.g. for the production of shampoo), such as cracked and incubated eggs deemed unfit for human consumption, are marketed in packs with a red band or label.

10.3 Methods for assessing egg classification

The criteria for distinguishing between grades and classes of eggs are based mainly on analytical criteria and as such are easy to monitor.

10.3.1 Candling

All eggs from packing stations are subject to candling and this test can reveal most of the criteria distinguishing Grades A, B and C. When candling, the egg is illuminated by a beam of light from one side of the egg, revealing its internal characteristics when viewed from the other side. The egg yolk of a high-quality egg will not be clearly revealed because of the thick egg white that surrounds it. The older the egg, the more the egg white breaks down, which allows the yolk to be seen more clearly. In addition, the air cell enlarges as moisture gradually evaporates through the membrane. Other defects that can be identified using this technique are cracks, blood spots and rots (Smith, 1991). Whereas mass candlers are used by the industry to allow many eggs to be observed at any one time, inspectors, when sampling batches for quality control purposes, use an individual candler that candles one egg at a time. In the UK, the EMI regularly downgrade approximately 5% of eggs that have passed the industrial candling process during their inspections. There have been many attempts to develop superior techniques to candling in order to monitor egg quality. These have included ultrasonics, spectro-photometry, and laser scanning; however, candling is still by far the most important test used to assess the quality of an intact egg. Although automated techniques have been developed, they are still at present in the minority; the trained eye is being used in most UK packing stations.

10.3.2 Detection of washing

Eggs classed as Grade A in the European Union may not be washed. The washing of eggs can be identified by the use of ultraviolet light because the water stains are clearly visible.

10.3.3 Weighing

All packaging stations have grading machines that are inspected regularly and checked for accuracy.

10.3.4 Inspections and audits

Other criteria such as 'best before' date and date of production are regulated by regular inspections and audits of premises. In the UK, this is carried out by the EMI.

10.4 Incubated clears

In 1985, the journal *Confectionery and Manufacture and Marketing* (Anon, 1985) reported "over 1 million eggs rejected from incubators are released each week from hatcheries and find their way into manufacturing for the food industry. These eggs have failed to hatch having been kept in incubators for more than 6 days at temperatures of 37°C". An increased emphasis on inspection and testing has ensured that, in 1995, the use of incubated eggs is no longer the major problem that it was considered to be in 1985. Legislation specific to hatching eggs requires them to be marked in such a way as to readily and clearly distinguish them as 'incubator clears'. The EMI has an enforcement role in this respect. The use of incubated eggs poses a potential health hazard as incubation provides conditions for the growth of *Salmonella* spp. In addition, as legislation has now banned the use of incubator clears in food (EEC 1907/90), these eggs are now of limited value. There is therefore a potential commercial benefit in fraudulently mislabelling incubated eggs as class A, B or C eggs and using them in the production of liquid egg. There has been much work carried out to identify poor quality eggs and incubator rejects in particular. Levels of β-hydroxybutyric acid (BHB), which is formed during embryonic development, have been used as an indicator of embryonic development in eggs (Salwin *et al.*, 1972; Uijttenboogaart *et al.*, 1986). Since there is little increase in BHB arising from the incubation of infertile eggs, the BHB assay is a specific test for fertilised incubator clears, in other words undeveloped fertilised eggs.

10.5 Liquid egg

There has been a substantial increase in the production of liquid egg in recent years. The increase may be due in part to consumer concerns about food safety. Liquid egg may in this respect be favoured by the consumer as it is required to be pasteurised in the production process. In addition, the increase in the use of egg products is a result of the increasing popularity of convenience and ready-made meals incorporating egg products and the decline in shell-egg consumption. In the EU, there is prescribed legislation to ensure the safety and quality of liquid egg manufactured in Member States.

10.5.1 Legislation

In 1993, Statutory Instrument No. 1520 (The Egg Products Regulations, 1993) came into force. These Regulations implemented, in part, the Council Directive EEC 437/89 (EEC 437/89) as amended, as well as emphasising national practices. The Regulations are far-reaching, covering aspects of the preparation and manufacture of egg products, as well as prescribing analytical limits and methods of analysis, to ensure the quality and safety of egg products. From 14 October 1993, these new Regulations revoke the Liquid Egg (Pasteurisation) Regulations 1963. The Regulations make provision for the preparation and manufacture of egg products. They state that only establishments approved by relevant food authorities can manufacture egg products (e.g. liquid egg) and that all whole egg or egg yolk must be pasteurised at 64.4°C for at least 2.5 min or by an equivalent pasteurisation process.

These Regulations prescribe a series of specifications and tests that must be carried out with each batch of egg product in order to ensure that these criteria are met. The egg products must conform to both chemical and microbiological limits to satisfy quality and safety criteria. The main exception is where egg products are used for the preparation of food at the establishment where they were obtained (i.e. immediate use, in which case neither pasteurisation or compliance with the analytical specifications is prescribed).

An important aspect of these Regulations is that, from 14 October 1993, the use of centrifuging (or crushing) is prohibited. Among other aspects covered, are details on storage, transportation, packing and marking of egg products.

10.5.1.1 Analytical limits. These are contained in Schedules 2–4 of Statutory Instrument 1520 and comprise the following:

1. Whole egg or yolks that are pasteurised at 64.4°C for 2.5 min must pass the α-amylase test (i.e. must have a Lovibond reading of >3 units).

2. β-hydroxybutyric acid (3 OH-butyric acid) concentration $\leqslant$10 mg/kg in the dry matter of the unmodified product.
3. A lactic acid concentration $\leqslant$1000 mg/kg of egg product dry matter in the untreated egg product.
4. A succinic acid concentration $\leqslant$25 mg/kg of egg product dry matter.
5. The quantity of egg-shell remains, egg membrane and any other particles in the egg product shall not exceed 100 mg/kg of egg product.
6. No *Salmonella* spp. in 25 g or 25 ml of egg product.
7. A mesophilic aerobic bacteria concentration of less than 10^5 in 1 g or 1 ml.
8. An *Enterobacteriaceae* concentration of less than 10^2 in 1 g or 1 ml.
9. No *Staphylococcus aureus* in 1 g of egg products.

10.5.2 Storage indicators

Lactic acid and succinic acid have been identified as indicators of microbial activity; lactic acid is associated mainly with contamination from handling and succinic with the microbiological activity within the egg. Microbiological tests are prescribed in the EU on the pasteurised product. By definition these tests cannot assess the quality of the eggs arriving at the processing plant prior to pasteurisation. Regulation EEC 437/89 includes limits for succinic acid (25 mg/kg DMB), and lactic acid (1000 mg/kg DMB).

10.5.3 Centrifugation

Regulation EEC 437/89 prohibits the use of centrifugation in the production of liquid egg. The centrifuge has traditionally been used as a means of processing damaged eggs and also as a means of recovering additional egg white from the shells obtained in the cracking process (Figure 10.2). When eggs are broken in the cracking process, the shells are usually fed into a large industrial centrifuge and spun down to produce a dry, manageable product and also to recover additional albumen from the shell. Prior to the introduction of UK Regulation Statutory Instrument 1520 in 1993 the additional recovered material would have, in most cases, been fed back into the production process and pasteurised with the rest of the liquid egg. Although it has been of considerable commercial value to producers, its use is now prohibited because of safety concerns over damaged eggs and poor quality eggs being processed and also because of the increased likelihood of microbiological contamination through contact between the contents of the egg with the external surface of the shell. The liquid recovered from the centrifuge can still be used for the production of animal feedstuffs and for industrial use. However, it is possible that producers could still use it for the production of liquid egg for human

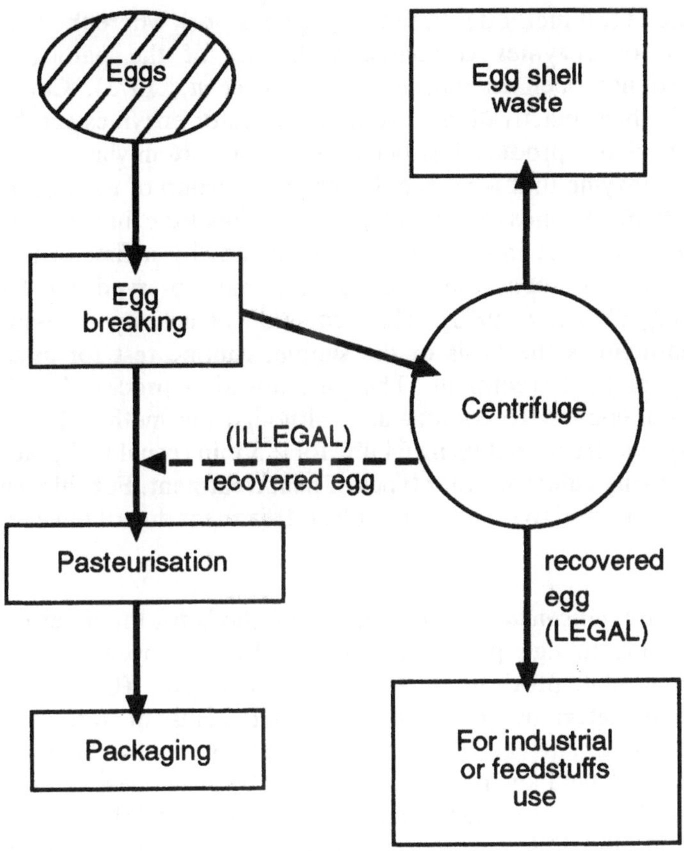

Figure 10.2 Diagram illustrating the use of centrifugation in the production of liquid egg.

consumption as there are no analytical methods available to detect its use. Again enforcement is carried out through regular audits and inspection, although, in this case, the agency concerned is the Environmental Health Department.

10.5.4 Methods of analysis

10.5.4.1 Detection of pasteurisation using the α-amylase test. This method is prescribed in the Regulations and involves incubating the sample at 44°C with starch. Any active amylase is able to break down the starch, so that the addition of iodine will not give an intensely blue coloured iodine/starch complex. When the enzyme has been denatured by a prior pasteurisation process, the starch is unaffected and the normal blue colour develops.

The use of this method is based on previous work where the relative heat resistance of enzymes compared with that of the pathogens to be eliminated has been established (Cotterill *et al.*, 1973). Consequently, testing the heat inactivation of suitable indicator enzymes can be used to determine if the product has been pasteurised. α-amylase is a naturally occurring enzyme that is present in the yolk portion of the egg. It belongs to a family of enzymes that hydrolyse the D-glycosidic linkage of the starch molecule. The heat inactivation of α-amylase, by a first order reaction mechanism, has been found to parallel that observed for *Salmonella senftenburg* (Murthy, 1970). This reduction of activity of α-amylase by heat treatment is the basis of the simple, routine test for assessing the adequacy of heat treatment. The pasteurisation process for liquid egg varies from country to country and, although the method is suitable for detecting heat treatment using 64.4°C for 2.5 min (usual UK practice), it is not necessarily valid for other types of heat treatment. For this reason, egg products subjected to other types of heat treatment do not have to pass this test.

10.5.4.2 Organic acid determination. Methods for the determination of organic acids in egg products have usually employed enzyme or gas chromatographic procedures. Gas chromatographic (GC) methods are capable of determining all three organic acids of interest (e.g. *β*-hydroxybutyric acid, succinic and lactic acid). There have been several GC methods published over the last 25 years (Staruszkiewicz *et al.*, 1970; Salwin *et al.*, 1972; Littman *et al.*, 1982; Littman, 1985).

The method of Staruszkiewicz incorporated a continuous extraction system, followed by derivatisation. The resulting propyl esters were separated on a packed column. A capillary GC method (Littman, 1985) using a celite column clean-up and formation of the methyl ester prior to separation on a capillary column was reported as being problematical (Stijve and Diserens, 1987). Although GC techniques are accurate and sensitive when determining the relatively low levels of *β*-hydroxybutyric acid prescribed in EC and UK legislation, they are time-consuming in terms of the numerous manipulations that are required to perform satisfactory GC analysis. Furthermore, although improvements have been made to the original methodology over the years, there is currently a lack of GC methods capable of determining satisfactorily the three organic acids of interest at the prescribed limits on a reproducible basis.

Enzyme procedures (Parry *et al.*, 1980; Elenbaas *et al.*, 1980; 1986) for the determination of these acids are available and have the advantage that they potentially offer a means of screening rapidly large numbers of egg samples. There are kits available for the determination of D- and L-lactic, succinic and D-*β*-hydroxybutyric acid after precipitation of proteins using Carrez solution or perchloric acid.

The β-hydroxybutryic acid assay is determined via its oxidation with β-hydroxybutyric acid dehydrogenase. The enzyme is specific for the D-isomer and catalyses the reversible reaction:

$$CH_3CH(OH)CH_2CO_2H + NAD \longleftrightarrow CH_3COCH_2CO_2H + NADH + H^+ \qquad (10.1)$$

where NAD = nicotinamide adenine dinucleotide; and NADH = reduced NAD. The concentration of NADH is determined by measuring the increase in absorbance at 340 nm. Similar enzyme procedures are available for the determination of succinic and L- and D-lactic acid. Methods for D- and L-lactic make use of the enzymes D-lactate dehydrogenase and L-lactate dehydrogenase, respectively.

Enzyme methods for the determination of succinic acid make use of the enzyme succinyl-CoA synthetase (SCS). This enzyme, in the presence of inosine–5–triphosphate (ITP) and coenzyme A, converts succinic acid to succinyl-CoA with a simultaneous formation of inosine–5–diphosphate, which eventually results in the oxidation of NADH to NAD. The change in absorbance is measured at 340 or 365 nm. Enzyme kits for the determination of these three organic acids are available commercially. They have advantages over GC techniques in terms of speed of the analysis (2–3 h) and the large number of samples that can be analysed in one batch. The enzyme procedures for succinic and D-β-hydroxybutyric acid suffer from a lack of sensitivity for the levels prescribed in the EC Directive.

Other methods developed for the detection of incubator rejects (i.e. fertile and infertile rejects) have used electrophoresis (Harwalkar, 1968; Csuka *et al.*, 1973) to measure the increase of reducing protein (Cattano, 1979). Both techniques are based on the detection of denaturation of proteins during incubation. These procedures are limited in terms of sensitivity and are not applicable to pasteurised products because of the heat-induced protein denaturation.

Methods using differential scanning calorimetry (DSC) have also been reported (Smith, 1964; Donovan *et al.*, 1975; 1976; Ma *et al.*, 1990; Raymond *et al.*, 1992) and are based on the formation of a more thermostable *S*-ovalbumin from ovalbumin during the incubation process. They are not applicable to pasteurised egg products. Raymond and co-workers (1992) suggest that 7% adulteration of egg white with incubator reject eggs can be detected using DSC.

10.5.4.3 Microbiological tests. EC Directive 437/89 (EC 437/89) prescribes limits for *Salmonella* spp., mesophilic aerobic bacteria (plate count at 30°C), Enterobacteriaceae and *Staphylococcus aureus*. British Standard methods are prescribed in Schedule 4 of the Statutory Instrument 1521:

- *Salmonella* spp.: BS 5763: Part 4, 1990 (ISO 6579)
- Mesophilic aerobic bacteria: BS 5763: Part 1, 1991 or BS 5763: Part 1, 1979 (ISO 4833)
- Enterobacteriaceae: BS 5763: Part 10, 1986 (ISO 7402), subject to additional instructions in Part IV of Schedule 4
- *Staphylococcus aureus*: BS 5763: Part 7, 1983 (ISO 6888).

10.6 Conclusions

Egg production is both a food safety and quality issue. Egg quality is now defined by EC and UK legislation. In many aspects of the prescribed legislation, where appropriate analytical methodology is unavailable, inspection and audit are the only effective ways of monitoring compliance.

References

Anon (1985) Danger from using egg incubator rejects in food processing. *Confectionery, Manufacture and Marketing* **22(5)**, 29–30.

BS (British Standard) 5763: Part 1, 1979, (ISO 4833) Methods for microbiological examination of food and animal feeding stuffs. Enumeration of micro-organism – colony count at 30°C. British Standards Institute, Milton Keynes.

BS (British Standard) 5763: Part 7, 1983, (ISO 6888) Methods for microbiological examination of food and animal feeding stuffs. Enumeration of *Staphylococcus aureus* by colony count technique. British Standards Institute, Milton Keynes.

BS (British Standard) 5763: Part 10, 1986, (ISO 7402) Methods for microbiological examination of food and animal feeding stuffs. Enumeration of *Enterobacteriaceae*. British Standards Institute, Milton Keynes.

BS (British Standard) 5763: Part 4, 1990, (ISO 6579) Methods for microbiological examination of food and feeding stuffs. Detection of *Salmonella*. British Standards Institute, Milton Keynes.

BS (British Standard) 5763: Part 1, 1991. Methods for microbiological examination of food and animal feeding stuffs. Enumeration of micro-organisms count technique at 30°C. British Standards Institute, Milton Keynes.

Cotterill, O.J., Glauert, J.L. and Krause, G.F. (1973) Thermal destruction curves for *salmonella* oranienburg in egg products, *Poultry Sci.* **52**, 570–72.

Csuka, J., Novy, J. and Jirosova, Z. (1973) Electrophoretic study of incubated fertile eggs. *Brit. Poult. Sci.* **14**, 203.

Daenens, P. and Leander, L. (1976) Lactic and succinic acid levels and refractive indices in the determination of age of eggs. *JAOAC* **59(3)**, 613–16.

Donovan, J.W., Mapes, C.J., Davis, J.G. *et al.* (1975) A differential scanning calorimetric study of the stability of egg white to heat denaturation. *J. Sci. Food Agric.* **26**, 73.

Elenbaas, H.L., Muuse, B.G., Haasnoot, B. *et al.* (1986) *J. Agric. Food Chem.*, **34**, 633.

EEC 437/89, Council Directive of 20 June 1989 on hygiene and health problems affecting the production and the placing on the market of egg products. OJ L 212, 22.7.89, pp. 87–100.

EEC 1907/90, Council Regulation (EEC) No. 1907/90 of 26 June 1990 on certain marketing standards for eggs. OJ L 173, 6.7.90, pp. 5–11.

EEC 1274/91, Commission Regulation (EEC) No. 1274 of 15 May 1991 introducing detailed rules for implementing Regulation (EEC) No. 1907/90 on certain marketing standards for eggs. OJ L 121, 11–24. 1991.

Harwalker, V.R. (1968) Electrophoresis differentiation of fresh and incubator rejects eggs. *J. Sci. Food Agric.* **34**, 663.
Heaney, R.K. and Curtis, R.F. (1976) Some further studies on incubator reject eggs. *J. Sci. Food Agric.* **27**, 1057.
Kessler, C., Sinell, H.J. and Weigner, J. (1990) Beurteillung des Frischezustandes von Huhnereiern in Abhangigkeit von der Gewichtsklasse. *Archiv. Fur Lebensmittelhygiene* **41(4)**, 81–85.
Littman, S. (1985) Zur chemisch-analytischen Untersuchung von Eiprodukten und Beurteilung ihres hygienischen Zustandes an hand des gehaltes an verderbsspezifischen organischen Sauren. *Deutsche Lebensmittel-Rundschau* **81**, 11.
Littman, S., Schulte, E. and Acker, L. (1982) Evaluation of the hygienic condition of egg products before pasteurisation by the pattern of organic acids. *Z. Lebensm Unters Forsch* **175**, 101–105.
Ma, C.Y., Sahasrabudhe, M.R., Poste, L.M. *et al.* (1990) Gamma irradiation of shell eggs. Internal sensory quality, physicochemical characteristics and functional properties. *Can. Inst. Food Sci. Technol. J.* **23**, 226.
Murthy, G. (1970) Thermal inactivation of α-amylase in liquid egg products. *J. Food Sci.* **35**, 352–56.
Parry, A.E.J., Robinson, D.S. and Wedzicha, B.L. (1980) An enzymic assay for β-hydroxybutyric acid in liquid whole egg. *J. Sci. Food Agric.* **31**, 905.
Reagan, J.G., York, L.R. and Dawson, L.E. (1971) Improved methods for the determination of certain organic acids in pasteurised and unpasteurised liquid and frozen whole egg. *J. Food Sci.* **36**, 351–54.
Raymond, D.E., Harwalker, V.R. and Ma, C.Y. (1992) Detection of incubator reject eggs by differential scanning calorimetry. *Food Res. Internat.* **25**, 31–35.
Robinson, D.S., Barnes, E.M., Taylor, J. (1975) Occurrance of β-hydroxybutyric acid in incubator reject eggs. *J. Sci. Food Agric.* **26**, 91.
Salwin, H., Staruszkiewicz, W.F. Jr., and Bond, J.F. (1972) Formation of β-hydroxybutyric acid in eggs during incubation. *J. AOAC* **55**, 458.
Smith, E.R. (1991) Overview of egg candling lights used by inspectors. *Dairy, Food and Environmental Sanitation* **11**, 640–42.
Smith, M.B. (1964) Studies of ovalbumin. 1. Denaturation by heat and the heterogeneity of ovalbumin. *Aust. J. Biol. Sci.* **17**, 261.
Staruszkiewicz, W.F., Bond, J.F. and Salwin, H. (1970) Quantitative gas chromatographic determination of β-hydroxybutyric acid with application to eggs. *J. Chromatogr.* **51**, 423–32.
Statutory Instrument 1520 (1993) The Egg Products Regulations 1993, HMSO.
Stive, T. and Diserens, J.M. (1987) Simplified method for the detection of organic acids in eggs and egg products. *Deutsche Lebensmittel-Rundschau* **83**, 2.
Stuke, T., Dickermann, D. and Hildenbrandt, G. (1987) Enzymeatisch und gaschromatographische Bestimmung organischer Säuren in Eiprodukten. *Archiv fur Lebensmittelhygiene* **38**, 93–120.
Uijttenboogaart, T.G., Steverink, A.T., Elenbaas, H.L. *et al.* (1986) Determination of incubator-reject eggs in egg and egg products by means of the content. 2. Levels of 3-HBA in different kinds of eggs and egg products and the development of 3-HBA during incubation. *J. Agric. Food Chem.* **34**, 667.

11 Other commodities

M.N. HALL

11.1 Introduction

The preceding chapters have reviewed the subject of food authenticity in relation to a wide range of products and issues. Aspects such as species or variety, geographical origin, age or vintage, process received, and admixture with base materials have all been considered. Indeed, any attribute of a food product or material could be the subject of an authenticity debate if specific claims are made regarding that property. Not surprisingly, therefore, the issues of authenticity are diverse and become more so with the introduction of new products to the marketplace and the development of new products and processes (Hall, 1993). This chapter seeks to explore further food groups, not already covered, and some of the issues that are not specific to particular product types. It also demonstrates by example the range of approaches that can be taken in authenticity testing.

Some food materials are covered by specific regulations; the majority of authenticity issues are, however, regulated through the labelling laws. In the UK, enforcement through the Food Safety Act 1990 under the section 'food not of the nature or substance or quality demanded' or the Trade Descriptions Act 1968 may be employed.

11.2 Tea

Most tea that is consumed in the UK is black tea and involves the stages of withering, leaf damage, fermentation and drying during its manufacture. The market for teas in the UK was valued at over £600 million in 1993 of which the majority were premium teas, that is, the leading brands (Mintel International, 1993a).

Tea is produced from the young leaves and shoot tips of the plant *Camellia sinensis*, which is believed to originate from China but is now grown in many geographical locations. There are two main varieties of the tea plant (*Camellia sinensis*, var. *sinensis* and *Camellia sinensis*, var. *assamica*) and these show quite distinct characteristics. Nevertheless, they hybridise freely to yield plants with characteristics that tend, to differing extents, toward one or other of the two varieties. In general, teas made from hybrids resembling the var. *assamica* type have a deep orange-brown colour and strong taste, whereas those of the var. *sinensis* type are lighter in colour and have more delicate scented flavours (Gill, 1992). Furthermore, the area in which the tea is grown can have a significant influence on its characteristics – by virtue of soil, climatic and processing influences. The properties of teas are therefore developed as a result of a combination of influences including hybrid type, geographical origin and processing method. However, particular hybrid types and processes are favoured in different areas so that teas are commonly referred to as typically 'Indian' or 'China' teas.

'China' teas have the properties of var. *sinensis* predominating and are manufactured using a rolling process (orthodox) to damage or bruise the leaves prior to fermentation. 'Indian' teas, on the other hand, are largely var. *assamica* in character and can be produced using either the orthodox method, a more drastic cutting process (cut, tear, curl – CTC) or some other method of manufacture. In addition, there are a multitude of specialised teas produced in specific areas, by particular processes or flavoured in various ways. Table 11.1 indicates the major teas from specific origins.

Table 11.1 Specific origins and characteristics of major teas[a–c]

Common name	Origin/type
Darjeeling	Tea from an area in the north-east of India in the Himalayan foothills. A high-priced tea considered by many connoisseurs as the finest speciality tea. The tea produces a pale liquor and has a delicate muscatel-like flavour.
Assam	Tea from a larger area in Bengal, one of India's most profilic areas. The teas have a good depth of flavour and richness of colour. Assam tea is often a significant component in 'pure Indian' blends.
Nilgiri	Less well known than Darjeeling or Assam these teas come from the mountains in the south-west of India. The teas are lighter than Assam but have a pleasant brightness and briskness of flavour.
Dimbula	One of the better known Ceylon teas from the hills of Sri Lanka. The teas have typical 'high-grown' characteristics.
Ouvah	From the eastern slopes of the central mountain of Sri Lanka. A full, rich-flavoured tea used in both blends and as a high-quality single tea.
Lapsang souchong	A full-bodied tea from China with a characteristic strong tarry taste.
Oolong	A fine tea from China or Taiwan with a peachy flavour and reddish liquor.

[a]Source: Wilson and Cifford (1992).
[b]East African teas are characterised by their intense colour, which is greatly valued in tea blends.
[c]Kenyan teas in particular are considered as being of good quality and are often sold as 'pure Kenyan' blends.

11.2.1 Standards, legislation and authenticity issues

Despite all of the different types and characteristics of teas, the industry has lived with few, if any, specific laws and regulations. In most countries, therefore, general food legislation is applied to tea; some countries, however, have adopted specific laws to govern the tea industry. For example, around the time of World War I, Belgium, Luxembourg and The Netherlands enacted the Benelux Tea Law, which defines what tea is. The specifications or standards for teas are now laid down in ISO 3720 (1986) and include the definition and limits of specified impurities.

As previously stated, the market for teas in the UK is extremely large and tea is still the most popular drink, accounting for 43% in volume terms of all drinks consumed. Speciality teas are increasingly popular with a value of £26 million in 1992; Earl Grey being the best known, although Assam, another speciality tea, is also doing well. This reflects the consumer attraction toward pure single-origin tea.

The major authenticity questions for teas are therefore:

- Is the material really tea as specified?
- Is the tea really the declared type from the correct origin?

Tea is a commodity and, as such, its price is based on its availability and quality. Availability can be affected by many factors such as the size of the crop (which is, in turn, affected by climate) and political influences. For example, in 1992, the drought in Africa and India caused tea prices to rise significantly. It is factors such as these that affect the availability from specific areas that can provide a considerable financial incentive for product adulteration or false description.

11.2.2 Methods of authenticity testing

The presence of material other than that which is truly tea is an issue similar in nature to instant coffee adulteration with non-bean material. A significant difference, however, is that tea in its normal state is not roasted or extracted and its component parts can be examined visually, for example in the assessment of the amount of stalk material present. A chemical test is also prescribed for measurement of fibre, which is indicative of inferior quality material being used.

There are no defined means by which teas can be authenticated in terms of type and origin. Variations in the sensory, chemical and physico-chemical properties, arising as a result of where or how teas are produced may, however, be useful indicators. Perhaps the first indication can come from the sensory character of the tea, which, as already shown in Table 11.1, varies significantly between types. The variations arising from other factors, such as season and climatic changes, however, make this approach tentative at best. Indeed, if adulteration is taking place it would almost certainly be at a level that is not perceptible, or with a tea with similar sensory characteristics.

The major chemical components of teas are polyphenolics in either their natural or oxidised forms in green tea and black tea, respectively. The composition of the finished tea material is a result of both the way the tea is processed and the chemical precursors present in the green leaf. Much work has been carried out to characterise teas chemically and to identify the most important phenolic components in terms of sensory character. Several classes of phenolic compounds have been isolated, from black tea, and structurally elucidated such as the theaflavins (Roberts, 1950), theafulvins (Bailey *et al.*, 1992), theacitrins (Powell *et al.*, 1994) and the less well chemically defined thearubigins. Even with all of the work carried out to date, much remains to be learnt about the complex phenolic chemistry of tea and the way in which individual compounds are generated and interact to influence character. It seems likely, however, that with a greater understanding and sufficient characterisation of non-volatile tea chemistry, this complex analytical pattern could be useful in authentication. This view is supported by the results of work carried out on green teas in which the level of total flavonols and total nitrogen were used to discriminate green teas produced from tea shoots that were either shaded or unshaded (Omori, 1983). Furthermore, the combination of sensory and chromatographic information and use of statistical or neural network approaches has been shown to enable good classification by country of origin (Tomlins and Gay, 1994).

The phenolic compounds give rise to the colour, mouthfeel and basic flavour of teas. Much of the more specific flavour character arises from the volatile aroma compounds produced as a result of the fermentation and firing stages. The withering stage can also influence significantly the generation of volatiles as can other treatments during processing (Takeo, 1984; Kobayashi *et al.*, 1985; Mahanta and Barnah, 1989). Certainly teas produced with zero, partial and full fermentation can be relatively easily distinguished on the basis of the differences in their volatile profiles. The use of volatile profiles could provide a means of distinguishing tea types so long as the differences from natural variation are taken into account. Varieties have been shown to be distinguishable by reference to the ratio of linalool, geraniol and 2–phenyl ethanol: *assamica* varieties were found to be high in linalool compared with *sinensis* varieties, which are high in geraniol, and hybrid varieties which have intermediate levels (Takeo, 1981). A further indicator was found to be the terpene index (the ratio of peak areas of gas chromatographic peaks from linalool to those of linalool and geraniol), which was shown to be able to differentiate teas of different varieties from different areas (Owuor *et al.*, 1987; and Takeo, 1983).

In common with most living systems, the composition of tea is influenced by the conditions in which it is grown. One of the most significant factors is the soil and the inorganic components found in the tea will, to some extent, reflect those of the soil, indeed, analysis of plant tissue is used as a method of assessing soil fertility. Tea is grown on a wide range of soil types with extensive differences in composition, structure and physical properties (Othieno, 1992). By measuring the levels of certain minerals in teas it may be possible to relate these back to the growing area and so authenticate teas on the basis of geographical origin. Such a method could be further developed to use the ratios of specific isotopes and match these to those found in the soil of the region or even estate.

The use of stable isotope ratios for assessing authenticity has been described in previous chapters of this book and could be applicable determining the geographical origin tea. Other approaches such as infra-red spectroscopy (NIRS) have also shown considerable promise. This form of spectroscopy is characterised by broad peaks, composed of overtone and combination absorption bands, the shape of which are influenced by physical structure and chemical components. NIRS has been used successfully to determine moisture and theaflavin levels and to predict the sensory quality of black tea (Hall *et al.*, 1988). More recently, the use of NIRS data for the determination of the geographical origin of orange juice has been successfully demonstrated using a combination of instrumental analysis and multivariate statistical analysis. A further approach could involve the use of several types of data such as that from volatile, non-volatile, inorganic analysis and sensory data in combination with multi-variate statistics or neural network interpretation.

11.3 Spices, herbs and condiments

The term 'spices and condiments' covers a wide range of specific materials but is also used as a general product description. According to ISO 676, the term applies to natural vegetable products or mixtures, without any extraneous matter, that are used for flavouring, seasoning, and imparting aroma to foods.

The total market for this type of product in 1993 in the UK was valued at over £60 million, of which herbs formed the largest sector. Sales of herbs were worth £16.1 million, followed by spices, worth £14.4 million, and pepper at £13.3 million. The market for other products, such as curry powders and various seasonings, was worth in total approximately £16 million (EIU, 1994). The pattern of consumption of spices has seen a change over recent years, with a slight reduction in sales of basic herbs and spices but an increase in the more exotic ones. This trend is almost certainly due to a combination of the home consumer using more convenience foods, and the greater interest in exotic and ethnic foods. The number of cookery programmes and books together with the range of foreign cuisine available in restaurants, and the increase in foreign travel are all likely to have contributed to this. The trend toward 'natural' and 'healthy' foods has also stimulated the interest in highly priced, premium products and the more unusual herbs and spices.

Spices and condiments can come from a wide range of plant types and from different parts of the plant, examples of which are shown in Table 11.2.

11.3.1 Authenticity issues

The inherent high value of certain spices and condiments increases the incentive for unscrupulous traders to seek ways of extending or substituting the expensive commodities with lower cost materials. The issues of authenticity therefore relate to adulteration with materials of inferior quality, such as immature material, other parts of the same plant, material from other plant types or with spent spice material from which the essential oils have already been removed. Powdered spices can also be adulterated by admixture with base materials, such as starches, colorants or even fine soil, stones or sawdust. A significant issue also concerns the adulteration of

Table 11.2 Origin of major types of herbs, spices and condiments[a]

Common name	Plant part derived from and plant type
Anise or aniseed	The fruit of *Pimpinella anisum* Linn.
Balm	The leaf of *Melissa officianalis* Linn.
Basil	The leaf of *Ocimum basilicum* Linn.
Bay	The leaf of *Laurus nobilis* Linn.
Black pepper	The fruit of *Piper nigrum* Linn.
Caper	The flower parts of *Capparis spinosa* Linn.
Caraway	The seeds of *Carum carvi* Linn.
Cardaman	The fruit and seed of Zingiberaceae family.
Cinnamon	The bark of *Cinnamomum zeylanicum* Blume.
Coriander	The leaf and the seed of *Coriandrum sativum* Linn.
Cumin	The seed of *Cuminum cyminum* Linn.
Dill	The seed of *Anthum graveolens* Linn.
Fennel	The seeds of *Foeniculum vulgare* P. Miller.
Garlic	The bulb of *Allium sativum* Linn.
Ginger	The rhizome of *Zingiber officinale* Roscoe.
Mace	The aril of *Myristica fragrans* Houttuyn.
Mint	The leaf of *Mentha* spp. Linn.
Nutmeg	The kernel of *Myristica fragrans* Houttuyn.
Rosemary	The leaf of *Rosmarinus officinalis* Linn.
Saffron	The stigma of *Crocus sativus* Linn.
Sage	The leaf of *Salvia officinalis* Linn.
Thyme	The leaf of *Thymus vulgaris* Linn.
Turmeric	The rhizome of *Curcuma longa* Linn.
Vanilla	The pod of *Vanilla pompona* Schneider.

[a]Source: Pruthi (1980).

the essential oils, which are extracted from the spices. These aspects are, however, covered within the next section of this chapter.

11.3.2 Methods of authenticity testing

The addition of mineral impurities, such as soil and fine stones, to spices and herbs is relatively easy to detect. An indication of this type of adulteration can be gained by the determination of ash content remaining after the organic matter is destroyed at high temperature. Another valuable technique is microscopy, which can be used to detect foreign plant material as well as a wide range of other extraneous matter. The use of polarised light microscopy is commonly used to reveal minerals, which show up as brightly coloured, shining objects when under the correct illumination.

An identification of the part and plant type from which particles originate can also be achieved using microscopic examination. For example the addition of fennel to anise can be detected by observation of structures from the fruit surface, which is smooth in the case of fennel but covered with numerous warty hairs in anise. Several works provide descriptions and diagrams of the structures specific to different plant species and plant parts. Perhaps one of the most thorough comes in the volumes by Winton and Winton, which, although old, remain extremely valuable texts (Winton and Winton, 1939). Physiological characteristics can also be used in the differentiation of certain spices such as cassias from different countries of origin and from true cinnamons. The amount of mucilagenous material is measured as a sediment after heating in a mixture of isopropyl alcohol and water (Stahl *et al.*, 1969).

The chemistry of the spices and herbs also differs and may indicate gross adulteration. As previously mentioned, near infra-red spectroscopy (NIRS) has been applied to various authenticity issues and has been shown by the author to differentiate spices. Further work is, however, required to determine the extent to which NIRS could detect partial adulteration.

Perhaps the most important aspect of any herb or spice, by definition of its purpose, are its flavour components. Analysis of the pattern and level of these compounds, together with identification of specific marker volatiles is a very direct method of detecting adulteration. Further discussion on these methods is, however, reserved for the section on flavours and essences.

11.4 Flavours and essences

Unlike many of the products discussed in this book, the direct purchase of flavours and essences by the consumer is relatively small. Most of the use of these materials, of which there is a large range, is as ingredients for use in formulated or processed products. Flavourings, particularly when natural materials are used, are often of high value, and the cost of them is inevitably passed onto the purchaser of the finished product.

11.4.1 Standards, legislation and authenticity issues

The EC framework directive for flavouring substances for use in foodstuffs (EEC, 1988) is designed to ensure that flavourings are safe and to facilitate trade within the European Union. The directive identifies four categories of flavourings: (i) flavouring substances (this is further divided into natural, nature-identical and artificial); (ii) flavouring preparations, (iii) process flavourings; and (iv) smoke flavourings.

A flavouring substance is a defined chemical compound that is obtained or produced:

- by appropriate physical, enzymatic or microbiological processes from material of animal or vegetable origin (natural)
- by chemical synthesis or isolation to produce a substance that is chemically the same as that found in natural material of animal or vegetable origin (nature-identical)
- by chemical synthesis to produce a substance that is not identical to that in natural material of animal or vegetable origin (artificial).

The directive also contains specific restrictions on the use of the term 'natural' such that the word natural or any other word with substantially the same meaning may only be used for flavourings in which the flavouring component contains only flavouring substances and/or flavouring preparations. Furthermore, if the word 'natural' is used with reference to a foodstuff or flavouring source then the flavouring component must have been isolated by appropriate physical, chemical or microbiological processes from the specified food or source.

12.4.2 Methods of authenticity testing

Several approaches have been put forward to ensure that flavouring materials are authentic and originate from the correct plant origin. Thin-layer chromatography was successfully used in the 1970s to separate extracted volatiles from cardamoms to differentiate greater cardamom from true cardamom (Sen *et al.*, 1977) and to reveal the addition of coumarin to vanilla flavouring (Sen Gupta *et al.*, 1973). Liquid chromatography has been used to indicate foreign components in vanilla extract and was shown through a collaborative trial to reveal adulteration (Potter, 1971).

In order for flavour compounds to be perceived, they must be volatile so that they can come into contact with the olfactory sensors in the nasal cavity. Their volatility also provides the means by which they can be separated and identified and a way by which some authentication can be achieved. Gas chromatography has been applied to the analysis of flavour volatiles for several years, for example in the testing of natural cinnamon oil for adulteration by synthetic cinnamon flavour (Zuercher *et al.*, 1974). The development of combined gas chromatography and mass spectroscopy allowed much greater confidence in the identification of volatile components and hence more definitive evidence of adulteration. It is often possible to use such instrumentation in a selective ion-monitoring mode so that specific indicators of a suspected adulterant can be detected. Not surprisingly, there has been much work published on the separation and identification of the flavour components. Some of the most recent approaches involve chiral chromatography – an approach particularly suited to distinguishing natural and synthetic materials. Chiral chromatography involves the separation of components based on their stereochemical properties (Tresl and Engel, 1985; Schreier *et al.*, 1991). The technique relies on the fact that compounds synthesised naturally are of one optical isomer, whereas chemical synthesis gives rise to a racemic mixture. The detection of the non-natural isomer gives clear evidence of the use of a synthetic compound.

Perhaps one of the most commonly adulterated flavouring materials of all is vanilla – an expensive natural product that can be synthesised chemically more cheaply. The application of stable isotope ratio measurements has been explored with considerable success in this area. The two main approaches to isotopic authentication methods are isotopic mass spectrometry and deuterium–NMR. Isotopic mass spectrometry can be used to give values for the ratios ^{13}C:^{12}C, ^{2}H:^{1}H or ^{18}O:^{16}O as total figures and therefore to detect crude forms of adulteration such as vanillin from non-bean sources (Anon, 1979; Bricout and Koziet, 1978). More sophisticated fraud can now involve materials enriched with ^{13}C to give ratios consistent with the authentic material and so escape detection using mass spectrometry. Furthermore, by using the correct mixture of vanillins derived from guaicol and lignins it is possible to produce ^{2}H:^{1}H total ratios similar to those of vanillin from the natural bean source. Experimental difficulties with the measurement of oxygen stable isotope ratios has limited the use of this approach. More recently the use of deuterium–NMR has made it possible to measure the ^{2}H:^{1}H ratio at specific sites on the molecule of interest and hence provides information about the origin of that molecule. In this way, it is often possible to determine the type of plant and the method of production of the flavour compound. Several papers have been written on the use of this technique, the factors influencing the site-specific fractionation of the isotopes and its specific application to flavour authentication (Guillou *et al.*, 1992).

11.5 Organic production of agricultural products

Consumer demand for materials produced using organic production methods is increasing and has stimulated a new market opportunity for farmers. The market price for agricultural products produced by such methods is generally higher and reflects the less-intensive utilisation of the land. There are several reasons why consumers choose to purchase 'organic produce' and these relate to perceived health and environmental benefits. Surveys have indicated that there is a commonly held view that the presence of any synthetic chemical residue on or in a food is harmful and that by buying 'organic produce' the intake of such chemicals can be avoided. Some consumers believe that organically produced food tastes better than conventionally produced food. A belief also exists that consumption of organically produced material confers health benefits other than those associated with safety. Perhaps more substantially, however, some consumers choose to subscribe to this particular type of agricultural practice, which they consider contributes to the protection of the environment and conservation of the countryside (Woodward, 1990).

There was a major increase in the consumption of 'organic food' during the early 1970s, particularly in products such as grains, pulses and bread. Most of this was imported material sold through relatively small specialised 'wholefood' shops. In the mid-1980s the larger supermarkets perceived a commercial advantage in marketing and selling organically produced food and offered fresh produce, meat and dairy products for sale. In 1993 the value of sales of 'organic food' in the UK was in excess of £100 million and had seen a growth of close to 500% over the previous 6 years. It is estimated that, although the growth rate has now slowed down, the market will be worth over £150 million by 1998 (Mintel International, 1993b) with the largest product sector being vegetables, worth over £45 million in 1992, followed by fruit, valued at around £20 million. Such was the size and pattern of market growth in the late 1980s and early 1990s that domestic production of organic produce could not keep up with the demand and a significant amount was imported (Wright, 1994).

11.5.1 Standards, legislation and authenticity issues

The control of products sold as 'organic' comes under a Council of the European Communities Regulation, which applies to the production, preparation, marketing, labelling, inspection and certification of such products. This Regulation (Council Regulation No. 2092/91) as amended and elaborated, stipulates that, in order for claims to be made that a food has been produced organically, prescribed principles of organic production at farm level must be satisfied. These principles restrict the amount and type of materials that may be applied to increase the fertility of the soil and/or to control pests diseases and weeds. Within the UK, standards have been produced by the UK Register of Organic Food Standards, commonly referred to as UKROFS (Council Regulation No. 2092/91). The European Regulation also details rules for registration, inspection and record-keeping for farms, processing and packing units. Furthermore, material imported from outside the EU must be grown and inspected to standards equivalent to those of the EU itself. The Regulation covers unprocessed crop products and processed products derived from them and in terms of labelling, classifies formulated products into three groups each with defined rules on the labelling permitted:

- organic: more than 95% organic agricultural ingredients
- partly organic: more than 50% but less than 95% organic agricultural ingredients
- non-organic: less than 50% organic agricultural ingredients.

For the reasons already mentioned, the consumer will pay a premium, often as much as 50–150%, for organically produced food. It is essential, therefore, that there can be an assurance that food labelled as such is truly 'organic' and has been produced using the appropriate method.

11.5.2 Methods of authenticity testing

Despite the opinion held by some people, there is little or no scientific evidence that fruit or vegetable materials produced organically taste different to their conventional equivalents (Schutz and Lorenz, 1976). Furthermore, such is the natural variability of food composition that it is extremely difficult to draw any conclusions from compositional data. However, in a survey of fruit and vegetables sold as organically grown, statistically significant differences were found in the levels of some nutrients and particularly in the amount of water present. The pattern of variation was extremely complex and showed little consistency between different crop types. Furthermore, when the water level was taken into account and nutrient levels were expressed on a dry matter basis, many of the apparent differences were found to disappear (Pither and Hall, 1990). There is some evidence that the level of nitrates in organically produced vegetables is lower than in those produced conventionally although, not surprisingly, there is considerable overlap of nitrate levels in the two groups (Rauter and Wolkerstorfer, 1982). Even the presence or absence of agrochemical residues cannot be taken as conclusive evidence of the agricultural practice employed (Anon, 1982). The majority of crops produced by either organic or conventional means do not contain detectable levels of pesticide residues and it is not unusual for residues to be found in organic products possibly as a result of spray drift (Collins and Naseef, 1993).

As previously noted, such is the natural variation of foods that it is difficult to envisage any direct chemical, physical or sensory method capable of distinguishing reliably between conventionally and organically produced materials. Even the combination of all of these types of data with the use of statistical analysis would seem unlikely to deal with such a task. It is therefore not surprising that the only method available and in use is that of an extensive and formalised system of documentation, inspection and certification.

11.6 Natural mineral water

The two major segments of the natural mineral water market in the UK are carbonated and still. Sales of bottled water over the period 1989–1994 have increased in volume by around 60% and in value by 37% to give a market worth over £300 million. The pattern of consumption of still and carbonated water has also changed, with still water now taking the greater share of market volume. This change in proportion is the result of a significant increase in purchases of still water and a slight decline in the carbonated water market. The origin of most natural mineral water in the UK was initially France who, it could be said, developed the market. In 1990, the imports of water fell significantly when Perrier was temporarily taken off the market because of a contamination issue involving benzene; the reduction of imports was even greater in the following year. In 1992, imports of natural mineral waters into the UK were valued at nearly £20 million, of which the majority were from France (EIU, 1993).

In a survey commissioned by Mintel to determine the consumer's attitude toward bottled water, the three most important features were found to be that it is natural, refreshing and healthy/good for you. These views are generally consistent among all age, class and geographical groups (Mintel International, 1994).

11.6.1 Standards, legislation and authenticity issues

In the Food Safety Act 1990, water is included as a food where it is used as a drink or as a food ingredient. Natural mineral waters are controlled in the UK through the Natural Mineral Waters Regulations 1985 No. 71 as amended, which in turn implement the European Council Directive No. 80/777/EEC. Legislation demands, among other things, that natural mineral waters originate from an underground source and are free from pollutants and from pathogenic microorganisms. The water must also have a consistent mineral composition, temperature and flow rate and be bottled at source. In order to meet the regulations, natural mineral water bottled for sale must also be labelled with information about the place where the source is exploited, the name of the spring, well or borehole and containers must have temper-evident closures. In addition, the method of carbonation or decarbonation must be specified to describe whether the water is naturally carbonated, carbonated with gas from the spring or artificially carbonated. Naturally carbonated natural mineral water is brought to the surface with carbon dioxide already dissolved in it and contains the same degree of carbonation as the water in the source. Carbonated natural mineral water is still when brought to the surface but is subsequently carbonated by gas injection.

All other water for drinking is controlled under the EU Drinking Water Standards Directive 80/778. Other terms are sometimes used to describe waters and the UK National Mineral Waters Association (NMWA, a part of the British Soft Drinks Association) has produced terms such as 'spring water' (coming from a single underground source but not recognised as a natural mineral water), 'table water' (all waters not from a single underground source), 'purified water' (water, coming from any source, that has had certain materials removed) and 'spa water' (spring water with a history of having therapeutic properties) to reduce confusion.

The questions of whether the water is from the designated source and or has been carbonated or decarbonated in the prescribed manner are therefore the most likely issues to be addressed in authenticity debates regarding water. An indication of the significance to the consumer is given by the fact that they will pay several hundred times as much for natural mineral water as for water directly from the tap.

11.6.2 Methods of authenticity testing

As natural mineral water is, by definition, from a designated source, authenticity can be established by comparison of the chemical and physical characteristics of a suspect sample against those of a verified standard sample. There are several parameters that can be considered as constant for water from a particular source (Andrews *et al.*, 1982). These include:

- the age of the precipitation, dissolved gases or dissolved elements
- the major elemental composition
- the isotopic composition
- the exsolved gases.

There is a very significant amount of data relating the geographical origin of water and the elemental isotopic ratios within it. The technique of stable isotope ratio analysis can therefore be a useful tool in the authentication of origin where a reference material is not readily available. The application of hydrogen and oxygen isotopes in this way follows the same principles as those referred to earlier in this book. Perhaps more speculative is the use of isotopic data on elements present in the water that originate from the source and can be used to identify authenticity.

However, inspection of bottling and extraction facilities, together with appropriate auditing of records and systems, is likely to be one of the most effective means of authentication.

11.7 Genetically engineered organisms

Genetic engineering involves the transfer of genes or other specific nucleotide sequences from one organism to a second, often unrelated organism. Organisms produced through genetic engineering are referred to as transgenic and pass on the acquired characteristics to successive generations. Genetic engineering has, to date, been applied successfully to a range of species bringing about changes in protein production capabilities, rate of maturation, enzyme activity, pest or disease resistance or susceptibility and nutritional composition (Jones, 1992).

11.7.1 Applications and use of genetic engineering

Genetic engineering has been applied to a wide range of issues to provide solutions to specific technological problems. Some of the developments are long-term but some have already been approved for use. All foods produced from crops, animals or microbes that have been genetically modified are regarded as novel in the UK and must be approved by the Advisory Committee on Novel Foods and Processes (ACNFP) prior to marketing. At the time of writing some eight products of genetic engineering have been approved for use in foods:

1. Baker's yeast, which breaks down sugar to carbon dioxide more quickly.
2. Brewer's yeast which is able to convert starch to alcohol. Conventionally, during brewing, amylase has to be added to break down the starch to sugar, which the yeast can then ferment to alcohol; with the modified yeast able to produce amylase, the addition of amylase is unnecessary.

3–5. Three microbes capable of producing bovine chymosin in fermentation systems. This enables the production of, for example, vegetarian cheese, which would normally be treated with chymosin extracted from bovine sources.

6. A tomato for producing pastes, with improved flavour and greater viscosity. Engineering involves the inclusion of 'antisense' polygalacturonase genes in tomato plants. This reduces the rate of cell wall degradation during ripening thus prolonging shelf-life by reducing softening of the fruit. In the USA the 'Flavr Savr'™ tomato has been approved and is now on sale at over 700 retail outlets across the USA.

7. A soybean that is tolerant to the herbicide glyphosate. Herbicide-tolerant crops are being developed to allow weeds to be controlled by use of herbicides that can be broken down quickly in the soil and which therefore have less impact on the environment. This is effectively making the plant resistant to a specific chemical rather than the chemical inactive to specific plants; it can thus provide much greater selectivity of action.
8. An oil-seed rape unable to self-pollinate. This is of significant agricultural importance in the control of crop attributes.

Work is progressing in many other areas, for example transgenic animals such as mice, rabbits, sheep, goats and pigs have been produced with inserted bovine, porcine or human growth hormone genes; the result being enhanced growth. Rabbits, goats and sheep have been engineered with the aim of pharmaceutical production in their milk. Mice have been engineered with a variety of genes in the study of particular diseases such as cancers. The insertion of multiple copies of native genes has also been successful such as for greater production of growth hormones. In plants there has been even more progress, such as in crop protection where the aim is to reduce the input of chemical control agents by enabling the plant to protect themselves against pests and diseases. Modification to increase shelf-life can reduce losses from spoilage and so make greater availability of fresh material possible. Nutritional composition and balance optimisation is enabling the production of methionine-rich proteins in soybean-based animals feeds; modifying the length or degree of saturation of fatty acids will allow valuable oils to be obtained from temperate crops; for example, laurate from oil-seed rape. The processing characteristics of products will be improved such as in potatoes where engineering will reduce fat uptake during frying or prevent browning. New sensory attributes could be introduced, for example genes for the sweet proteins thaumatin and monellin have been introduced into potatoes to give sweet-tasting tubers; colour alterations of fruits could also be possible. Product diversification such as pharmaceutical production in common crops could allow agricultural-scale production of drugs to ensure adequate supplies.

11.7.2 Labelling of genetically engineering materials

From the previous paragraph it can be seen that the developments and potential benefits of genetic engineering are enormous but controls and information are essential. The issue of genetic engineering is one that attracts much public interest and could present ethical concerns for certain groups of the population. For example, the transfer of genetic material from certain animals into other organisms is unacceptable to followers of particular religions. Furthermore, some members of the public have a general anxiety about the use of genetic engineering and consider it dangerous. In Australia, public outcry followed the discovery that experimental transgenic pigs had entered the human food chain; the pigs were carrying additional copies of the porcine growth hormone gene.

In the UK, following the report of the Polkinghorne Committee, the Food Advisory Committee decided that foods containing copies of ethically sensitive genes should be labelled. They announced in 1994 that:

- this labelling should be a statutory requirement
- it should take the form of 'contains copies of X genes'
- identity-preserved foods (e.g. whole fruits) should carry the label next to the name
- an ingredient in a mixture should carry the label next to its listing in the ingredients list
- ingredients not normally requiring listing would nevertheless require listing if they contain copies of ethically sensitive genes
- materials processed to the extent that the DNA is no longer present (e.g. refined sugar) would not need to be labelled under the current guidelines, although a European Regulation on novel food may affect this.

However, forthcoming European legislation may broaden the scope of these labelling requirements.

11.7.3 Methods of genetic engineering

Methods of gene transfer fall into two broad categories: the so-called 'vector-mediated transfers' and the 'vectorless transfers'. Vectors are DNA segments that allow another piece of DNA to be replicated. Vector-mediated transfer relies on the use, for example, of pathogenic agents, which, by naturally infecting the recipient organism insert part or all of their own DNA into the host DNA. Any disease-causing genes normally found in such vectors are replaced by the gene to be engineered into the recipient, so that the vector carries the transgenes into the host genome. A good examples of this is the use of the tumour-inducing plasmid of the plant pathogen *Agrobacterium tumefaciens* to introduce foreign genes into dicotyledonous plants. The wild type of the pathogen induces the formation of galls on plant but the vector form is rendered non-pathogenic by the replacement of key plasmid-borne genes with those to be transferred.

Vectorless transfer of DNA relies on physical methods. Examples include: micro-injection of DNA directly into cells, firing high-velocity DNA-coated micro-particles directly at cells and electrically inducing permeability of cell membranes to DNA in a surrounding solution. Vectorless transfer has proved to be particularly useful for engineering organisms for which there is no natural or practicable vector, for example, monocotyledonous plants that are not susceptible to *A. tumefaciens*.

It is usual that, when the actual coding region of the gene of interest is inserted, other DNA sequences are co-transferred into the recipient. These sequences will include vector sequences, as mentioned above, and sequences involved in the control of expression of the coding sequences, mainly promoter and terminator regions. For example, genes inserted into plants are frequently linked to a promoter from cauliflower mosaic virus. Similarly, in animals, the mouse metallothionein promoter is frequently used. In some instances, specific inducible promoters are used. These are sequences induced by external factors, such as heat, physical wounding or organ-specific factors. The latter facilitates organ-specific expression of genes in plants so that a gene for an insecticidal protein may be present in all cells of a plant but only expressed (e.g. by toxin production) in those organs susceptible to attack by the insect.

11.7.4 Authenticity issues

As previously mentioned, genetic engineering is a highly emotive subject and one in which there is great consumer interest and, in some cases, concern. This factor, together with some of the ethical issues, makes it essential that consumers can choose whether or not they purchase and eat food produced in this way. The issues are therefore twofold: is the food, or does the food contain, a product of genetic engineering and is any declared modification the actual and only modification present?

11.7.5 Methods of authenticity testing

The first difficulty associated with determining whether a food is composed of or contains genetically engineered material comes in the diversity of food types, processes and storage regimes. Each will present its own difficulties for the extraction of DNA in an appropriate condition for testing. Secondly, although in some instances the specific transgene (and therefore nucleotide sequence) will be known, in many cases it will not be defined. Thirdly the genetic modification may be an inserted gene, a modified native gene, or further copies of a native gene.

Despite the significant technological developments, referred to earlier, the availability and use of genetically modified materials and products derived from them is in its early days. There are therefore no current definitive methods for their detection. Despite the diversity of problems, the principles underlying the range of appropriate approaches will almost certainly be consistent: the use of gene probes to assess the presence in foods of specific nucleotide sequences. These approaches could involve the use of probes to sequences known to be present in the transgenic food.

Alternatively, they could probe for sequences commonly employed in biotechnology, such as vectors, promoters and other regulator sequences to provide generic indicators of transgenic material. In addition, the use of restriction fragment length polymorphism (RFLP) maps would allow the transgenic and conventional materials to be distinguished (Tanksley *et al.*, 1989).

Probing of the DNA present in food, by whichever means, requires the extraction of DNA so that the probe can be applied. A DNA or gene probe is merely a segment of a single strand of the DNA double helix, with a known or specific nucleotide sequence, that will bind to its complementary sequence. The probe also has some easily detected label attached to it, such as an enzyme, fluorescent molecule or radioactive atom. If a probe to a specific target sequence, such as a transgenic marker, comes into contact with that target it will hybridise (join together with the target) to reform the double helix. Subsequent detection of the probe label will indicate the transgenic nature of the material.

11.8 Irradiated products

Food irradiation is a preservation technique that can not only destroy microorganisms but can also inhibit sprouting in vegetables, retard ripening and induce sterility in pests and parasites, thus providing a means of population control. In some instances, irradiation can also improve the technological properties of foods.

11.8.1 Methods of irradiating food

Irradiation of foods is a treatment in which food is exposed to short-wave, high-energy radiation. The source of this ionising radiation can be either radioactive isotopes, such as Cobalt 60 and Caesium 137, which emit gamma rays, or electrochemical devices that generate X-rays or high-energy electrons. The choice of which type of radiation/source is used depends upon the type of food to be treated and other economic considerations; see Table 11.3

Several units of measurement are used with regard to irradiation treatments, the most important of which in legislative terms is the Gray (Gy); this is the unit of energy that characterises the dose of radiation absorbed by the food.

$$1 \text{ Gy} = 1 \text{ Joule/kg or } 1000 \text{ Joules/kg} = 1 \text{ kGy} \qquad (11.1)$$

The radiation dose is commonly classified into three levels. Low doses (up to 1 kGy – Racidation), medium doses (1–10 kGy – Radurisation) and high doses (10·50 kGy) may all be used depending upon the particular

Table 11.3 Characteristics of radiation types and sources[a]

Source	Type of radiation	Properties
Radioactive isotopes	Gamma rays	Good mechanical reliability and faults can be easily repaired High penetration power, allowing deeper treatment and application to larger packs
Electrochemical device	High-energy electrons	Electron accelerators are more likely to break down and require trained staff to operate them. The machines can be switched off when not in use and can be integrated into a production line The dose rate is very high and therefore allows high throughput; it can, however, be varied for the type of pack Electrons have limited penetration (5–8 cm)
Electrochemical device	X-rays	Provide a compromise between penetration depth and initial investment, although running costs are high

[a]Source: Fink and Rehmann (1994).

Table 11.4 Applications of different radiation doses[a]

Radiation dose (kGy)	Type of food	Effect of irradiation
Low dose		
0.05–0.15	Potatoes, onions, garlic, etc.	Sprout inhibition
0.2–0.5	Fresh fruits and vegetables	Delay of ripening
0.2–1.0	Cereals, dried fruit, meat, etc.	Insect/parasite deinfestation
Medium dose		
1.5–3.0	Fresh fish, strawberries, bread, etc.	Delay of mould growth
2.0–5.0	Fresh/frozen meat and fish, eggs, etc.	Spoilage decontamination
2.0–7.0	Grapes, dehydrated vegetables, etc.	Increase juice yield, reduce cooking time
High dose		
30–50	Poultry, meat, seafood, etc.	Sterilisation, including spores
10–50	Nuts, spices, special preparations, etc.	Decontamination to replace use of chemicals

[a]Sources: Robins (1991); Fink and Rehmann (1994).

application and purpose, as shown in Table 11.4. Irradiation can also be used in combination with other preservation protocols and, although there has been limited use of such combination procedures so far, many believe that it is in this way that irradiation will be at its most useful.

During irradiation, packaged food is either put into its container prior to irradiation or the packaging is presterilised and the irradiated food and container are brought together aseptically. In either case, the effect of the irradiation treatment on the packaging material is of great importance. The material must be able to withstand the radiation and not produce chemicals that migrate into the food or interact with it. Accordingly, the type of packaging that is to be exposed to irradiation processes needs to be fully tested and approved. The USA Food and Drug Administration has approved a range of single-layer films for use in food irradiation with acceptable radiation doses.

11.8.2 Standards, legislation and authenticity issues

The approval and control of food irradiation varies greatly from country to country. In the USA, radiation is defined as a food additive under the Food, Drug and Cosmetic Act and is regulated as such (Urbain, 1986). The issue of consumer concern has undoubtedly restricted the uptake of the technology and some of the European member states await a directive from the European Commission before allowing irradiated foods to be marketed; over 35 countries worldwide, however, permit irradiation. In Europe, a proposal was put forward in 1988, to come into force in 1993, to permit irradiation in a specified range of products at specified maximum doses and with provisions for controlling irradiation facilities. In 1991, regulations were introduced into the UK that prescribed a system of licensing for food irradiation facilities (The Food (Control of Irradiation) Regulations, 1991). The regulation also approved maximum doses for general types of foods:

- bulbs and tubers 0.2 kGy
- vegetables, cereals 1.0 kGy
- fruit 2.0 kGy
- fish and shellfish 3.0 kGy
- poultry 7.0 kGy
- spices and condiments 10.0 kGy

Irradiated foods cannot be identified by the consumer and therefore accurate labelling is essential to allow consumers to choose whether they eat food treated in this way or not. The International Consultative Group on Food Irradiation (ICGFI) has recommended labelling, together with public information and education campaigns. In the UK the Food Labelling Regulations 1984 as amended (1990) demand that food that has been treated in this way, or which contains irradiated ingredients, must be labelled to indicate this. Some materials are exempt from this but the remainder must be marked either as 'irradiated' or 'treated with ionising radiation' (Painter, 1992).

The issues of authentication are therefore whether a food not marked as such has been irradiated and whether any treatment is within the approved restrictions applying to that dose and pack combination.

11.8.3 Methods of authenticity testing

A wide range of methods have been developed or proposed for the detection of irradiation treatments, and no single method is appropriate for all food types. Four basic categories of test exist: DNA methods; biological and microbiological methods; physical methods; and chemical methods (Raffi and Belliardo, 1991).

Irradiation causes several different forms of DNA damage, such as base modification, damage of the sugar, single- and double-strand breaks and even crosslinking between bases and between DNA and protein. DNA modification can therefore be detected by use of high-performance liquid chromatography (HPLC) or by gas chromatography linked to mass spectrometry (GC–MS) to measure the breakdown products of the purine and pyrimidine bases (Robins, 1991). Other methods, such as fluorimetry and immunochemistry, have also been used in this context. One specific indicator of base changes is dihydrothymidine, which is produced only by irradiation treatment and has been detected using an immunochemical test (Deeble *et al.*, 1994). Strand breaks have also been used to detect irradiation but cannot be considered as induced only by irradiation; the use of mitochondrial DNA has, however, been suggested to be specific to ionising radiation (Bergaentzle *et al.*, 1994). One procedure that shows considerable promise is microgel electrophoresis. This technique was proposed by Cerda (Cerda *et al.*, 1992) and relies upon smaller DNA strand segments being more mobile in an electrophoretic gel. The nuclei of lysed cells are embedded in agarose on a microscope slide and exposed to an electric current. By observation of the pattern or comet of DNA streaming from the nuclei irradiated products can be detected. Figure 11.1 shows the different patterns derived from chicken treated with 0 and 5 kGy (Haine and Jones, 1994).

Microbiological methods rely on the population kinetics and highly variable sensitivity of different microbial species to ionising radiation. The population of microorganisms within an irradiated food will be biased toward ionisation-resistant species; for example as a general rule, Gram-negative bacteria are more sensitive than Gram-positive ones. Furthermore, as enzymes are not significantly deactivated by irradiation, a comparison of enzyme activity and viable counts can also give an indication of an irradiation treatment, as can the ratio between viable and non-viable bacteria. Several workers have progressed these approaches and have proposed screening methods based upon them (Jones *et al.*, 1994; Rowe and Towner, 1994).

One of the more fully evaluated methods of detection is that of luminescence measurement (Sanderson *et al.*, 1994). This physical characteristic relies upon energy, stored up within the food matrix during irradiation, being released upon thermal or optical stimulation. The

(a)

(b)

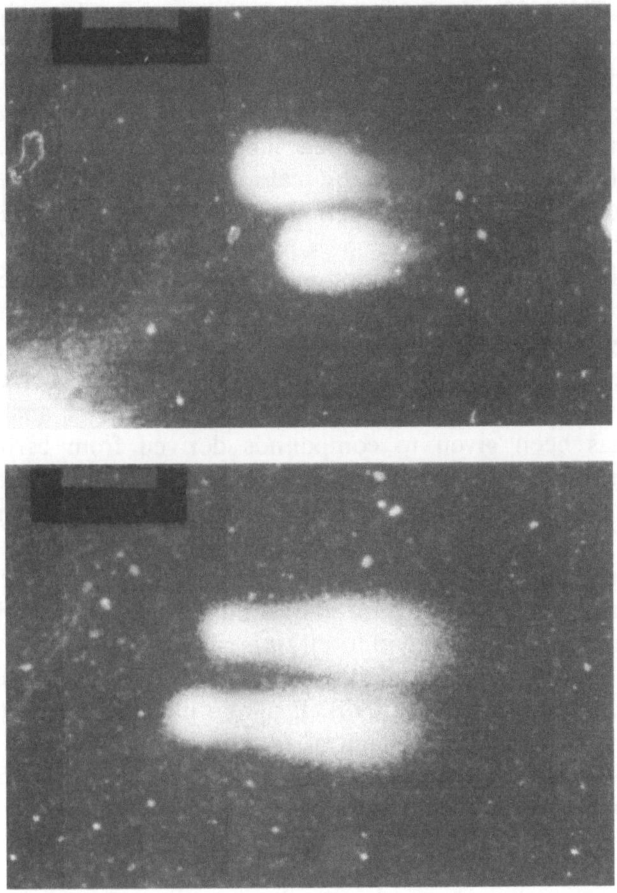

Figure 11.1 DNA stream from the nuclei of irradiated chicken cells treated with: (a) 0 kGy; and (b) 5 kGy.

released energy or luminescent emission is measured and can be used to determine irradiation. In order to be successful, the test sample must have a structure capable of trapping the energy; soft food materials therefore cannot be tested directly. However, many foods contain or have materials suitable for this technique associated with them, such as soil in vegetables, grit in crustacea and shells on shellfish.

Electron spin resonance spectroscopy (ESR) also relies upon entrapment of irradiation-induced free electrons. Free electrons in the form of free radicals or in their isolated state can be formed during irradiation and, in materials containing rigid structures such as bone or shell, can remain available for detection by ESR. This technique has therefore been applied to a range of materials including meat and poultry, shellfish and crustacea

(Stewart *et al.*, 1992). In bone, the free-electron signal can be detected following irradiation at doses as low as 0.2 kGy. Work has indicated that a residual signal can also be derived from crystalline cellulose in materials, such as the seeds of raspberries (Goodman *et al.*, 1994). Other physical methods that have been researched include freezing-point depression measured by differential scanning calorimetry (Dublin *et al.*, 1991), viscosity measurements (Nurnberger *et al.*, 1991) and changes in conductivity (Schertz, 1991).

Chemical indicators of irradiation result from the effect of the treatment on the proteins, carbohydrates and lipids present in foods. In the case of protein degradation, the only marker that has been studied to any significant extent is ortho-tyrosine, however, its presence in non-irradiated materials calls it into question as an effective target for analysis. Little attention has been given to compounds derived from carbohydrates; whereas the volatile compounds arising from lipid breakdown have been the subject of more interest. Alkanes, alkenes, aldehydes and cyclobutanones have been identified as the major radiolytic products from lipids. Upon irradiation, the fatty acids in foods give rise to their respective 2-alkylcyclobutanones. Owing to the common occurrence of palmitic, stearic, oleic and linoleic acid in foods, it is the cyclobutanones from these fatty acids that have been analysed and used as specific indicators of irradiation (Stevenson *et al.*, 1990). Progress has been made toward the development of an immunochemical technique, which would provide a much simpler means of detection and so provide a screening method for foods containing a significant amount of fat (Hamilton *et al.*, 1994).

References

Andrews, J.N., Burgess, W.G., Edmunds, W.M. *et al.* (1982) The thermal springs of Bath. *Nature* **298(5872)**, 339–43.

Anon (1979) Isotope 'label' assures truth in vanilla extract labelling. *Food Product Devel.* **13(3)**, 16.

Anon (1982) Organic and conventional products – the difference is in the viewpoint. *Ernaehrung* **6(12)**, 616.

Bailey, R.G., Nursten, H.E. and McDowell, I. (1992) Isolation and analysis of a polymeric thearubigin fraction from tea. *J. Sci. Food Agric.* **59**, 365–75.

Bergaentzle, M., Hasselmann, C. and Marchioni, E. (1994) Detection of irradiation foods by the mitochondrial DNA method. *Food Sci. Technol. Today* **8(2)**, 111–13.

Bricout, J. and Koziet, J. (1978) Characterisation of synthetic substances in food flavours by isotopic analysis. In: *Flavour of Food and Beverages* (Eds G. Charalambous, and G.E. Inglett). Academic Press, New York, pp. 199–208.

Cerda, H., Hofsten, B.V. and Johanson, K.J. (1992) Identification of irradiated food by microelectrophoresis of DNA from single cells. In: *Recent Advances in Detection of Irradiated Food.* BCR information. EUR 14315EN. Commission of the European Communities, 382–86.

Collins, M. and Nassif, W. (1993) Pesticide residues in organically and conventionally grown fruit and vegetables in New South Wales, 1990–91. *Food Austral.* **45(9)**, 429–31.

Deeble, D.J., Christiansen, J.F., Jones, M. *et al.* (1994) Detection of irradiated food based on DNA base changes. *Food Sci. Technol. Today* **8(2)**, 96–98.
Dubini, B., Leonardi, M., Messina, G. *et al.* (1991) Use of differential scanning calorimetry (DSC) for the identification of irradiated poultry meat. In: *Potential New Methods of Detection of Irradiated Food*. BCR information, EUR 13331EN. Commission of the European Communities, Luxembourg, 144–47.
EEC (1979) Council Directive No. 79/112 (EEC). OJ No L 033; 08.02.79 and as amended.
EEC (1980) Council Directive No. 80/777 (EEC). OJ No L 229.08.80 and as amended.
EEC (1988) Council Directive No. 88/388/EEC. OJ No I 229/11; 30.08.80.
EEC (1991) Council Regulation No. 2092/91 (EEC) (1991). OJ No L 198; 22.07.91.
EU Drinking Water Standards Directive 80/778.
Fink, A. and Rehmann, D. (1994) Research priorities relating to food irradiation. *Study Report No. 3, Flair, EUR 15017EN*. European Commission, Directorate-General XIII, Luxembourg.
Food Labelling Regulations (1984), SI 1984: No. 1305 as amended. HMSO, London.
Food Labelling (Amendment) (Irradiated Food) Regulations (1990), SI 1990: No. 2489. HMSO, London.
Food (Control of Irradiation) Regulations (1991), SI 1990: No. 2490. HMSO, London.
Food Safety Act (1990). HMSO, London.
Gill, M. (1992) Speciality and herbal teas. In: *Tea – Cultivation to consumption* (Eds K.C. Wilson and M.N. Clifford). Chapman and Hall, London, pp. 513–33.
Goodman, B.A., Deighton, N. and Glidewell, S.M. (1994) Detection of irradiated foods of plant origin by electron paramagnetic resonance spectroscopy. *Food Sci. Technol. Today* **8(2)**, 110–11.
Guillou, C., Remaud, G. and Martin, G.J. (1992) Applications of NMR to the characterisation and authentication of foods and beverages. *Trends in Food Sci. Technol.* **3**, 197–201.
Haine, H.E. and Jones, J.L. (1994) Microgel electrophoresis of DNA as a method to detect irradiated foods. *Food Sci. Technol. Today* **8(2)**, 103–104.
Hall, M.N. (1993) Authenticity issues for the food industry. In: *Food Technology International Europe 1993*, Sterling Publishing Group Plc., Hong Kong, pp. 193–95.
Hall, M.N., Robertson, A. and Scotter, C. (1988) Near infra-red prediction of quality, theaflavin and moisture content of black tea. *Food Chem.* **27** 61–75.
Hamilton, L., Elliott, C.T., Boyd, D.R. *et al.* (1994) The use of cyclobutanones in the development of an enzyme-linked immunosorbent assay (ELISA) for the detection of irradiated foods. *Food Sci. Technol. Today* **8(2)**, 100–101.
ISO 3720 (1986) *Black tea – definition and basic requirements*. International Standards Organisation.
ISO 4676 (1982) *Spices and Condiments – nomenclature. First list, trilingual edition*. International Standards Organisation.
Jones, J.L. (1992) Genetic engineering of crops: its relevance to the food industry. *Trends in Food Sci. Technol.* **3(3)**, 54–59.
Jones, K., MacPhee, S., Stuckey, T. (1994) The direct epifluorescent filter technique (DEFT): a screening method for the detection of irradiated foods. *Food Sci. Technol. Today* **8(2)**, 105–106.
Kobayashi, A., Tachiyama, K., Kawakami, K. *et al.* (1985) Effect of solar withering and turn-over treatment during indoor withering on the formation of pouchong tea aroma. *Argric. Biol. Chem.* **49**, 1655–60.
Mahanta, P.K. and Baruah, S. (1989) Relationship between process of withering and aroma characteristics of black tea. *J. Sci. Food Agric.* **46**, 461–68.
Mintel International (1993a) Tea. In: *Market Intelligence*. Mintel International, London.
Mintel International (1993b) Vegetarian and Organic Food. In: *Market Intelligence*. Mintel International, London.
Mintel International (1994) Bottled Water. In: *Market Intelligence*. Mintel International, London.
Natural Mineral Water Regulations (1985) SI 1985: No. 71. HMOS, London.
Nurnberger, E., Heide, L. and Bogl, K.W. (1991) Viscosity measurements – A comprehensive overview of the method and its applicability for the identification of

irradiated foodstuffs. In: *Potential New Methods of Detection of Irradiated Food*. BCR information, EUR 13331EN. Commission of the European Communities. 121–43.

Omori, K. (1983) Discrimination between Gyokuru and Sen-cha. *Tea Res. J.* **58**, 28–35.

Othieno, C.O. (1992) Soils. In: *Tea – Cultivation to Consumption* (Eds K.C. Wilson and M.N. Clifford), Chapman and Hall, London, pp. 137–70.

Owuor, P.O., Takeo, T., Horita, H. *et al.* (1987) Differentiation of clonal tea by terpene index. *J. Sci. Food Agric.* **40**, 341–45.

Painter, A.A. (1992) *Butterworths Food Law*. Butterworths, London.

Pither, R.J. and Hall, M.N. (1990) Analytical survey of the nutritional composition of organically grown fruit and vegetables. *Technical Memorandum No. 597*. Campden Food and Drink Research Association.

Potter, R.H. (1971) Non-vanillin vanilla volatiles in vanilla extract. *J. Assoc. Official Analytical Chemists* **54(1)**, 39–41.

Powell, C., Clifford, M.N., Opic, S.C. and Gibson, C.I. (1994) Fractionation of thearubigins: Isolation of theacitrins. In: *What's in a Cuppa? – Recent Advances in the Chemistry and Biochemistry of Tea*. SCI Lecture Papers Series No. 0027.

Pruthi, J.S. (1980) Spices and condiments: chemistry, microbiology, technology. In: *Advances in Food Research, Suppl.* **4**. Academic Press, London.

Raffi, J.J. and Belliardo, J.-J. (Eds) (1991) *Potential New Methods of Detection of Irradiated Food*. BCR information, EUR 13331EN. Commission of the European Communities.

Rauter, W. and Wolkerstorfer, W. (1982) Nitrate in vegetables. *Zeitschrift fuer Lebensmittel Untersuchung und Forschung* **175(2)**, 122–24.

Roberts, E.A.H. (1950) The phenolic substances of of manufactured tea. II. Their origin as enzyme oxidation products in fermentation. *J. Sci. Food Agric.* **9**, 212–16.

Robins, D. (1991) *The Preservation of Food by Irradiation – a factual guide to the process and its effect on food*. CWA Consultants, Food Technology Department, IBC Technical Services Ltd., London.

Rowe, T.F. and Towner, K.J. (1994) Potential of non-radioactive probes for the retrospective detection of bacteria present in foodstuffs before irradiation treatment. *Food Sci. Technol. Today* **8(2)**, 107–108.

Sanderson, D.C.W., Carmichael, L.A., Ni Riain, S. *et al.* (1994) Luminescence studies to identify irradiated food. *Food Sci. Technol. Today* **8(2)**, 93–96.

Schertz, H. (1991) Detection of irradiated potatoes by conductivity measurements. In: *Potential New Methods of Detection of Irradiated Food*. BCR information, EUR 13331EN. Commission of the European Communities, pp. 152–55.

Schreier, P., Bernreuther, A., Krammer, G. *et al.* (1991) Chirospecific capillary gas chromatography (HRGC) and on-line HRGC-isotope ratio mass spectrometry – the state of the art in flavour analysis. In: *Authentication and Quality Assessment of Food Products*. European seminar, Nantes, France, October 1991 (Organised by Eurofins Laboratories, Nantes, France).

Schutz, H.G. and Lorenz, O.A. (1976) Consumer preferences for vegetables grown under commercial and organic conditions. *J. Food Sci.* **41**, 70–73.

Sen, A.R., Sen Gupta, P., Sardar, P.K. *et al.* (1977) Thin-layer chromatographic detection of *Amomum subulatum* Roxob. In: *Elletaria cardamomum* Maton. *J. Assoc. Official Analytical Chemists* **60(2)**, 307–308.

Sen Gupta, P., Sen, A.R., Bose, A. and Mathew, T.V. (1973) Detection of coumarin in vanilla flavoured food by TLC. *Res. India* **18**, 96.

Stahl, W.H., Skarzynski, J.N. and Voelker, W.A. (1969) Differentiation of certain types of cassias and cinnamons by measurement of mucilagenous character. *J. Assoc. Official Analytical Chemists* **52(4)**, 741–44.

Stevenson, M.H., Crone, A.V.J. and Hamilton, J.T.G. (1990) Irradiation detection *Nature* **334**, 202–203.

Stewart, E.M., Stevenson, M.H. and Gray, R. (1992) Detection of irradiation in scampi tails – effects of sample preparation, irradiation dose and storage on ESR response in the cuticle. *Internat. J. Food Science Technol.* **27**, 125–32.

Takeo, T. (1981) Variation in amounts of linalool and geraniol produced in tea shoots by mechanical injury. *Phytochem.* **20**, 2149–51.

Takeo, T. (1983) Effect of clonal specificity of the monoterpene alcohol composition of tea shoots on black tea aroma profile. *J. Agric. Res. Quart.* **17**, 120–24.

Takeo, T. (1984) Withering effect on the aroma formation found in oolong tea manufacture. *Agric. Biol. Chem.* **48**, 1083–85.

Tanksley, S.D., Young, N.D., Paterson, A.H. (1989) RFLP mapping in plant breeding: new tools for an old science. *Biotechnol.* **7**, 259–66.

Tomlins, K.I. and Gay, C. (1994) Prediction of quality and origin of black tea and pine resin samples from chromatographic and sensory information: evaluation of neural networks. *Food Chem.* **50**, 157–65.

Trade Descriptions Act (1968). HMSO, London.

Tressl, R. and Engel, K.H. (1985) Biogenesis of chiral aroma components and their analytical characterisation at trace level. In: *Progress in Flavour Research 1984*, (Ed. J. Adda). Elsevier, Amsterdam.

Urbain, W.M. (1986) *Food Irradiation. Food Science and Technolgy.* A series of monographs, Academic Press, London.

Wilson, K.C. and Cifford, M.N. (1992) *Tea – Cultivation to Consumption*. Chapman and Hall, London.

Winton, A.L. and Winton, K.B. (1939) *The Structure and Composition of Foods*. John Wiley & Sons, Chichester.

Woodward, L. (1990) Measurement of quality of organic produce. In: *Symposium Proceedings: Organic food production. A growing business*? Campden Food and Drink Research Association, London.

Wright, S. (1994) *Handbook of Organic Food Processing and Production*. Blackie, Glasgow.

Zuercher, K. Hadorn, H. and Strack, C. (1974) Evaluation of cinnamon quality by gas chromatographic analysis of the essential oil. *Mitteilungen-aus-dem-Gebiete-der-Lebensmitteluntersuchung-und-Hygiene* **65(4)**, 440–52.

12 Milk and milk products

M.J. DENNIS

Milk has been an important traded commodity since historical times and hence may well have been one of the earliest commodities to be the subject of extension or adulteration. This has been recognised by enforcement authorities where legislation on milk has been in place for many years.

The authenticity of milk can, perhaps, be seen as a microcosm of food authenticity issues. When considering how the authenticity of milk has developed as a science, it may be possible to draw parallels with newer traded commodities with regard to how questions of authenticity may develop. Holeman (1984) describes how dairy products were among the earliest to be subject to control. The earliest adulterations of milk were made with 'a variety of unsophisticated yet hazardous and fraudulent adulterants'. A mixture of 1 gallon water, 2 oz sugar, 1½ oz salt and a little caramel was added to 4 gallons of milk in order to 'fool the lactometer'. Chalk, dextrins and plain water were other common early adulterants. Many of the tests that were developed in this early period were subsequently deleted from the association's 'Official Methods of Analysis' – presumably because formerly commonly used adulterants were no longer being encountered and because methods became outdated. Thus, in the future, analysts should be prepared to discard methods that are no longer pertinent to current issues.

Another point of particular interest from this early period is to note how changing consumer preferences can affect commodities. The fat content of milk has traditionally been one of its most important characteristics. Gravimetric methods were therefore developed at the turn of the century to ensure that fat was not being skimmed off milk. Now the perception of skimmed and semi-skimmed milk as being healthier than full-fat milk means that these products are in demand in their own right. With proper labelling they have become both accepted and, in some cases, preferred.

12.1 Authenticity issues

There are several important issues that are relevant to milk quality that are not strictly authenticity issues. The presence of bacterial contamination and antimicrobial drugs represent two cases in point. The official *Journal of the European Community* (EEC, 1991) lays down several methods of analysis relating to these quality issues, as well as to authenticity.

There are several ways in which the authenticity of milk may be challenged. The addition of water is one of the oldest and most obvious. Since milk is used as the raw material for other products such as cheese, there is also the potential to use waste from these processes (e.g. whey) as an extender. Other potential extenders may arise from other species. These might include milk from different species of animal to the inclusion of vegetable oils to replace butter fat.

There may be safety issues attached to this type of adulteration. Many people are allergic to cows' milk and therefore choose milk, cheese and yoghurt from other animals such as sheep and goats. If a significant quantity of cows' milk is added to these products without proper labelling, health problems can result. Speciality cheeses are made in several parts of the world. Under European legislation (EEC, 1992a; 1992b) it is possible to protect food names on a geographical or traditional recipe basis. This provision has been particularly useful for cheeses. As yet, analytical methods to determine origin are in their infancy and documentary evidence remains the best approach. Finally, milk and milk products undergo several industrial processes and treatments and are labelled as such. Methods are in place to ensure that these treatments have been properly carried out.

12.2 Analytical methods for milk and milk product authenticity

12.2.1 Added water

A standard methodology for determining added water in milk is now well established. The test measures the depression of the freezing point of milk and considerable data is available for authentic ranges from several countries. A cryoscopic method is detailed in 91/180/EC (EEC, 91). Rhom *et al.* (1992) found a mean value of −0.5257°C ± 0.00659°C for Austrian milk.

According to Harding (1990), the Hortvet technique is often used in the UK; this gives values about 0.017°C higher than the scale used in many continental countries. A mean value of −0.543°C was found for UK milk. A depression of only −0.530°C would be expected by chance from only six in 1000 samples. The difference between the mean value and this lower value would represent only 2.5% added water. The reason that this range is so small is that, for a cow to produce milk at all, the osmotic pressure of the milk must balance that of the blood and this can only vary within a small range. The slight differences that occur may be caused by differences in environment, season, diet and stage of lactation of the animals. Similar data is available for milk from other species. Typical ranges for sheep have been found between −0.550 and −0.590°C. (Pappas *et al.*, 1994).

12.2.2 Process control

While the depression of freezing point represents a valuable method for determining added water as an adulterant it might also be considered a method for ensuring process control. Modern heat treatments may use the direct injection of steam to heat milk quickly to high temperatures. This ensures microbiological stability while minimising unwanted chemical changes. Naturally this adds water to the product, which has then to be removed. Measurement of the depression of freezing point ensures that sufficient water has been removed from the product to return it to its original composition.

In general, there are rather fewer methods to establish the correct processing of commodities than there are to measure composition. Milk provides a good example of a commodity where the required process of thermal treatment has been established for some time. It is important to have methods that ensure that thermal treatment of milk has been correctly carried out because of the potential for microbial spoilage of under-treated product and because over-heated product may have a flavour that is unacceptable to some consumers. Council Directive 92/46/EEC (EEC, 1992c) describes the requirement for marketing raw and heat-treated milk. Pasteurised milk must be heated to 71.7°C for 15 seconds or an equivalent pasteurisation process to produce an equivalent effect. (UHT milk must be heated to 135°C for 1 s.)

Commission Decision 91/180/EEC describes the official methods for testing heat-treated milk. Milk contains the two enzymes, alkaline phosphatase and peroxidase. The phosphatase enzyme liberates phenol from disodium phenyl phosphate, which is then measured colorimetrically. Release of less than 4 μg phenol/ml milk is considered a negative reaction. The peroxidase enzyme decomposes hydrogen peroxide. Again, the atomic oxygen liberated is measured colorimetrically. Pasteurised milk must show a negative reaction to the phosphatase test and a positive reaction to the peroxidase test. If the peroxidase test is negative the product must be labelled as 'high-temperature pasteurised'.

It is quite common that when a food commodity is used to manufacture added value products, some parts of the commodity may remain for which there is only a limited market. Extracting value from such residues is seen as a key to ensuring a profitable business. Very often it is possible to use these materials in food products with appropriate labelling but there may be temptations to use them as extenders of quality products. The dairy industry provides a good example.

Whey, left over from cheese-making, is a low-value product which is often included in desserts with appropriate labelling. However, it represents a potentially profitable extender of milk products. In Brazil, the temptation is to add this material to pasteurised milk. Wolfschoon-Pombo and Furtado (1989) describe a simple method for determining this adulteration. The method measures casein-bound phosphorus and protein nitrogen content. The method was used to show that four out of nine pasteurised Brazilian milks contained up to 60% whey. These authors point out that, in Europe, whey is more likely to be added to non-fat dried milk powder. Imbert *et al.* (1993) described a procedure that is appropriate for both raw milk and milk powder. The procedure involves the measurement by HPLC of casein macropeptides produced from rennin in the cheese-making process.

Milk powder itself may be a potential adulterant of raw and pasteurised milk. Resmini *et al.* (1992) described the analysis of furosine by HPLC. This compound is formed by the heat treatment of milk and is a measure of the extent of the Maillard reaction. Typical raw milks contain a maximum level of furosine of 60 mg/l and pasturised milk 70–80 mg/l. For skimmed milk powder, values may range from 650–5000 mg/l. Of 124 Italian milk samples analysed, 26 (21%) from 17 different producers were deemed to contain milk powder by virtue of their furosine concentration.

12.2.3 The species of origin of milk

Sheep, goat and buffalo milk are all used to make cheese. Sheep and goat's milk are also used for making yoghurt. There can be times of the year when these milks are in short supply. It may then be tempting for producers to maintain their product volume by using cows' milk as an extender. There is now available extensive literature on research into methods of analysis for detecting mixtures of milk from different species in raw and further processed milk products. The two preferred methods are immunology and electrophoresis.

Some of the problems that may be encountered when using immunological methods have been alluded to in the chapter on meat species authenticity. Many of the same considerations will apply to milk. Indeed, Hernandez *et al.* (1994) have reviewed antibody based methods for both these issues in the same publication.

Several different experimental formats have been applied to this problem. Aranda *et al.* (1993) have described an immunodotting technique that is rapid (2 h) and will detect 0.5% goats' or cows' milk in sheeps' milk. It can also be used for determining their presence in ewes' milk cheese. Garcia *et al.* (1993) preferred a sandwich ELISA format, which achieved a similar detection limit. Rolland et al. (1993) managed to detect 0.125% cows' milk in sheeps' milk and 0.5% in cheese. However, perhaps the most impressive demonstration of the application of immunological techniques was provided by Levieux and Venien (1994). By using a monoclonal antibody against cows β-lactaglobulin, one part cows' milk in 100 000 parts goats' or ewes' milk could be detected.

Electrophoretic techniques also have a considerable following for this determination. Moio *et al.* (1990) describe an isoelectric focusing procedure. Again the detection limit for bovine milk in milk from other species was 0.5%. The Commission of the European Communities, as described in Commission Regulation (690/92/EEC) established an isoelectric focusing reference method for the detection of cows' milk casein in cheeses made from ewes' milk. This regulation permits the use of other methods. However, these must have a detection limit lower than 0.5%, they must not be subject to false-positive results, and they must have the required sensitivity after the usual long ripening periods that are typical of this type of cheese.

A comparison of electrophoresis (PAGE), isoelectric focusing (IEF) and radial immunodiffusion has been carried out by Amigo *et al.* (1992). The PAGE and IEF methods were in good quantitative agreement. The immunodiffusion method agreed qualitatively with the other methods in about 90% of the cases studied. Unfortunately, this simple and rapid procedure did not detect three samples containing goats' milk and two samples containing cows' milk in the 52 samples analysed. In this survey, only 36.5% of samples proved to be genuine ewes' milk cheeses, 52% contained goats' milk and 9.6% contained cows' milk. This is disappointing because these cheeses are of considerable economic importance to the farmers from this region of Portugal.

Electrophoretic and immunological techniques are likely to be the most important methods for species characterisation of milk and milk products but other techniques have also been described. Lopez-Fandino *et al.* (1993) and Urbanke *et al.* (1992) have both described the use of HPLC for this purpose. As yet this approach does not seem to offer any advantages for milk species' identification. However, Hewedy and Smith (1989) found fast protein liquid chromatography (FPLC) the best method for detecting soy 'milk' in pasteurised bovine milk. Urea gel electrophoresis was not promising but SDS–PAGE detected 5% soy 'milk'. However, FPLC was more sensitive and was able to detect 1% soy 'milk' in cows' milk. (It should be noted that the term 'soya milk' was deemed to infringe the Dairy Designation Regulations (1989/87/EEC). Although the product remains available, the name has been phased out.)

Considerable research has gone into determining the composition of the fat fraction from various milk products. Matter (1992) described the fatty acid composition of several meat and cheese products. Iverson and Sheppard (1989) found the lauric:capric acid (C12:C10) ratio useful for determining the presence of cows' milk in sheeps' or goats' milk cheese. This method is unfortunately not useful for detecting mixtures of sheeps' and goats' milk where ratios are similar. The ratio of butyric acid:caproic acid (C4:C6) was considered a valuable indicator of adulteration by Iwaida *et al.* (1979), although it was conceded that the method was insufficient on its own and should therefore be used in conjunction with other methods of analysis such as sterol composition.

The fatty acid composition is perhaps more useful for detecting the presence of fat that is not of milk origin. Fox and co-workers (1988) described the use of this approach for detecting vegetable fats in butter and cheese. The ratio of butyric:oleic acids (C4:C18:1) was used to evaluate the sample and could detect the addition of 10% partially hydrogenated vegetable fat. The sensitivity of the test was limited by the variability in composition of the natural products.

An attempt was subsequently made to take into account some of this variability by making an allowance for the season during which the cheese might be produced. This would allow the amount of vegetable fat detectable to be reduced from 10.5% to 8.2% (Fox *et al.* 1989). While this may be achievable theoretically, it seems to be a rather impractical approach.

However, these authors report success in analysing cheese as an ingredient of pizza. There are, as yet, few methods for determining the authenticity of a component of a prepared food product. In this case, it was possible to physically remove the cheese from the top of the frozen pizza and wash it with distilled water to remove tomato sauce etc. before analysis.

The use of ratios of dairy components provides a means of limiting the variability of the data. As discussed in the introduction, the more data available to combine into a model of the authentic foodstuff, the easier it becomes to detect the non-authentic product. Villanueva and co-workers (1988) described the application of a linear discriminant model to the triglyceride composition of milk fats. Triglycerides were analysed using gas chromatography. Using this model, non-milk fat of animal origin was detected in 100% of cases at the 10% adulteration level and 97% of cases at the 5% adulteration level.

Lercker *et al.* (1992) confirmed the value of triacylglycerol patterns in assessing the authenticity of butterfat. This gave a clear indication of the presence of tallow or lard. The inclusion of analytical data on diglycerides and 'minor fatty acids' (C14 and C15) improved the ability to detect adulteration.

Of course, as always, the ability to detect a component that should not be present provides the clearest demonstration of adulteration. As early as 1971, Keeney *et al.* (1971) reported a colorimetric procedure for determining vitamin E. This procedure was able to identify 2–3% corn oil or soybean oil in ice cream fat extracts. It was noted that the procedure could not be used for coconut oil because this was a poor source of the vitamin.

More recently, sterol analysis has been used for the same purpose. The major component of the sterol fraction in butter is cholesterol (99.7%).

Thus the presence of vegetable oil sterols is a good indicator of adulteration. Younes and Soliman (1987) found that β-sitosterol permitted the identification of cotton seed oil in butter fat.

Clearly the best method of analysis for assessing whether non-milk fats are present in a dairy product will depend on the source of the adulterant. Collomb and Spahni (1991) have reviewed in some detail the potential adulteration of milk products. In summary they find that:

- sterol analysis is the most sensitive method for detecting vegetable fat
- triglyceride analysis is the most universal method since it permits the identification of animal body fats
- fatty acid analysis is the least sensitive because of variations resulting from the influence of fodder
- hydrogenated fats may be detected using infrared spectroscopy to identify changes in the *cis/trans* ratio of unsaturated double bonds
- synthetic fats can be detected from products (ketones, mono and diglycerides) of the process of transesterification
- refined fats may be identified by the presence of typical hydrocarbons (e.g. 3,5 cholestadiene).

12.3 Effects of adulteration on the market for milk products

The producer who seeks to obtain a short-term gain by extending his or her product with an adulterant risks the possibility of legal action. If this approach is taken on bulk commodities such as milk, it is likely that this will be the only deterrent that they will face because there will always be a demand for basic commodities. However, those who manufacture a speciality product may well depress the market through their actions. When this occurs, it can be very difficult to retain the goodwill of the consumer and all those producers who are in this market are affected. An early example of this occurred in the USA in 1870 when 'filled' cheese was made with cheaper fat (usually lard). Sales of these cheeses to Britain were particularly important and exports accounted for about one-half the production quantity. However, sales of this inferior product resulted in exports falling to one-quarter of output and, by 1890, Wisconsin's reputation for fine cheese was lost. It took 20 years for this to be recovered (Small, 1983).

Ultimately, maintaining the quality, reliability and authenticity of food products can only be to the benefit of both consumers and producers.

References

Amigo, L., Ramos, M., Calhau, L. *et al.* (1992) Comparison of electrophoresis, isoelectric focusing, and immunodiffusion in determinations of cows' and goats' milk in Serra da Estrela cheeses. *Lait* **72**, 95–101.

Aranda, P., Sanchez, L., Perez, M.D. *et al.* (1993) Rapid immunoenzymatic method for detecting adulteration in ewes' milk. *Food Control* **4(2)**, 101–104.

Collomb, M. and Spahni, M. (1991) Adulteration of milk products. Review of the analytical criteria for the detection of vegetable and animal fats in milk fat. *Trav. Chim. Aliment. Hyg.* **82**, 615–62.

EEC (1987) Dairy Designation Regulations 1989/87/EEC.

EEC (1991) Council Directive 91/180.

EEC (1992a) EC Regulation 2081/92.

EEC (1992b) EC Regulation 2082/92.

EEC (1992c) Council Directive 92/46/EEC.

EEC (1992d) Commission Regulation 690/92/EEC.

Fox, J.R., Duthie, H. and Wulff, S (1988) Precision and sensitivity of a test for vegetable fat adulteration of milk fat. *J. Dairy Sci.* **71**, 574–81.

Fox, J.R., Duthie, H. and Wulff, S. (1989) Effect of the seasonal variation of fatty acids in milk fat on the sensitivity of a test for vegetable fat adulteration. *J. Dairy Sci.* **72**, 1981–85.

Garcia, T., Martin, R., Morales, P. *et al.* (1993) Sandwich ELISA for detection of caprine milk in ovine milk. *Milchwissenschaft* **48**, 10.

Harding, F. (1990) Milk adulteration – freezing point depression. *J. Soc. Dairy Technol.* **43(3)**, 61.

Hernandez, P.E., Martin, R., Garcia, T. *et al.* (1994) Antibody based analytical methods for meat species determination and detecting adulteration of milk. *Food and Agric. Immunol.* **6**, 95–104.

Hewedey, M.M. and Smith, C.J. (1989) Detection of soy milk in pasteurised bovine milk. *Food Hydrocolloids* **3(5)**, 399–405.

Holeman (1984) Food adulteration detection: 100 years of progress in AOAC methodology. *J. Assoc. Off. Anal. Chem.* **67(6)**, 1029–34.

Imbert, A., Nicolas, M. and Grenier, J. (1993) Detection of milk in whey's of enzymatic origin by identification of their caseinomacropeptides. *Sciences des aliments* **13**, 545–58.

Iverson, J.L. and Sheppard, A.J. (1989) Detection of adulteration in cow, goat, and sheep cheeses utilising gas-liquid chromatographic fatty acid data. *J. Dairy Sci.* **72**, 1707–12.

Iwaida, M., Ito, Y., Tonogai, Y. *et al.* (1979) Detection of adulterated milk fats by gas liquid chromatographic determination of butyrate and caproate. *J. Food Hyg. Soc.* **20(5)**, 329–31.

Keeney, M., Bachman, K.C., Tikriti, H.H. *et al.* (1971) Rapid vitamin E method for detecting adulteration of dairy products with non-coconut vegetable oils. *J. Dairy Sci.* **54(11)**, 1702–703.

Lercker, G., Frega, N., Boccio, F. *et al.* (1992) Bovine milk fat, triacylglycerol gas chromatographic pattern and bovine butterfat genuineness. *Scienza e technica lattiero-casearia* B43B(2), 95–110.

Levieux, D. and Venien, A. (1994) Rapid, sensitive two-site ELISA for detection of cows' milk in goats' or ewes' milk using monoclonal antibodies. *J. Dairy Res.* **61**, 91–99.

Lopez-Fandino, R., Acedo, M.I. and Ramos, M. (1993) Comparative study of HPLC of caseinomacropeptides from cows', ewes' and goats' milk. *J. Inst. Dairy Res.* **60**, 117–21.

Matter, L. (1992) Determination of the animal origin of dairy products and raw and cooked meats by GC analysis of the fatty acid methyl esters (FAME) obtained by transesterification. *J. High Res. Chromatogr.* **15**, 514–16.

Moio, L., Sasso, M.L. and Chianese, L. (1990) Rapid detection of bovine milk in ovine, caprine and water buffalo milk or cheese by gel isoelectric focusing on Phastsystem™. *Ital. J. Food Sci.* **3**, 185–90.

Pappas, C.P., Voutsinas, L.P. and Kondyli, E. (1994) Determination of added water in sheep milk by measurements of the freezing point and acidity. *Milchwissenschaft* **49**, 6, 309–12.

Resmini, P., Pellegrino, L., Masotti, F. *et al.* (1992) Detection of reconstituted milk powder in raw and in pasteurised milk by direct HPLC of furosine. *Scienza e technica lattiero-casearia* **43(3)**, 169–86.

Rhom, H., Solkner, J. and Pleschberger, C. (1992) The freezing point of Austrian milk. 3. Variations in the freezing point of individual milk. *Ernahrung* **16(3)**, 135–40.

Rolland, M.-P., Bitri, L. and Besancon, P. (1993) Polyclonal antibodies with predetermined specificity against bovine α_{s1}-casein: application to the detection of bovine milk in ovine milk and cheese. *J. Dairy Res.* **60**, 413–20.

Small, E. (1983) Defining cheddar cheese. *Daily Record*, October, pp. 192–96.

Urbanke, W., Luf, W. and Brandl, E. (1992) Use of HPLC for control of the adulteration of milk and milk products of different species. *Zeitschrift fur Lebensmittel Untersuchung und Forschung* **195**, 137–42.

Villanueva, B., de la Reguera, P.F. and Pinto, M. (1988) Application of linear discriminant analysis to the detection and quantification of adulteration of milk fat. *Agro Sur* **16(1)**, 47–52.

Wolfschoon-Pombo, A.F. and Furtado, M.A.M. (1989) Detection of adulteration of pasteurised milk with whey by determination of the casein-bound phosphorus and protein nitrogen content. *Z. Lebensm Unters Forsch* **188**, 16–21.

Younes, N.A. and Soliman, M.A. (1987) Adulteration of butterfat: sterol composition. *Grasas y Aceites* **38(6)** 372–74.

Index

Page numbers appearing in **bold** refer to figures and page numbers appearing in *italic* refer to tables.

MIX
Papier aus verantwortungsvollen Quellen
Paper from responsible sources
FSC® C105338

FSC
www.fsc.org

If you have any concerns about our products,
you can contact us on
ProductSafety@springernature.com

In case Publisher is established outside the EU,
the EU authorized representative is:
Springer Nature Customer Service Center GmbH
Europaplatz 3, 69115 Heidelberg, Germany

Printed by Libri Plureos GmbH
in Hamburg, Germany